T0401540

RENEWABLE ENERGY: RESEARCH, DEVELOPMENT AND POLICIES

BIOGAS

PRODUCTION AND PROPERTIES

RENEWABLE ENERGY: RESEARCH, DEVELOPMENT AND POLICIES

Additional books and e-books in this series can be found on Nova's website under the Series tab.

RENEWABLE ENERGY: RESEARCH, DEVELOPMENT AND POLICIES

BIOGAS

PRODUCTION AND PROPERTIES

JAMES G. SPEIGHT, PHD

Library of Congress Cataloging-in-Publication Data

ISBN: 978-1-53615-278-4

Published by Nova Science Publishers, Inc. † New York

CONTENTS

PREFACE

As the need for the replacement of fossil fuels as energy sources continues to grow and the depletion of the resources of natural gas continues, gaseous products from other sources (such as biomass) will continue to grow in importance. As this depletion of fossil fuel resources continues, biomass will play an increasingly important role in the future global energy infrastructure for the generation of power and heat, but also in the production of gaseous and liquid products for the production of fuels and chemicals and fuels. The dominant gas-from-biomass technologies will be anaerobic digestion, gas from landfills, pyrolysis gas and biomass gasification. Also, within the crude oil refinery system, gasification processes where biomass is co-gasified with crude oil residua are likely to supersede the other biomass-to-gas processes for the production of gases as intermediates for high-efficient power production or for the synthesis of chemicals and fuels.

Biomass from arable land and from grass land in various countries is already, to some extent, is successfully used for biogas production and, within the biogas production and utilization scenarios, energy crops are currently at the center of interest. There is also, however, the fear that the growth of crops for energy production will compete with the growth of crops for food production. At present the cultivation of energy crops takes place in competition with food production. Therefore, ways must be found to produce food on arable land and to use the surplus and residual substances in biogas plants. Furthermore, in the discussion on the production of gases from biomass (other forms of organic feedstocks, such as organic waste as contained in municipal solid waste will also be of importance). However, it is important to understand that the composition of the produced gas is very dependent on the type of gasification process and especially the gasification temperature.

Gas can be produced from raw materials such as agricultural waste (including manure), municipal waste, industrial sewage, and many other forms of organic waste. Thus, the term *biogas* has been expanded herein to include the gases known as *landfill gas* and *bio-synthesis gas* (also often referred to as bio-syngas). In the latter case, the prefix *bio* is used to differentiate the gas from the synthesis gas produced from the fossil fuels: crude oil residua, tar sand bitumen, coal, and oil shale. Both of these bio-gases are produced by the breakdown (due to chemical reactions and microbes) of biodegradable waste within a landfill (hence the term landfill gas) or in a gasification process. Also, by way of further expansion of the definition of biogas, gaseous products from the pyrolysis of biomass and to gases produced by the gasification of biomass are also included. It is worthy of note that this point that the gasification of biomass leads to two categories of gaseous products. Hence the need to include in this book gases produced by the gasification of biomass as a falling under the umbrella of biogas.

Within the context of this book, the term *biogas* is used to designate the mixture of different gases produced by the breakdown of organic matter in the absence of oxygen, usually referred to the anaerobic digestion of organic matter including manure, sewage sludge, municipal solid waste, biodegradable waste or any other biodegradable feedstock, under anaerobic conditions. Typically, biogas is comprised predominantly of methane, and carbon dioxide with, depending upon the source of the gas, smaller amounts of carbon monoxide, water, hydrogen sulfide, nitrogen, hydrogen, mercaptan derivatives, and oxygen.

Throughout this book wherever a temperature is given in degrees Centigrade, the corresponding temperature in degrees Fahrenheit is also presented. Also, as has been the habit in many books, because the temperatures involved in process are often high, foe convenience the Fahrenheit temperatures are rounded upward or downward to the nearest 5 degrees.

This book presents to the reader an overview, with some degree of detail, of the complexities of the production and utilization of biogas. The book can be used as a reference source and checklist for all the considerations and actions necessary for the production of biogas. Furthermore, in this book, gases, when produced from a biomass sources, are collectively referred to as biogas and the prefix *bio* is often added to the name as a means of identifying the source of the gas – for example, bio-producer gas, bio-synthesis gas or bio-syngas). It is the purpose of this book to describe biogas in terms of production and composition followed by descriptions of the methods of production.

The book is designed to be suitable for undergraduate students, graduate students, technicians, professionals, and managers who are working with biogas or who are interested in entering the area of biogas technology. Each chapter includes a list of references that will guide the reader to more detailed information. In addition, a detailed Glossary is so included to assist the reader with any unknown or difficult terminology.

Dr. James G. Speight, PhD
CD&W Inc., Laramie,
Wyoming 82070, USA

PART 1: RAW MATERIALS

Chapter 1

BIOGAS

1. INTRODUCTION

As the need for the replacement of fossil fuels as energy sources continues to grow, gases products from other sources (such as biomass) will continue to grow in importance. In fact, biomass will play an increasingly important role in the future global energy infrastructure for the generation of power and heat, but also in the production of gaseous and liquid products for the production of fuels and chemicals and fuels. Biomass is a source of alternative energy insofar as it can be used for fuel substitution and is (unlike the fossil fuels sources) a non-depleted resource which can be renewed and o annual basis.

The dominant biomass conversion technologies will be anaerobic digestion, gas from landfills, pyrolysis gas, and gasification of biomass, with gasification processes likely to supersede anaerobic digestion for the production of gases as intermediates for high-efficient power production or for the synthesis of chemicals and fuels. Isolation of the gas produced from the organic waste in a landfill will also be of extremely high interest. Furthermore, in the discussion on the production of gases from biomass (other forms of organic feedstocks, such as organic waste are also included here), it is important to understand that the composition of the produced gas is very dependent on the type of gasification process and especially the gasification temperature (Table 1.1) and the predominance of the various reactions (Table 1.2) (Speight, 2008, 2013a, 2013b, 2014a, 2014b; Luque and Speight, 2015).

Biomass from arable land and from grass land in various countries is already, to some extent, less successfully used for biogas production and, within the biogas production and utilization scenarios, energy crops are currently at the center of interest. And, there is also the fear that the growth of crops for energy production will compete with the growth of crops for food production. At present the cultivation of energy crops takes place in competition with food production. Therefore, ways must be found to produce food on arable land and to use the surplus and residual substances in biogas plants. Therefore it is necessary to consider other processes by which gas can be produced from biological sources. One particular process is the gasification of biomass.

Table 1.1. Composition of Various Biogas Samples

Constituents	Household waste	Wastewater treatment plant sludge	Agricultural waste
Methane, % v/v	50 – 60	60 - 75	60 - 75
Carbon dioxide, % v/v	38 – 34	33 - 19	33 - 19
Nitrogen, % v/v	5 – 0	1 - 0	1 - 0
Oxygen, % v/v	0 – 1	< 0.5	< 0.5
Water, % v/v	6	6	6
Hydrogen sulfide, mg/m^3	100 - 900	1000 - 4000	3000 - 10 000
Ammonia, mg/m^3	-	-	50 - 100
Aromatics mg/m^3	0 - 200	-	-
Properties:			
Density*	0.93		0.85
Wobbe index**	6.9		8.1

*Natural gas: 0.57.

**Natural gas: 14.9.

Table 1.2. Reactions that Occur During Gasification of a Carbonaceous Feedstock

$2C + O_2 \rightarrow 2CO$

$C + O_2 \rightarrow CO_2$

$C + CO_2 \rightarrow 2CO$

$CO + H_2O \rightarrow CO_2 + H_2$ (shift reaction)

$C + H_2O \rightarrow CO + H_2$ (water gas reaction)

$C + 2H_2 \rightarrow CH_4$

$2H_2 + O_2 \rightarrow 2H_2O$

$CO + 2H2 \rightarrow CH_3OH$

$CO + 3H2 \rightarrow CH_4 + H_2O$ (methanation reaction)

$CO_2 + 4H_2 \rightarrow CH_4 + 2H_2O$

$C + 2H_2O \rightarrow 2H_2 + CO_2$

$2C + H_2 \rightarrow C_2H_2$

$CH_4 + 2H_2O \rightarrow CO_2 + 4H_2$

Within the context of this book, the term *biogas* is used to designate the mixture of different gases produced by the breakdown of organic matter in the absence of oxygen, usually referred to the anaerobic digestion of organic matter including manure, sewage sludge, municipal solid waste, biodegradable waste or any other biodegradable feedstock, under anaerobic conditions. Typically, biogas is comprised predominantly of methane (CH_4) and carbon dioxide (CO_2) with, depending upon the source of the gas, smaller amounts of carbon monoxide (CO), water (H_2O, as vapor), hydrogen sulfide (H_2S), nitrogen (N_2), hydrogen (H_2), mercaptan derivatives (RSH, also called thiol derivatives) and oxygen (O_2) (Table 1.1) (Angelidaki and Ellegaard, 2003; Regueiro et al., 2012; Ramírez et al., 2015; Recebli et al., 2015; Mustafa et al., 2016).

Gas can be produced from raw materials such as agricultural waste (including manure), municipal waste, industrial sewage, and many other forms of organic waste (Dublein and Steinhauer, 2008). Thus, the term *biogas* has been expanded herein to include the gases known as *landfill gas*, *pyrolysis gas*, and *bio-synthesis gas* (also often referred to as bio-syngas). In the latter case, the prefix *bio* is used to differentiate the gas from the synthesis gas produced from the fossil fuels: crude oil residua, tar sand bitumen, coal, and oil shale. Both of these bio-gases are produced by the breakdown (due to chemical reactions and microbes) of biodegradable waste within a landfill (hence the term landfill gas) or in a gasification process. Also, by way of further expansion of the definition of biogas, gaseous products from the pyrolysis of biomass and to gases produced by the gasification of biomass are also included. It is worthy of note that this point that the gasification of biomass leads to two categories of gaseous products. Hence the need to include in this book gases produced by the gasification of biomass as a falling under the umbrella of biogas.

The first category is the product gas, which is produced at relatively low temperature gasification (below 1000°C, 1830°F) and containing carbon monoxide, hydrogen, methane, higher molecular weight alkanes (C_nH_{2n+2}), some alkene (olefin) derivatives (C_nH_{2n}), and some aromatic hydrocarbon derivatives, such as benzene (C_6H_6), toluene ($C_6H_5CH_3$), and other generally complex mixtures known collectively as tar (besides CO_2 and H_2O) – the bio-syngas components H_2 and CO typically contain only approximately 50% of the energy in the gas, while the remainder is contained in CH_4 and higher (aromatic) hydrocarbons.

The second category is the bio-synthesis gas which is produced at a higher temperature (above 1200°C, 2190°F) in the absence of a catalyst (thermal gasification) or presence of a catalyst (catalytic gasification) and, under these conditions, the organic feedstock biomass is undergoes near-complete conversion into carbon monoxide and hydrogen (with some carbon dioxide and water). The gases produced in either of these categories need additional assiduous gas cleaning and conditioning to afford a gaseous product with the correct composition and specifications for the final application as a chemical feedstock (Chapter 5).

In terms of nomenclature, within the gasification industry (saw commercial birth with the gasification of coal for heat, power, and chemicals) (Speight, 2013a, 2013b), gas produced by any gasification process has been described by a variety of names: *producer gas, town gas, generator gas*, and *synthesis gas*. With the advent of, and growing interesting in, the gasification of biomass other terms such as *wood gas* as well as the use of the prefix bio have been used to describe the gas produced from a biomass source. All of these gases, when produced from biomass have fallen under the collective umbrella of *biogas*, though biogas more typically refers to gas produced via microbes in anaerobic digestion. In the context of biomass gasification using air-aspirated gasifiers, the term *producer gas* is also a common term since the other terms have implications that do not necessarily apply to the gas produced by biomass gasification. On occasion, biogas is also called producer gas, especially when the gas originate from a gasification process.

By way of clarification, producer gas is the mixture of gases produced by the gasification of organic material (such as biomass) at relatively low temperatures (700 to 1000°C, 1292 to 1830°F). Producer gas is composed of carbon monoxide (CO), hydrogen (H_2), carbon dioxide (CO_2) and typically a range of hydrocarbons such as methane (CH_4) with nitrogen from the air. Producer gas can be burned as a fuel gas such as in a boiler for heat or in an internal combustion gas engine for electricity generation or combined heat and power (CHP). The composition of the gas can be modified by manipulation of gasification parameters. On the other hand, synthesis gas (syngas) is a mixture of carbon monoxide (CO) and hydrogen (H_2), which is the product of high temperature steam or oxygen gasification of organic material such as biomass. Following clean-up of the gas stream to remove any impurities such as tars, syngas can be used to produce organic molecules such as synthetic natural gas [SNG-methane (CH_4)] or liquid biofuels such as synthetic diesel (via Fischer-Tropsch synthesis).

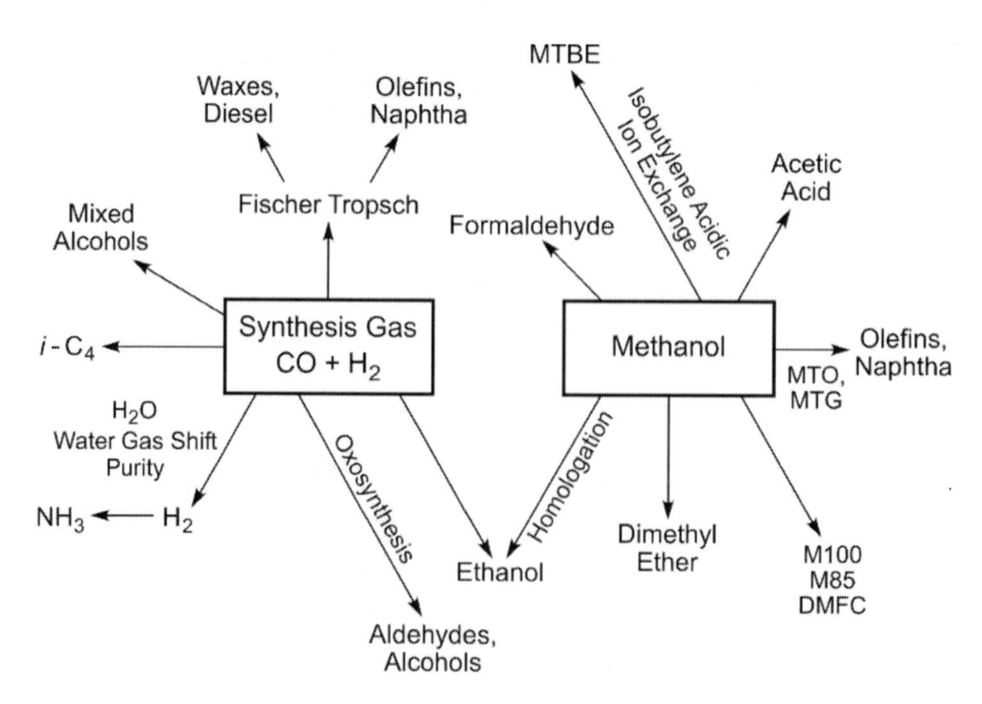

Figure 1.1. Routes to Chemicals from Synthesis Gas and Methanol.

Substitute natural gas (SNG) from biomass can also be considered a member of the biogas community. Substitute natural gas contains predominantly methane and resembles natural gas which, after cleaning (Chapter 7) is high purity (>95% v/v) methane. It is worth remembering here that methane is methane, whatever the source. Any differences in the behavior of the gas mixture that contains methane are caused by the types and amounts of non-methane contaminants. Substitute natural gas can be produced synthesized from synthesis gas but is typically present in the product gas from the gasification of a variety of carbonaceous feedstocks (Figure 1.1) (Speight, 2014b) – the product gas already contains significant amounts of methane.

In this text, all such gases, when produced from a biomass sources, are collectively referred to as biogas and the prefix *bio* is often added to the name as a means of identifying the source of the gas – for example, bio-producer gas, bio-synthesis gas or bio-syngas).

2. BIOGAS

Biogas is a combustible mixture of gases and consists mainly of methane (CH_4) and carbon dioxide (CO_2) and is formed from the anaerobic bacterial decomposition of organic compounds, i.e., without oxygen.

Table 1.3. General Properties of Methane

Chemical formula	CH_4
Molar mass	16.04 $g \cdot mol^{-1}$
Appearance	Colorless gas
Odor	Odorless*
Density	0.656 g/L (gas, 25°C, 1 atm/14.7 psi) 0.716 g/L (gas, 0°C/32°F, 1 atm/14.7 psi)
Melting point	-182.5°C; -296.4°F
Boiling point	-162°C; -260°F
Solubility in water	22.7 mg/L
Solubility	Soluble in ethanol, diethyl ether, benzene, toluene, methanol, acetone
Molecular shape	Tetrahedron
Flash point	−188 °C (-306.4°F)
Autoignition temperature	537°C (999°F)
Explosive limits	4.4 to 17% v/v in air

*Natural gas which is predominantly methane has an odor because of the addition of odiferous compounds (such as t-butyl thiol, also known as t-butyl mercaptan that has a skunk-like odor) that is added by the seller so that inherently dangerous gas leaks can be detected immediately.

The gases formed are the waste products of the respiration of these decomposer microorganisms and the composition of the gases depends on the substance that is being decomposed. If the material consists of mainly carbohydrates, such as glucose and other simple sugars and high-molecular compounds (polymers) such as cellulose and hemicellulose, the methane production is low. However, if the fat content is high, the methane production is likewise high.

Methane – a colorless and odorless gas with a boiling point of -162°C (-260°F) and it burns with a blue flame – is the major combustible constituents of biogas. Methane is also the main constituent (77 to 90% v/v) of natural gas. Chemically, methane belongs to the alkane series of hydrocarbons and is the simplest possible member of this series (C_nH_{2n+2}) form of these. At normal temperature and pressure, methane has a density of on the order of 0.66 to 0.72 g/L in the gas phase or 0.42 g/L in the liquid phase (Table 1.3). Due to carbon dioxide being somewhat heavier (Table 1.4), biogas has a slightly higher density than methane and is on the order of 1.15 g/liter. If biogas is mixed with 10 to 20% v/v air, there is the high probability of an explosion – explosive air is the name often applied to such a mixture.

Table 1.4. General Properties of Carbon Dioxide

Chemical formula	CO_2
Molar mass	44.01 g·mol^{-1}
Appearance	Colorless gas
Odor	Low concentrations: barely detectable High concentrations: sharp; acidic
Density	1.84 g/L (NTP*); 1.96 g/L (STP*)
Melting point	-56.6°C; -69.8°F
Solubility in water	1.45 g/L at 25°C (77°F)

*NTP: 0°C (32°F) and 1 atm (14.696 psi); STP: 15.6°C (60°F) and 1 atm (14.696 psi).

2.1. History

Historically, biogas was produced as a natural phenomenon and was not a designated industry. As already stated, biogas is produced naturally in swamps, bogs, rice paddies, and in the sediment at the bottom of lakes and oceans where anaerobic conditions prevail. Methane is also created in the rumen of ruminant animals (such as cows, sheep, deer, camels, and lamas).

It has been known from several centuries that combustible gas is generated when organic waste is allowed to rot in huge piles. It is also likely that Arab scientists new of such a gas in the 10th Century and, for example, it is documented that in the 17th Century, Van Helmont recorded that decaying organic material produced flammable gases. In 1776, Volta resolved that there was a direct connection between how much organic material was used and how much gas the material produced. That this combustible gas is methane was established by the work conducted independently by John Dalton and Humphrey Davy in the first decade of the 19th Century (Tietjen 1975). In 1868, it was observed that the formation of methane during the decomposition of organic matter was through a microbiological process and in the 1890s, microbes responsible for the release of hydrogen, acetic acid, and butyric acid during methane fermentation of cellulose were isolated. It was also observed that methane was formed due to micro-organism-mediated reaction between hydrogen and carbon dioxide (McCarty et al., 1982). Later, in 1910, it was also reported that fermentation of complex materials occurs through oxidation-reduction reactions to form hydrogen, carbon dioxide, and acetic acid and that hydrogen then reacts with carbon dioxide to form methane (McCarty et al., 1982).

One of the oldest biogas systems is the septic tank, which has been used for the treatment of wastewater since the end of the 19th Century and is still used for isolated land areas (such as isolated communities, farms or ranches) where there is no sewage removal system. This led to the design and construction of biogas systems and plants that started as early as the mid-19th Century. Also, in the 1890s, the Englishman Donald Cameron constructed a special septic tank, from which the gas was collected and used for street lighting. In some cases, the biogas was mixed with coal gas – the more popular gas for street lighting at that time.

In the 1920s, the construction of biogas plants for wastewater treatment started in Denmark and, in such cases, the gas was initially used to heat the digester tank and the main purpose was therefore not to extract energy from the gas but to decompose organic matter in the wastewater and thus reduce amount of sludge (a product of the treatment process) as well as to stabilize the sludge. In the following period and until shortly after the Second World War, there was a substantial growth in the biogas industry, particularly in Germany, Britain and France, and the technology also gradually found its way into agriculture and, during that period, energy production from the gas superseded all other uses.

In the 1950s, development of biogas source was at a low ebb because of the plentiful supply of cheap crude oil and natural gas. In addition, coal also played a role in the lessening of interest in the production and use of biogas. However, there was an awakening in the interest in biogas in the mid-1970s because of the following the oil crisis in 1973. In the decades since the 1970s, research and development programs related to biogas have been initiated with the aim of testing and constructing different types of biogas plants using various sources (animal manure in one such source) for the production of biogas. As a result, there are (currently) many biogas facilities installed at sewage treatment plants. In addition, communal biogas plants of various sizes have been constructed to treat manure and produce gas at many livestock farms. These biogas plants also supplement the manure feedstock by accepting organic waste from the food industry and slaughterhouses, whereby the energy from the waste is extracted and the nutrients recycled to the agricultural sector. In addition, on a world-wide basis, the interest in biogas has increased and there are on-farm facilities and a number of biogas plants associated with landfill sites and with different industries that produce waste water that has a high content of organic constituents.

But with the political agreement in many countries promoting energy policies related to green energy and on the realization that the economics of biogas production (and, hence, electricity production) are now favorable, the sector is making strong headway into various energy scenarios.

2.2. Origin

Biogas originates from the metabolic activities of bacteria in the process of biodegradation of organic material under anaerobic conditions (conditions without air and sometime referred to as fermentation). Thus, the natural generation of biogas is an important part of the biogeochemical carbon cycle, which is a pathway by which a chemical substance moves through the biosphere and lithosphere, atmosphere, and hydrosphere of the Earth. In addition to natural systems, biogas is produced in different anthropogenic environments such as landfills, waste water treatment plants (WWTP), and biowaste digesters during anaerobic degradation of organic material.

Table 1.5. Simplified Illustration of the Generation of Biogas

Substrate	Stage 1 products	Stage 2 products	Stage 3 products*
Organic waste	Carbon dioxide		Carbon dioxide
	Acetic acid		Methane
	Propionic acid	Carbon dioxide	Carbon dioxide
		Acetic acid	Methane
		Carbon dioxide	Carbon dioxide
	Butyric acid	Acetic acid	Methane
		Carbon dioxide	Carbon dioxide
		Acetic acid	Methane
Stage 1 catalyst: fermentive bacteria			
Stage 2 catalyst: acetogenic bacteria			
Stage 3 catalyst: methanogenic bacteria			
* Hydrogen is also produced in minor quantities in each stage			

Biogas is typically produced by anaerobic digestion (also called anaerobic fermentation) (Table 1.5), although other processes are known, which involves the activities of three different bacterial communities. In the biogeochemical cycle, methanogens (methane-producing bacteria) are the final link in a chain of micro-organisms which degrade organic material and return the decomposition products to the environment. It is in this process that biogas is generated and the constituents of biogas – methane, hydrogen, and carbon monoxide – can be combusted or oxidized with oxygen and it is this energy release which allows biogas to be used as a fuel. Biogas can be used as a low-cost fuel in any country for any heating purpose, such as cooking, and it can also be utilized in modern waste management facilities where it can be used to run any type of heat engine, to generate either mechanical or electrical power (Bove and Lunghi, 2005; Budzianowski, 2011; Niemczewska, 2012).

The process of biogas-production depends on various parameters (Chapter 3, Chapter 4) (Weiland, 2010; Nasir et al., 2012). For example, changes in ambient temperature can have a negative effect on bacterial activity. The microbes that produce biogas microbes consist of a large group of complex and differently acting microbe species, notably the methane-producing bacteria. The whole biogas-process can be divided into four steps: hydrolysis, acidification, acetogenesis, and methane formation. As a result of these four steps, biogas typically from different digcstible sources contains constituents that can range considerably – up to 70% v/v methane and as much as 45% v/v carbon dioxide when the methane contact is lower. Depending on the source, biogas can also contain, e.g., nitrogen, hydrogen sulfide, halogenated compounds and organic silicon compounds.

Biogas is considered a carbon dioxide-neutral biofuel and, if used as vehicle fuel, emits lower amounts of nitrogen oxide, hydrocarbon and carbon monoxide emissions than gasoline engines or diesel engines. Interest in the use of biogas as vehicle fuel and in fuel cells has also increased (Chapter 8). To be utilized for these purposes, in particular, but also increasingly for other purposes, a high methane concentration is required with minimum no-methane constituents and, consequently, the raw biogas has to be cleaned (upgraded). The number of technologies for upgrading biogas has increased rapidly in recent years (Chapter 7) along with knowledge about the use of biogas as a fuel for various vehicles (Chapter 8).

When biogas is used as an energy source, hydrogen sulfide and halogenated compounds, commonly found in biogas from landfills, can cause damage (such as corrosion) to engines. Also, depending on the particular application, biogas may need to be upgraded so that no compounds harmful to engines are combusted and released to the atmosphere as obnoxious gases, such as sulfur oxides and nitrogen oxides.

2.3. Composition

The composition of biogas is highly variable. For example, biogas from sewage digesters usually contains from 55 to 65% v/v methane, 35 to 45% v/v carbon dioxide, and <1% nitrogen whereas biogas from organic waste digesters usually contains from 60 to 70% methane, 30 to 40% carbon dioxide and <1% v/v nitrogen, while in biogas from landfills typically from 45 to 55% v/v methane, 30 to 40% v/v carbon dioxide, and 5 to 15% v/v nitrogen. Besides the main components, biogas also contains hydrogen sulfide and other sulfide compounds, aromatic compounds, and halogenated compounds, and siloxane derivatives. The latter compounds (i.e., halogenated compounds, and siloxane derivatives) are more common in landfill biogas than in biogas from the anaerobic digestion of manure (unless the livestock have been on a very mysterious diet).

Although the amounts of trace compounds are low compared to methane, they can have environmental impacts such as stratospheric ozone depletion, the greenhouse effect and/or the reduction in local air quality. In addition, some compounds cause engine damage leading to engine failure if the gas is used as an energy source. Many volatile organic compounds (VOCs) with high vapor pressure and low solubility, which can occur in biogas from some sources, are harmful to the environment and/or to human health. For example, aromatic derivatives, heterocyclic compounds, ketone derivatives, aliphatic derivatives, terpene derivatives, alcohol derivatives, and halogenated aliphatic derivatives, for example, occur in particular in landfill gas (Allen et al., 1997, Spiegel et al., 1997, Eklund et al., 1998, Shin et al., 2002, Jaffrin et al., 2003). Also, many toxic volatile organic compounds are formed from household waste which includes cleaning compounds, plastics, synthetic textiles, coatings, pesticide derivatives, and pharmaceutical derivatives, (Reinhart 1993).

The predominant sulfur compound in biogases is hydrogen sulfide (H_2S) but other sulfur compounds such as sulfide derivatives (-S-, including hydrogen sulfide, disulfide derivatives (-S-S-), and thiol derivatives (-SH, also called mercaptans) can also occur in biogas. In the presence of water, sulfur compounds can cause corrosion to compressors, gas storage tanks and engines and, thus, these compounds need to be removed before biogas can be utilized as energy. Under anaerobic conditions hydrogen sulfide and other sulfide compounds originate along several different pathways (Wilber and Murray 1990). For example, methane thiol (CH_3SH) and dimethyl sulfide (DMS, CH_3SCH_3) are formed by the degradation of sulfur-containing amino acids such as cysteine ($HO_2CCH(NH_2)CH_2SH$) that may be present in manure:

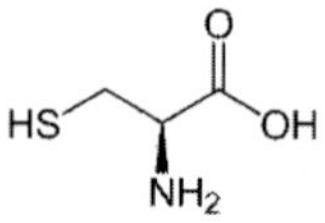

HS, OH, O, NH_2

Cysteine

Methanethiol (CH_3SH) and dimethyl sulfide (DMS, CH_3SCH_3) and also produced by the anaerobic methylation of sulfide derivatives. When dimethyl sulfide is reduced by the methanogenic bacteria, methane and methane thiol are formed.

$$CH_3SCH_3 + [2H] \rightarrow CH_3SH + CH_4$$

Methanethiol later forms methane, carbon dioxide and hydrogen sulfide (Lomans et al., 2002). Thus:

$$CH_3SH + [2H] + [2O] \rightarrow CH_4 + CO_2 + H_2S$$

Aromatic and chlorinated compounds are widely used in industry as solvents, and fluorinated compounds have been used as refrigerating aggregates, foaming agents, solvents and propellants (Scheutz et al., 2004) and, unless environmental laws are strictly enforced, levels of alkanes and aromatic compounds as well as those of halogenated and oxygenated compounds can occur in landfills and the amounts of these compounds in a landfill are dependent on the composition and stage of decomposition of waste (Allen et al., 1997; Jaffrin et al., 2003).

Halogenated compounds are often found in landfill gases and only rarely in biogases produced from sewage sludge or organic wastes. If this type of biogas is used for energy production, constituents containing organochloride derivatives can contribute to corrosion in combustion engines while in certain combustion conditions the formation of dioxins and furans is also possible (Allen et al., 1997). 1,2-Dioxin is a heterocyclic, organic compound with the chemical formula ($C_4H_4O_2$, 2-dioxin and 1,4-dioxin and their derivatives) are hazardous to human health.

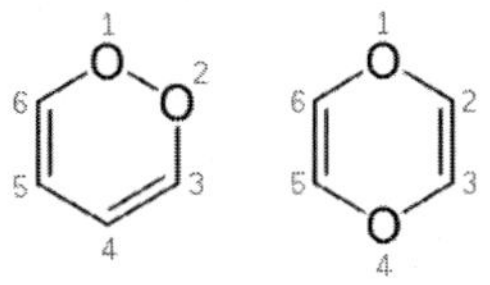

1,2-Dioxin 1,4-Dioxin

The subgroups of silicone derivatives containing silicon-oxygen (Si-O-Si) bonds with organic radicals bonded to Si are particularly troublesome in biogas. The organic radicals can include methyl (CH_3), ethyl (C_2H_5) as well as other organic functional groups and. Moreover, the structure of a siloxane can be linear or cyclic. Most siloxane derivatives have high vapor pressure and low water solubility, and they also have a high Henry's law constant. Henry’s law is a gas law which states that the amount of dissolved gas is proportional to its partial pressure in the gas phase. A high Henry’s law constant indicate siloxane derivatives can move easily from water to air. In waste water digesters and landfills, siloxane derivatives are volatilized into biogas.

Organic silicon compounds present in biogas are oxidized during biogas combustion into microcrystalline silicon dioxide (SiO_2), a residue with chemical and physical properties similar to glass. Silicon dioxide collects in deposits on valves, cylinder walls, and liners, causing abrasion and blockage of pistons, cylinder heads, and valves. In gas turbines, siloxane deposits usually form on the nozzles and blades, causing erosion of the turbine blades and subsequently lowering operating efficiency. Moreover, the glass-like residues of silicon dioxide can de-activate the surface of the catalyst in an emission control system (Schweigkofler and Niessner 1999). In addition, there is a correlation between the increasing emissions of carbon monoxide and the build-up of silicate-based deposits from siloxane combustion in generator engines. Some organic silicon compounds can also appear in the engine oil after the combustion process – in this case, the engine oil needs to be changed more frequently and the engine will need attention.

Along with the most common silicon compounds in biogases, biogases may also contain organic silicon compounds other than siloxane derivatives that cause similar detrimental effects in the combustion process. For example methoxy trimethyl silane, tetramethylsilane, trimethyl fluorosilane, and trimethyl propoxysilane have been detected in waste water digester biogases.

The use of siloxane derivatives is increasing, for example, in household industrial cleaning products because volatile methyl siloxane derivatives (VMSs) solvents are aroma free and widely available. They can also originate from the hydrolysis of polydimethylsiloxane (PDMS), which is an organosilicon compound that is the most widely used silicon compound in household and industrial applications. It is used, for example, as a softener and wetting agent in textile manufacturing, as a component of many surface treatment formulations, and in many domestic products such as shampoos, deodorants, and gels. Most siloxane derivatives are very volatile and decompose in the atmosphere into silanol derivatives, which are eventually oxidized into carbon dioxide. These compounds are insoluble in water and have a high adsorption coefficient. Some siloxane derivatives end up in waste water and are adsorbed onto the extracellular polymeric substances (EPS) of sludge flocks.

During anaerobic digestion, siloxane derivatives are volatilized from the sludge during anaerobic digestion and end up in biogas (Dewil et al., 2006). Silicone derivatives may occur in digester sludge having been added to the digester as anti-foaming agents, where they can biodegrade into siloxane derivatives (Dewil et al., 2006). Organic silicon compounds end up in landfills from sources such as shampoo bottles and other containers in which some of the product remains, through landfilling of waste water treatment sludge, and from packaging, and construction materials. The degradation of high molecular silico-organic compounds in landfills may also form volatile silicon compounds (Schweigkofler and Niessner 1999) and, because of the widespread use, siloxane derivatives are commonly found in air, water, sediment, sludge, and biota, and the variation in concentrations can be high.

In chemical terms, biogas which is primarily methane (CH_4) and carbon dioxide (CO_2) and may, depending upon the source and the production process, also contain varying amounts of hydrogen sulfide (H_2S), moisture (H_2O), and siloxane derivatives which contain the silicon-oxygen bond. Thus:

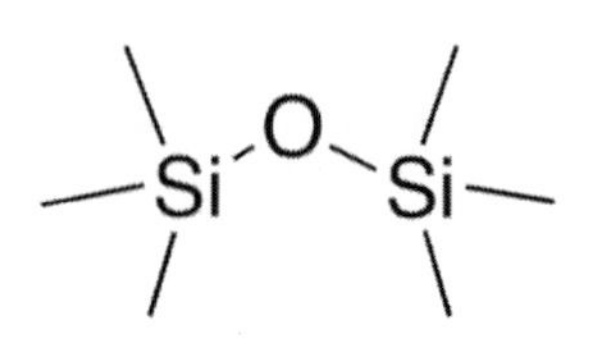

By way of further clarification, a siloxane is a functional group in organosilicon chemistry and the parent siloxane derivatives include the oligomeric and polymeric hydride derivatives which have the formula $H(OSiH_2)_nOH$ and $(OSiH_2)_n$, respectively. Siloxane derivatives also include branched compounds, the defining feature of which is that each pair of silicon centers is separated by one oxygen atom. The siloxane functional group forms the backbone of silicone polymers of which an example is polydimethylsiloxane:

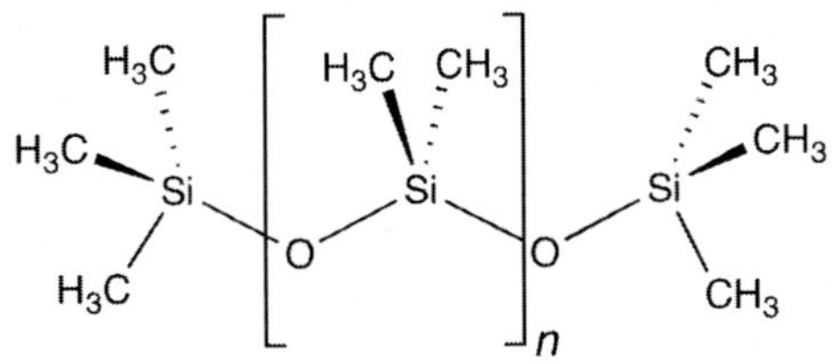

In biogas, the gases methane, hydrogen, and carbon monoxide (CO) can be combusted or thermally-oxidized with oxygen which releases energy:

$$CH_4 + 2O_2 \rightarrow CO_2 + 2H_2O + \text{energy}$$
$$2H_2 + O_2 \rightarrow 2H_2O + \text{energy}$$
$$2CO + O_2 \rightarrow 2CO_2 + \text{energy}$$

This energy release from the above reactions allows biogas to be used as a fuel – t can be used for any heating purpose, such as cooking and can also be used in a gas engine to convert the energy in the gas into electricity and heat. As already noted, the composition of biogas varies depending upon the substrate composition, as well as the conditions within the anaerobic reactor (temperature, pH, and substrate concentration) and, hence, only ranges of composition can be given (Table 1.6).

The origin of the siloxane derivatives in biogas is due to the anaerobic decomposition of silicon-based constituents that are commonly found in soaps and detergents. During combustion of biogas containing siloxane derivatives, silicon is released and can combine with free oxygen or other elements in the combustion gas leading to deposits that contain predominantly silica (SiO_2) or silicate derivatives (Si_xO_y) and can also contain calcium, sulfur, zinc, and phosphorus.

High levels of methane are produced when manure is stored under anaerobic conditions. In addition, during storage and when manure has been applied to the land, nitrous oxide (N_2O) is also produced as a byproduct of the denitrification process – nitrous oxide is 320 times more aggressive as a greenhouse gas than carbon dioxide and methane 25 times more than carbon dioxide.

Table 1.6. Ranges of Composition of Biogas

	Landfill gas	Anaerobic digester gas	Natural gas
Density	1.3	1.1	0.82
Relative density (air = 1.0)	1.1	0.9	0.63
Wobbe index	18	27	55
Methane number*	>130	>135	73
Methane, % v/v	35-65	60-70	85-92
Heavy hydrocarbons	0	0	9
Hydrogen, % v/v	0-3	0	-
Carbon dioxide, % v/v	15-40	30-40	0.2-1.5
Nitrogen, % v/v	5-40	-	0.3-1.0
Oxygen, % v/v	0-5	-	-
Hydrogen sulfide, ppm v/v	0-100	0-4000	1.1-5.9
Ammonia, ppm v/v	5	100	-
Total chlorine, ppm v/v	20-200	0-5	-

*A measure of the gas resistance to knocking in an internal combustion engine; see Malenshek, M., and Olen, D.B. 2009. Methane Number Testing of Alternative Gaseous Fuels. Fuel, 88: 650-656.

3. Landfill Gas

Landfill gas (LFG) is a natural byproduct of the decomposition of organic material in landfills. Landfill gas is composed of methane (the primary component of natural gas) and carbon dioxide as well as non-methane organic compounds (Table 1.7). Landfill gas is a type of biogas generated from the decomposition of the biodegradable organic fraction (BOF) of municipal solid waste by microorganisms. This generally occurs under semi-controlled conditions in a landfill; its constituents depend on the composition and age of the waste. The landfill gas is produced by wet organic waste decomposing under anaerobic conditions in a biogas. The waste is covered and mechanically compressed by the weight of the material that is deposited above. This material prevents oxygen exposure thus allowing anaerobic microbes to thrive.

The gas builds up and is slowly released into the atmosphere if the site has not been engineered to capture the gas and is released in an uncontrolled way leading to hazardous situations – the gas can be explosive when it escapes from the landfill and mixes with oxygen. The lower explosive limit of methane is 5% v/v in air and upper explosive limit is 15% v/v methane in air. However as a cautionary note, methane (as a greenhouse gas) is twenty eight times more potent than carbon dioxide. Large variations may exist in the composition of landfill gas due to differences in sources of municipal solid waste and operating conditions at the landfill. The three main gas constituents – methane (CH_4), carbon dioxide (CO_2), and hydrogen sulfide (H_2S)] are used to characterize landfill gas (Table 6.1).

Table 1.7. Some Physical Characteristics of the Constituents of Landfill Biogas

Characteristics	Unit	CO_2	CH_4*	H_2S**
Molecular weight	G	44.1	16.04	34.08
Boiling point (14.7 psi)	°C	-164.0	-161.61	-59.6
Freezing point at (14.7 psi)	°C	-78.0	-182.5	-82.9
Relative density (15°C, 59°F) (air = 1.0)		1.53	0.555	1.189
Density at 0°C, 32°F	Kg/m^3	1.85	0.719	1.539
Flammable limits in air	% v/	-	5.3-14	4.3-45
Solubility in water	Kg/m^3	4.0	24.0	3.4

*Combustion: $CH_4 + 2O_2 \rightarrow CO_2 + 2H_2O$.

**Combustion: $H_2S + 2O_2 \rightarrow SO_3 + H_2O$.

Typically landfill gas is composed of 45 to 60% v/v methane, 40 to 60% v/v carbon dioxide, 0 to 1.0% v/v hydrogen sulfide, 0 to 0.2% v/v hydrogen (H_2), trace amounts of nitrogen (N_2), low molecular weight hydrocarbons (dry volume basis) and water vapor (saturated) (Kreith and Goswami, 2007). The production rates for landfill gas vary spatially within the waste mass in the landfill and its movement is by diffusion (concentration) and convection (pressure). Landfill gas moves through the path of least resistance and several concerns are noted.

Landfill gas contains methane which is highly combustible, making it a potential fire and explosion hazard in the landfill environment or in adjacent structures. Also, landfill gas is capable of migrating significant distances through soil, increasing the risk of explosion and exposure; as a result, serious accidents (injury, loss of life, property damage) may occur when landfill conditions favor gas migration. As landfill gas is produced, the pressure gradient upward may create cracks and disrupt the geo-membrane in the landfill cover. Methane, a component of landfill gas, is an asphyxiant in high concentrations and is *greenhouse gas* that is 23 times more potent than carbon dioxide as a contributor to global warming, (v) migrating gas may cause adverse effects such as stress to vegetation, by lowering the oxygen content of soil gas available in the root zone – plant roots require sufficient oxygen to carry out normal respiration processes, and methane acts as an asphyxiant to roots. The gas generated at landfills and vented to the atmosphere emits nuisance odors (such as the odor of hydrogen sulfide) and septic soil. Emissions of non-methane organic compounds in landfill gas contribute to the degradation of local air quality. Vinyl chloride has been found to be present in substantial concentrations in landfill gas, resulting in health and safety concerns. Finally, uncontrolled landfill gas is a loss of potential resources; instead, it can be a satisfactory fuel for boilers, gas turbines, internal combustion engines, or upgraded to pipeline quality gas and distributed to consumers (Chapter 8).

In most countries government regulations require control of landfill gas and require gas management systems as a component of the landfill cover insofar as owner/operators of municipal solid waste landfill sites must ensure that the concentration of methane generated by the facility does not exceed 25% of the lower explosion limit (LEL) for methane in facility structures (excluding gas control or recovery system components) or the lower explosion limit at the facility property boundary.

The lower explosion limit is the lowest percent by volume of a mixture of explosive gases in air that will propagate a flame at 25°C (77°F) and atmospheric pressure. Methane is explosive when present in the range 5 to 15% v/v in air. At methane concentrations in excess of 15% v/v, the gas mixture will not explode as the gaseous mixture is considered 'rich' from a methane perspective. This 15% threshold is the upper explosive limit (UEL) defined as the maximum concentration of a gas, above which the substance will not explode when exposed to an ignition source. The explosive hazard range occurs between the lower explosive limit and the upper explosive limit. Methane concentrations below the upper explosive limit, fire may still be possible and asphyxiation will occur. A sudden dilution of methane can bring the mixture back within an explosive range. Also, landfill gas which escapes into the atmosphere may significantly contribute to environmental effects. In addition, volatile organic compounds (VOCs) in landfill gas contribute to the formation of photochemical smog.

Briefly, smog is a type of severe air pollution – the word *smog* originated in the mid-20th Century as a blending of the words smoke and fog to refer to smoky fog, its opacity, and odor. This kind of visible air pollution is composed of nitrogen oxides, sulfur oxides, ozone, smoke, and particulate matter. Man-made (anthropogenic) smog is derived from coal combustion emissions, vehicular emissions, industrial emissions, forest and agricultural fires and photochemical reactions of these emissions. Photochemical smog, is a type of air pollution derived from vehicular emission from internal combustion engines and from industrial fumes. These pollutants react in the atmosphere with sunlight to form secondary pollutants that also combine with the primary emissions to form the photochemical smog. The developing smog is usually toxic to humans and can cause severe sickness, a shortened life span, or premature death.

Landfill gas has a promising future as a synthetic fuel and a source of renewable energy; it is produced from the anaerobic digestion of the biodegradable organic constituents in solid waste and comprises methane, carbon dioxide, water vapor and lesser amounts of other gases (Chapter 7). Landfill gas production rates vary spatially within the landfill and mathematical modeling software is used to provide potential estimates from waste in place.

Gas migration takes place by diffusion and convection and landfill gas moves through the path of least resistance, uncontrolled gas migration is undesirable as it can result in explosions, fire, odor, physical disruption/damage to landfill cover and toxic vapor emissions. Given such a backdrop passive and active gas collection systems are usually installed. At larger landfills the gas could be processed and used for energy production; it may also generate revenue to help offset closure costs. Bioreactor landfills provide an innovative way to achieve rapid degradation and waste stabilization and enhanced landfill gas generation within a relatively short time however they require careful monitoring and control for optimum performance. Over the long term they provide considerable environmental protection, cost savings and revenue potential to help defray operational costs. Landfill mining also provides an opportunity to recover carbonaceous material that could be put to incineration with energy recovery in a waste to energy plant.

4. SYNTHESIS GAS

Synthesis gas, a mixture composed primarily of carbon monoxide and hydrogen but also water, carbon dioxide, nitrogen and methane, has been produced on a commercial scale since the early part of the 20th Century (Speight, 20134; Luque and Speight, 2015). The gas can be used for producing a liquid hydrocarbon mixture in the gasoline range using Fischer-Tropsch technology.

Gasification to produce synthesis gas can proceed from just about any organic material, including biomass and plastic waste. The resulting synthesis gas burns cleanly into water vapor and carbon dioxide.

$$C + O_2 \rightarrow CO_2$$
$$CO_2 + C \rightarrow CO_2$$
$$C + H_2O \rightarrow CO + H_2$$

Alternatively, the synthesis gas may be converted efficiently to methane by the Sabatier reaction or to a diesel-like synthetic fuel via the Fischer Tropsch reaction.

The Sabatier reaction (or the Sabatier process) involves the reaction of hydrogen with carbon dioxide at elevated temperatures (300 to 400°C, 570 to 750°F) and pressures in the presence of a nickel catalyst to produce methane and water:

$$CO_2 + 4H_2 \rightarrow CH_4 + 4H_2O$$

On the other hand, the Fischer-Tropsch reaction (the Fischer–Tropsch process) involves a series of chemical reactions that produce a variety of hydrocarbons, ideally having the formula (C_nH_{2n+2}) (Table 1.8). Thus:

$$(2n + 1)H_2 + nCO \rightarrow C_nH_{2n+2} + nH_2O$$

In this equation, n is typically 10 to 20. The formation of methane ($n = 1$) is unwanted and most of the alkane derivatives produced tend to be straight-chain, suitable as diesel fuel. In addition to alkane formation, competing reactions give small amounts of alkene derivatives, as well as alcohol derivatives and other oxygenated products.

In principle, synthesis gas can be produced from any hydrocarbon feedstock Luque and Speight, 2015). These include: natural gas, naphtha, residual oil, petroleum coke, coal, biomass, and more pertinent to the present context, from biogas, in which case it is often referred to as bio-synthesis or bio-syngas. Synthesis gas, from whatever the source, is combustible and often used as a fuel source or as an intermediate for the production of other chemicals.

Gasification processes are used to convert a carbon-containing (carbonaceous) material into a synthesis gas (Chapter 6) which has a relatively low calorific value, ranging from 100 to 300 Btu/ft^3. The uses of bio-synthesis gas include a feedstock for petrochemical processes and in gas-to-liquid processes, which employ Fisher-Tropsch chemistry (Chapter 6) to produce liquid fuels, as feedstock for chemical synthesis, as well as being used in the production of fuel additives, including diethyl ether and methyl t-butyl ether (MTBE), acetic acid and its anhydride, synthesis gas could also make an important contribution to chemical synthesis through conversion to methanol (Figure 1.1).

The composition of the products from the synthesis gas production processes from biomass is varied insofar as the gas composition varies with the type of biomass feedstock and the gasification system employed (Speight, 2013; Luque and Speight, 2015; Speight, 2014a, 2014b). Furthermore, the quality of gaseous product(s) must be improved by removal of any pollutants such as particulate matter and sulfur compounds before further use (Chapter 7), particularly when the intended use is a water gas shift or methanation reaction.

5. Upgrading

Raw biogas produced from digestion is approximately 60% v/v methane and 29% v/v carbon dioxide with trace elements of hydrogen sulfide and is inadequate for immediate use because of the corrosive nature of hydrogen sulfide. The property of biogas depends largely on the quantity of combustible methane while the amount of carbon dioxide has no impact on the biogas property due to its lacking combustibility. Organic substances that are typically found in waste include protein, carbohydrate, and fat compound, both in the forms of solid and solution.

Biogas cleaning (Chapter 7), which is the first step in the biogas upgrading process, has gained increased attention due to the recognition that as conventional fossil fuel sources of gas are being depleted the sources of biogas – the various forms of biomass (Chapter 2) – offer an unlimited supply of feedstocks for the production of the gas. It is therefore important to have an optimized upgrading process in terms yielding high methane content in the upgraded gas. It is also very important to minimize, or if possible avoid, emissions of methane during the upgrading process, since methane has a greenhouse gas effect that is twenty three times greater than the greenhouse gas effect of carbon dioxide.

Biogas can be cleaned in a conventional gas processing facility with the recognition that the main difference in the composition between biogas and natural gas relates to the carbon dioxide content. Carbon dioxide is one of the main components of biogas, while natural gas contains lower amounts carbon dioxide. In addition, natural gas also contains higher levels of hydrocarbons other than methane. These differences result in a lower energy content of biogas per unit volume compared to natural gas and by separating carbon dioxide from the biogas in an upgrading process, the energy content of upgraded biogas becomes comparable to natural gas.

Table 1.7. Carbon Chain Groups of the Range Which Can Be Produced by the of Fischer-Tropsch Reaction

Carbon number	Group name
C1-C2	Synthetic natural gas (SNG)
C3-C4	Liquefied petroleum gas (LPG)
C5-C7	Low-boiling naphtha
C8-C10	High-boiling naphtha
C11-C16	Middle distillate
C17-C30*	Low-melting wax
C31-C60	High-melting wax

*The C17 n-alkane (n-heptadecane) is the first member of the alkane series that is not fully liquid under ambient conditions (melting point: 21°C (70°F).

Biogas from anaerobic digestion and landfills consists primarily of methane (CH_4) carbon dioxide (CO_2). Trace components that are often present in biogas are water vapor, hydrogen sulfide, siloxanes, hydrocarbons, ammonia, oxygen, carbon monoxide and nitrogen. The end result is that the biogas for use must be very clean to reach pipeline quality and must be of the correct composition for the gas distribution network to accept – any carbon dioxide, water, and hydrogen sulfide.

In order to transfer biogas into biomethane, two major steps are performed: (i) a cleaning process to remove the trace components and (ii) an upgrading process to adjust the calorific value. Upgrading is generally performed in order to meet the standards for use as vehicle fuel or for injection in the natural gas grid. Different methods for biogas cleaning and upgrading are used. They differ in functioning, the necessary quality conditions of the incoming gas, the efficiency and their operational bottlenecks. Condensation methods (demisters, cyclone separators or moisture traps) and drying methods (adsorption or absorption) are used to remove water in combination with foam and dust.

A number of techniques have been developed to remove hydrogen sulfide (H_2S) from natural gas streams which are also applicable to biogas streams (Chapter 7) (Kohl and Nielsen, 1997; Mokhatab et al., 2006; Speight, 2014a; Miltner et al., 2017; Speight, 2019). As a preliminary step, provided it does not interfere with the digestion process, air dosing and addition of iron chloride into the digester tank that can remove hydrogen sulfide during the digestion period. Also, techniques such as adsorption on iron oxide pellets and absorption in liquids remove hydrogen sulfide after digestion. Subsequently, trace components such as siloxane derivatives, hydrocarbon derivatives, ammonia, oxygen, carbon monoxide, and nitrogen will require extra removal steps. Finally, methane must be separated from carbon dioxide using pressure swing adsorption, membrane separation, physical or chemical carbon dioxide-absorption (Table 1.9) (Chapter 7) (Kohl and Nielsen, 1997; Mokhatab et al., 2006; Speight, 2014a; Miltner et al., 2017; Speight, 2019).

Biogas upgrading processes need to be designed to avoid or minimize emissions. For example, in absorption processes some of the methane can be absorbed to the absorption liquid and be released into the air with the exhaust gas. Hydrogen sulfide, as well as other reduced sulfur compounds and halogenated compounds in the exhaust gas, can have environmental and health risks if released into the air untreated. The absorption liquid used in the process should also be treated, for example with other waste waters.

In general, there are four main methods of upgrading (also called gas processing and gas refining): (i) water washing, (ii) pressure swing absorption, (iii) absorption, and amine gas treating (Table 1.9) (Chapter 7) (Kohl and Nielsen, 1997; Mokhatab et al., 2006; Speight, 2014a; Miltner et al., 2017; Speight, 2019). In addition to these processes, the use of membrane separation technology for biogas upgrading is increasing. The most prevalent method is water washing where high pressure gas flows into a column where the carbon dioxide and other trace elements are scrubbed by cascading water running counter-flow to the gas. This arrangement could deliver methane (98% v/v purity).

Briefly, in the context of this chapter, the process whereby pollutants from the gas stream are dissolved into a solvent liquid stream by mass transfer involves physical absorption. The difference in concentration of the solute between the gaseous and liquid phases is the driving force for mass transfer – this is chemical absorption if a chemical reaction between the pollutant from the gas stream and the component of the liquid phase occurs and the reaction can be reversible or irreversible (Kohl and Nielsen, 1997; Mokhatab et al., 2006; Speight, 2014a; Miltner et al., 2017; Speight, 2019). The solubility of gases into water is dependent on several factors, such as pressure, temperature, liquid/gas ratio etc. According to Henry's law there is a linear relationship between the partial pressure of a gas and its concentration of the gas in dilute solution.

There are several absorber designs, but the primary function of all is to increase the area of contact between the liquid and the gas phases under conditions favoring mass transfer (Kohl and Nielsen, 1997; Mokhatab et al., 2006; Speight, 2014a; Miltner et al., 2017; Speight, 2019). This is usually achieved by bubblizing the gas (i.e., dividing the gas into small bubbles) in a continuous liquid phase, spreading the liquid into films that flow through a continuous gas phase or forming the liquid into small drops in a continuous gas phase (Kohl and Nielsen, 1997). A solvent to absorption process in which the organic contaminant is as soluble as possible should be chosen to decrease the size of the absorption equipment and increase the effectiveness of the process. Liquid and gas should be distributed evenly in the absorption column to avoid gas channeling, which causes a decrease in the removal efficiency of gas components (Kohl and Nielsen, 1997; Hunter and Oyama, 2000; Mokhatab et al., 2006; Speight, 2014a; Miltner et al., 2017; Speight, 2019).

Table 1.8. Summary of Processes Used for Cleaning Biogas

Separation concept	Process	Principle
Adsorption	Pressure Swing adsorption	Adsorption of CO_2 using a molecular sieve
Physical absorption	Pressurized water wash	Dissolution of CO_2
	Selexol, Rectisol, Purisol processes	Dissolution of CO_2 in a specialized solvent
Chemical absorption	Alkanolamine wash	Chemical reaction of CO_2 with the alkanolamine
Membrane separation	Polymer membrane (dry separation)	Membrane permeability of H_2S and CO_2 is higher than CH_4
Cryogenic process	Low temperature process	Phase transformation of CO_2 to liquid
		CH_4 remains gaseous

A counter-current flow column has a high absorption efficiency because it provides a long contact time between the vapor and liquid phases. This type of column is often used in biogas purification systems, as it is considered to be a simple, efficient and versatile method capable of treating corrosive compounds at relatively low cost. The biogas is usually pressurized and channeled to the bottom of the counter-current absorption column and the solvent is sprayed into the column from the top. The cleaned biomethane is taken from the top of the column after which the biomethane is dried. The used solvent can be regenerated in the desorption column at low pressure and re-used in the absorption process. In addition, packed columns can be used which can also have a co-current flow and the tower can be vertical or horizontal. Other absorption designs include trays or open vessels (Kohl and Nielsen, 1997; Hunter and Oyama, 2000; Mokhatab et al., 2006; Speight, 2014a; Miltner et al., 2017; Speight, 2019).

In the biogas upgrading scheme, water is the most common solvent used and, depending on the site, pure water or treated water from a waste water treatment plant can be used with flow-through or with water circulation. Water washing is an effective upgrading process for biogas produced from different sources – carbon dioxide and hydrogen sulfide are more soluble in water than methane and thus it is possible to produce high quality methane gas. Other impurities such as ammonia, sulfur dioxide, hydrogen chloride, hydrogen fluoride, chlorine, aldehydes, organic acids, alcohols, silicon tetrafluoride, and silicon tetrachloride have also been removed by means of water scrubbing (Kohl and Nielsen 1997, Mokhatab et al., 2006; Speight, 2014a; Miltner et al., 2017; Speight, 2019).

The commonly used organic solvents include polyethylene glycol derivatives and alkanolamine derivatives (Kohl and Nielsen, 1997; Mokhatab et al., 2006; Speight, 2014a; Miltner et al., 2017; Speight, 2019). . Carbon dioxide and hydrogen sulfide are more soluble in organic solvents than in water, and thus the upgrading process is more efficient. During the absorption process some of methane may also dissolve into the absorption liquid. As a consequence, in many plants, a flash tank is used to recover methane from the absorbing liquid at a low-to-intermediate pressure (30 to 60 psi) thereby decreasing the potential losses of the bio-methane.

Finally, the methane content in any gas reject stream, such as in the water from a water scrubber, or in any other stream leaving the upgrading plant should be minimized. Several techniques for biogas upgrading exist today and they are continually being improved. In parallel, new techniques for gas cleaning are under development (Chapter 7) (Kohl and Nielsen, 1997; Mokhatab et al., 2006; Speight, 2014a; Miltner et al., 2017; Speight, 2019). These new developments can also lead to other advantages such as lower methane emission which is important from both an economical and environmental perspective. Generally, methane in biogas can be concentrated via a biogas upgrader to the same standards as natural gas, which itself has to go through a cleaning process, and is called (for source identification purposes) *biomethane* (Ryckebosch et al., 2011)

Throughout the gas cleaning operation, it is essential to minimize any losses of methane not only for economic reasons but also for environmental reasons. Thus the release of methane to the atmosphere should be minimized by treating the off-gas, air or water streams leaving the plant even though the methane cannot be utilized. In short, although methane is a valuable renewable energy source, it is also a harmful greenhouse gas if emitted into the atmosphere. In any biogas cleaning plant, the gas produced is generally used for on-site energy production. Also, in several countries methane-rich gas from landfills is not used for energy production, it must be collected and disposed of in an acceptable manner to prevent release of methane (a potent greenhouse gas) into the atmosphere.

REFERENCES

Allen, M. R., Braithwaite, A. & Hills, C. C. 1997. Trace Organic Compounds in Landfill Gas at Seven U. K. Waste Disposal Sites. *Environ. Sci. Technol*, 31: 1054-1061.

Angelidaki, I. and Ellegaard, L. 2003. Co-digestion of Manure and Organic Wastes in Centralized Biogas Plants. *Appl. Biochem. Biotechnol*, 109(1-3): 95-105.

Bove R., and Lunghi P. 2005. Electric Power Generation from Biogas Using Traditional and Innovative Technologies. *Energy Conversion & Management*, 47, 1391- 1401, 2005.

Budzianowski W. 2011. Sustainable Biogas Energy in Poland: Prospects and Challenges. *Renewable and Sustainable Energy Reviews*, 2011.

Dewil, R., Appels, L., and Baeyens, J. 2006. Energy Use of Biogas Hampered by the Presence of Siloxanes. *Energy Convers. Manage,* 47:1711-1722.

Dublein, D. and Steinhauer, A. 2008. *Biogas from Waste and Renewable Resources.* Wiley-VCH, John Wiley & Sons Inc., Hoboken, New Jersey.

Eklund, B., Anderson, E. P., Walker, B. L. & Burrows, D. B. 1998. Characterization of landfill gas composition at the fresh kills municipal solid-waste landfill. *Environ. Sci. Technol,* 32: 2233-2237.

Hunter, P., and Oyama, S. T. 2000. *Control of Volatile Organic Compound Emissions. Conventional and Emerging Technologies*. John Wiley & Sons Inc., Hoboken, New Jersey.

Jaffrin, A., Bentounes, N., Joan, A. M. & Makhlouf, S. 2003. Landfill Biogas for Heating Greenhouses and Providing Carbon Dioxide Supplement for Plant Growth. *Biosyst. Eng,* 86: 113-123.

Kohl, A., and Nielsen, R. 1997. *Gas Purification.* 5th Edition. Gulf Publishing Company, Houston, Texas.

Lomans, B. P., Pol, A., and Op den Camp, H. J. M. 2002. Microbial Cycling of Volatile Organic Sulfur Compounds in Anoxic Environments. *Water Sci. Technol*, 45(10): 55-60.

Luque, R., and Speight, J. G. (Editors). 2015. *Gasification for Synthetic Fuel Production: Fundamentals, Processes, and Applications*. Woodhead Publishing, Elsevier, Cambridge, United Kingdom.

McCarty, P. L. 1982 One Hundred Years of Anaerobic Treatment. In: *Anaerobic Digestion, Proceedings. The Second International Symposium on Anaerobic Digestion 1981*. Hughes, D, E., Stafford, D., A., Wheatley, B. I. (Editor). Elsevier Biomedical, Amsterdam, Netherlands. Page 3-22.

Miltner, M., Makaruk, A., and Karasek, M. 2017. Review on Available Biogas Upgrading Technologies And Innovations Towards Advanced Solutions. *Journal of Cleaner Production,* 161(10): 1329-1337.

Mokhatab, S., Poe, W. A., and Speight, J. G. 2006. *Handbook of Natural Gas Transmission and Processing*. Elsevier, Amsterdam, Netherlands.

Mustafa, M. Y., Calay, T. K., and Román, E. 2016. Biogas from Organic Waste – A Case Study. *Procedia Engineering,* 146: 310-317.

Nasir, I. M., Mohd, Ghazi. T. I. and Omar, R. 2012. Anaerobic Digestion Technology in Livestock Manure Treatment for Biogas Production: *A Review. Eng. Life Sci.*, 12(3): 258-69.

Niemczewska, J. 2012. Characteristics of Utilization of Biogas Technology. *Nafta-Gaz*, LXVIII(5): 293-297.

Persson M., Wellinger A. 2006. *Biogas Upgrading and Utilization, IEA Bioenergy*. International Energy Agency, London, United Kingdom.

Ramírez M., Gómez J. M. and Cantero D. 2015. Biogas: Sources, Purification and Uses. In: *Hydrogen and Other Technologies*. U. C. Sharma, S. Kumar, R. Prasad (Editors). Studium Press LLC, New Delhi, India. Chapter: 13, page 296-323.

Rauch, R. 2001. Biomass Gasification to produce Synthesis Gas for Fuel Cells, Liquid Fuels and Chemicals, IEA Bioenergy Agreement, Task 33: *Thermal Gasification of Biomass.*

Recebli, Z., Selimli, S., Ozkaymak, M., and Gonc, O. 2015. Biogas Production from Animal Manure. *Journal of Engineering Science and Technology*, 10(6): 722-729.

Regueiro, L., Carballa, M., Alvarez, J. A. and Lema, J. M. 2012. Enhanced Methane Production from Pig Manure Anaerobic Digestion Using Fish and Biodiesel Wastes as Co-Substrates. *Bioresour. Technol*, 123: 507-513.

Reinhart, D. R 1993. A review of recent studies on the sources of hazardous compounds emitted from solid waste landfills: A U.S. experience. *Waste Manage. Res*, 11: 257-268.

Röshe, L.; John, P., and Reitmeier, R. 2003. Organic Silicon Compounds. In: *Ullmann's Encyclopedia of Industrial Chemistry*. John Wiley and Sons, Hoboken, New Jersey.

Ryckebosch E., Drouillon, M., Vervaeren H. 2011. Techniques for Transformation of Biogas to Biomethane. *Biomass & Bioenergy*, 35(5): 1633-1645.

Scheutz, C., Mosbaek, H., and Kjeldsen, P. 2004. Attenuation of Methane and Volatile Organic Compounds in Landfill Soil Covers. *J. Environ. Qual*, 33: 61- 71.

Schweigkofler, M., and Niessner, R. 2001. Removal of Siloxanes in Biogases. *J. Hazard. Mater.* B, 83: 183-196.

Shin, H-C., Park, J. W., Park, K. & Song, H-C. 2002. Removal characteristics of trace compounds of landfill gas by activated carbon adsorption. *Environ. Pollut,* 119: 227-236.

Speight, J. G. 2008. *Synthetic Fuels Handbook: Properties, Processes, and Performance*. McGraw-Hill, New York.

Speight, J. G. 2013a. *The Chemistry and Technology of Coal.* 3rd Edition. CRC-Taylor and Francis Group, Boca Raton, Florida.

Speight, J. G. 2013b. *Coal-Fired Power Generation Handbook*. Scrivener Publishing, Beverly, Massachusetts.

Speight, J. G. 2014a. *The Chemistry and Technology of Petroleum* 4th Edition. CRC-Taylor and Francis Group, Boca Raton, Florida.

Speight, J. G. 2014b. *Gasification of Unconventional Feedstocks*. Gulf Professional Publishing Company, Elsevier, Oxford, United Kingdom.

Speight, J. G. 2019. *Natural Gas: A Basic Handbook.* 2nd Edition. GPC Books, Elsevier, Gulf Publishing Company, Elsevier, Cambridge, Massachusetts.

Spiegel, R. J., Preston, J. L. & Trocciola J. C. 1997. Test results for fuel-cell operation on landfill gas. *Energy*, 22: 777-786.

Tietjen, C. 1975. From Bio-dung to Biogas – A Historical Review of the European Experience. In: Energy, agriculture, and waste management. *Proceedings of the 1975 Cornell Agricultural Waste Management Conference*. W. J. Jewell (Editor). Ann Arbor Science, Ann Arbor, Michigan. Page 207-260.

Weiland, P. 2010. Biogas Production: Current State and Perspectives. *Appl. Microbiol. Biotechnol,* 85(4): 849-60.

Wilber, C., and Murray, C. 1990. Odor Source Evaluation. *Biocycle,* 31(3): 68-72.

Chapter 2

BIOMASS

1. INTRODUCTION

Reducing national dependence of any country on the fossil fuels is of critical importance for long-term national security and continued economic growth for many countries. Supplementing fossil fuel consumption with renewable biomass resources is a first step towards this goal. The realignment of the chemical industry from one of petrochemical refining to a bio-refinery concept is, given time, feasible has become a national goal of many countries that use the fossil fuels as source of energy, particularly the countries that rely on imported crude oil and natural gas as the major sources of energy (Speight, 2008, 2011a). However, clearly defined goals are necessary for increasing the use of biomass-derived feedstocks in the areas of energy production and industrial chemical production and it is important to keep the goal in perspective. In this context, the increased use of biofuels should be viewed as one of a range of possible measures for achieving self-sufficiency in energy in the future, rather than a panacea for immediate replacement of fossil fuels (Crocker and Crofcheck, 2006).

Also, in the case of crops grown for energy production, there is the need for caution to ensure that the use of crops for energy does not cause a scarcity in terms of those same crops that are need form food. In add, whatever the crop, a potentially large resource that can be used for energy production, is discarded. For example, the straw associated with the wheat crop in often ploughed back into the soil, even though only a small proportion is needed to maintain the necessary level of organic matter for future crop growth. Thus, a huge renewable resource is not being usefully exploited since, for example, wheat straw contains a range of potentially useful chemicals. These chemical constituents include: (i) cellulose (Figure 2.1) and related compounds which can be used for the production bioethanol, biogas, and/or bio-oil, (ii) silica compounds which can be recovered and used as filter materials such as those necessary for water purification, and (iii) long-chain lipid derivatives (sometimes used less correctly to mean fats but does include other non-fat chemicals) (Figure 2.2). The chemicals specified in the three categories (above) can, after treatment, be used in cosmetics or in other specialty chemicals (Speight, 2019a).

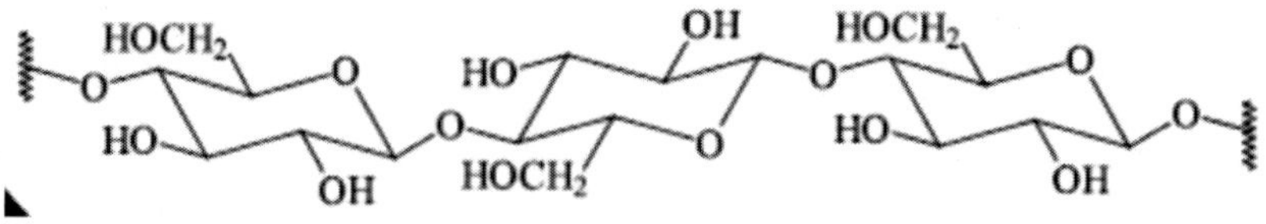

Figure 2.1. Generalized structure of cellulose.

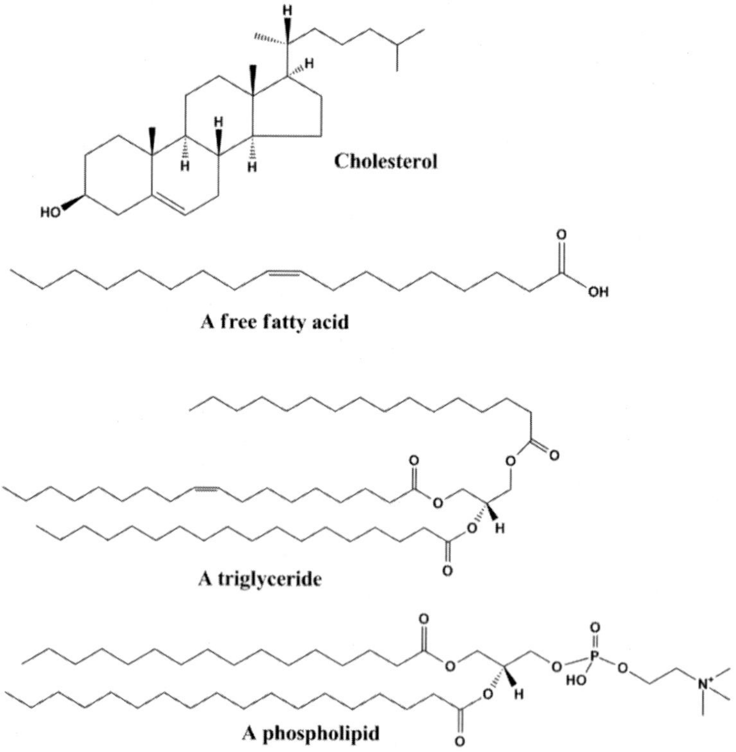

Figure 2.2. Chemical structures of Some Common Lipid Derivatives.

Biomass is the plant material derived from the reaction between carbon dioxide in the air, water and sunlight, via photosynthesis, to produce carbohydrates that form the building blocks of biomass (Wright et al., 2006). Typically photosynthesis is the process that converts the available energy from sunlight to chemical energy which is stored in the chemical bonds of the structural components of biomass. When biomass is processed either by chemical or biological methods, the carbon-containing constituents of the biomass are oxidized to produce carbon dioxide and water. The process is cyclical, since the released carbon dioxide is then available to produce new biomass. In the current context, the term *biomass* refers to plant material such as energy crops grown specifically to be used as fuel or as other plant material such as (i) fast-growing trees, (ii) switch grass, (iii) agricultural residues and by-products, such as straw, sugarcane fiber, and rice hulls, and (iv) residues from forestry, construction, and other wood-processing industries (McKendry, 2002a, 2002b; NREL 2003). Some of these materials may even have been advertently or inadvertently diverted to landfills which, under the appropriate circumstances could result in the production of biogas (Chapter 1).

Thus, the term *biomass* is used to describe any material of recent biological origin, including plant materials such as trees, grasses, agricultural crops, and even animal manure. Other biomass components, which are generally present in minor amounts, include triglycerides, sterols, alkaloids, resins, terpenes, terpenoids and waxes. This includes everything from *primary sources* of crops and residues harvested/collected directly from the land, to *secondary sources* such as sawmill residuals, to *tertiary sources* of post-consumer residuals that often end up in landfills. A *fourth source*, although not usually categorized as such, includes the gases that result from anaerobic digestion of animal manures or organic materials in landfills (Wright et al., 2006).

The most important biomass energy sources are wood and wood wastes, agricultural crops and their waste byproducts, municipal solid waste, animal wastes, waste from food processing, and aquatic plants and algae. As a result of these varying sources, biomass feedstocks and fuels exhibit a wide range of physical, chemical, and agricultural/process engineering properties.

2. HISTORY

The discoveries that led to the harnessing of fire – undoubtedly one of the most important discoveries in the history of mankind – led to the combustion of wood as a source of heat and energy and contributed to the commencement of birth of civilization as we know it. As a result, wood remained the most widely used raw material for many centuries, not only to burn fires, but also as building material. In fact, solid biofuels such as wood, various types of grasses, and dried cattle manure have been used since man learned to control fire.

The invention of the steam engine allowed mankind to obtain mechanic energy from the combustion of wood, whereas up to the 18th Century wind and water where the only mechanic energy sources available, thanks to wind and water mills. During the Industrial Revolution wood started to become scarce owing to the massive deforestation carried out to produce energy. In addition, liquid fuels from biological sources was used since the early days of the automobile industry. However, when crude oil began being cheaply extracted from the formations in the Earth (thanks to the major oil reserves discovered in Pennsylvania and Texas), the use of fuels from crude oil expanded rapidly after which World War I became the first truly mechanized war in history.

Before World War II, biofuels were seen as providing an alternative to imported oil in countries such as Germany, which sold a blend of gasoline with alcohol fermented from potatoes under the name Reichskraftsprit. After the war, cheap Middle Eastern oil lessened interest in biofuels and the continued search for reserves of coal, crude oil, and natural gas indicates an abundant supply of these fossil fuels. In fact, the depletion with time and use of the fossil fuel resources was not even a consideration.

In the latter part of the 20th Century, the realization that the fossil fuel resources could be depleted became a reality – although the exact time of depletion when there are no more (or insufficient) supplies of these fossil fuel resources to meet the demand is subject to much error and debate (Speight and Islam, 2016). Then, with the so-called oil shocks of 1973 and 1979, there was an increase in interest – by various levels of government – in biomass as a source of fuels and chemicals. This interest diminished after the counter-shock of 1986 that made oil prices cheaper again. It is only recently that the possible disappearance of fossil fuels and the pollution produced by the combustion has led to the search for alternate energy sources – and this has been realized with the *re-discovery* that the various forms of biomass should be given serious consideration as sources of energy (Speight, 2011a).

In fact, in the 21st Century, factors such as: (i) the considerable variation in the oil price market, (ii) the general rise oil prices, (iii) concerns over the potential real and imagined) peak in crude oil and natural gas production, (iv) greenhouse gas emissions (global warming), and (v) instability among the oil-producing nations (Speight, 2011b) has created renewed interest in fuels from biomass.

3. Biomass Feedstocks

The supply of crude oil, the basic feedstock for the production of fuels and petrochemicals, is finite and the dominant position of crude oil will eventually become unsustainable as supply/demand issues begin to diminish and any economic advantage over other alternative feedstocks is negated. This situation will be mitigated to some extent by the exploitation of more technically challenging fossil fuel resources and the introduction of new technologies for the production of fuels and chemicals from coal, tar sand bitumen, and heavy oil (Speight, 2008, 2013, 2019b). However, turning to these resources in place of conventional crude oil may have serious environmental consequences. Having made such a statement, biomass is not such an environmentally energy source and the production of fuels from biomass will also be subject to environmental controls. Nevertheless, the continued use of fossil resources at current rates without any or little control of the emissions will have serious and irreversible consequences for the global climate. Thus, the petroleum and petrochemicals industries are is coming under increasing pressure not only to compete effectively with global competitors utilizing more advantaged hydrocarbon feedstocks but also to ensure that its processes and products comply with increasingly stringent environmental legislation.

Biomass is a sustainable feedstock for chemicals and energy products that could potentially enhance the energy independence of the world. As an energy source that is highly productive, renewable, carbon-neutral, and easy to store and transport, biomass has drawn worldwide attention recently. Biomass offers important advantages as a combustion feedstock due to the high volatility of the fuel and the high reactivity of both the fuel and the resulting char. However, it should be noted that in comparison with fossil fuels such as coal, crude oil, natural gas, heavy oil, and tar sand bitumen, biomass contains much less carbon and more oxygen and has a low heating value (Speight, 2011a). With higher oxygen content than fossil fuels, biomass feedstocks have fundamentally lower energy content.

The most important biomass energy sources are wood and wood wastes, agricultural crops and their waste byproducts, municipal solid waste (MSW), animal wastes, waste from food processing, and aquatic plants and algae. The average majority of biomass energy is produced from wood and wood wastes (64%), followed by municipal solid waste (24%), agricultural waste (5%), and landfill gases (5%) (Demirbas, 2001a).

For a given production line, the comparison of the feedstocks must include at least one of the following includes several issues (Gnansounou et al., 2005): (i) chemical composition of the biomass, (ii) emission of greenhouse gases, acidifying gases and ozone depletion gases, and (iii) absorption of minerals to water and soil. The production of chemicals from renewable plant-based feedstocks utilizing state-of-the-art conversion technologies presents an opportunity to maintain competitive advantage and contribute to the attainment of national environmental targets. Bioprocessing routes have a number of compelling advantages over conventional petrochemicals production; however, it is only in the last decade that rapid progress in biotechnology has facilitated the commercialization of a number of plant-based chemical processes. It is widely recognized that further significant production of plant-based chemicals will only be economically viable in highly integrated and efficient production complexes producing a diverse range of chemical products. This biorefinery concept is analogous to conventional oil refineries and petrochemical complexes that have evolved over many years to maximize process synergies, energy integration and feedstock utilization to drive down production costs.

Plants offer a unique and diverse feedstock for a variety of products, including fuel products and petrochemical products. Plant biomass can be gasified to produce synthesis gas; a basic chemical feedstock and also a source of hydrogen for a future hydrogen economy. In addition, the specific components of plants such as carbohydrates, vegetable oils, plant fiber and complex organic molecules known as primary and secondary metabolites can be utilized to produce a range of valuable monomers, chemical intermediates, pharmaceuticals and materials:

3.1. Categories

For example, *primary biomass feedstocks* refer to the biomass that is harvested or collected from the field or forest where it is grown. Examples of primary biomass feedstocks currently being used for bioenergy include grains and oilseed crops used for transportation fuel production, plus some crop residues (such as orchard trimmings and nut hulls) and some residues from logging and forest operations that are currently used for heat and power production. In the future it is anticipated that a larger proportion of the residues inherently generated from food crop harvesting, as well as a larger proportion of the residues generated from ongoing logging and forest operations, will be used for bioenergy (Smith, 2006). Additionally, as the bioenergy industry develops, both woody and herbaceous perennial crops will be planted and harvested specifically for bioenergy and product end-uses.

Secondary biomass feedstocks differ from primary biomass feedstocks in that the secondary feedstocks are a by-product of processing of the primary feedstocks. By *processing*, it is meant that there is substantial physical or chemical breakdown of the primary biomass and production of by-products; *processors* may be factories or animals. Field processes such as harvesting, bundling, chipping or pressing do not cause a biomass resource that was produced by photosynthesis (e.g., tree tops and limbs) to be classified as secondary biomass.

Specific examples of secondary biomass includes sawdust from sawmills, black liquor (which is a byproduct of paper making), and cheese whey (which is a by-product of cheese making processes). Manures from concentrated animal feeding operations are collectable secondary biomass resources. Vegetable oils used for biodiesel that are derived directly from the processing of oilseeds for various uses are also a secondary biomass resource.

Tertiary biomass feedstocks includes post-consumer residues and wastes, such as fats, greases, oils, construction and demolition wood debris, other waste wood from the urban environments, as well as packaging wastes, municipal solid wastes, and landfill gases. A category *other wood waste from the urban environment* includes trimmings from urban trees, which technically fits the definition of primary biomass. However, because this material is normally handled as a waste stream along with other post-consumer wastes from urban environments (and included in those statistics), it makes the most sense to consider it to be part of the tertiary biomass stream. Domestic waste products include paper, containers, tin cans, aluminum cans, and food scraps, as well as sewage. Industrial waste products include paper, wood, and metal scraps, as well as agricultural waste products. Biodegradable wastes, such as paper fines and industrial bio-sludge, into mixed alcohol fuels (e.g., isopropanol, isobutanol, iso-pentanol).

The proper categorization of *fats and greases* may be debatable since those are byproducts of the reduction of animal biomass into component parts. However, most fats and greases, and some oils, are not available for bioenergy use until after they become a post-consumer waste stream, it seems appropriate for them to be included in the tertiary biomass category. Vegetable oils derived from processing of plant components and used directly for bioenergy (e.g., soybean oil used in biodiesel) would be a secondary biomass resource, though amounts being used for bioenergy are most likely to be tracked together with fats, greases and waste oils.

Many different types of biomass can be grown for the express purpose of energy production. Crops that have been used for energy include: sugar cane, corn, sugar beets, grains, elephant grass, kelp (seaweed) and many others. There are two main factors which determine whether a crop is suitable for energy use. Good energy crops have a very high yield of dry material per unit of land (dry tonnes/hectare). A high yield reduces land requirements and lowers the cost of producing energy from biomass. Similarly, the amount of energy which can be produced from a biomass crop must be less than the amount of energy required to grow the crop. In some circumstances like the heavily mechanized corn farms in the U.S. Midwest, the amount of ethanol which can be recovered from the corn is barely larger than the fuel required for tractors, fertilizers, and processing.

However, it is the inherent properties of the biomass source that determines both the choice of conversion process and any subsequent processing difficulties that may arise. Equally, the choice of biomass source is influenced by the form in which the energy is required and it is the interplay between these two aspects that enables flexibility to be introduced into the use of biomass as an energy source. Dependent on the energy conversion process selected, particular material properties become important during subsequent processing.

The main material properties of interest, during subsequent processing as an energy source, relate to: (i) moisture content – intrinsic and extrinsic, (ii) calorific value, (iii) proportions of fixed carbon and volatiles, (iv) mineral matter content, manifested as the yield of combustion ash, (v) alkali metal content, and (vi) the cellulose/lignin ratio. For dry biomass conversion processes, the first five properties are of interest, while for wet biomass conversion processes, the first and last properties are of prime concern (McKendry 2002a, 2002b).

The simplest, cheapest and most common method of obtaining energy from biomass is direct combustion. Any organic material with a water content low enough to allow for sustained combustion can be burned to produce energy. The heat of combustion can be used to provide space or process heat, water heating or, through the use of a steam turbine, electricity. In the developing world, many types of biomass such as manure and agricultural wastes are burned for cooking and heating.

Thus, almost all crops, whether grown for food, animal feed, fiber or any other purpose, result in some form of organic residues after their primary use has been fulfilled. These organic residues, as well as animal wastes (excrement) can be used for energy production through direct combustion or biochemical conversion. Current worldwide production of crop residues is very large; but an increased scale of use for fuel may have significant environmental impacts, the most serious being those of lost soil fertility and soil erosion.

Most crop residues are returned to the soil, and the humus resulting from their decomposition helps maintain soil nutrients, soil porosity, water infiltration and storage, as well as reducing soil erosion. Crop residues typically contain 40% of the nitrogen, 80% of the potassium, and 10% of the phosphorous applied to the soil in the form of fertilizer. If these residues are subjected to direct combustion for energy, only a small percentage of the nutrients is left in the ash. Similarly, soil erosion will increase. Estimates for the United States indicate that 22% of crop residues could be removed, providing energy equivalent to 5% of the needs of the United States.

Table 2.1. General Categorization of Renewable and Non-renewable Energy Sources

Energy Source	Type	Sub-type	Process	Products	
Renewable	Biomass	Biological	Microbial	Alcohol	
				Biogas	
		Thermal	Briquetting	Solid fuel	
			Gasification	Bio-syngas	
			Pyrolysis	Biogas	
				Bio-oil	
				Bio-char	
	Geothermal			Heat	
	Hydro			Power	
	Hydrogen			Various	
	Ocean			Heat/power	
	Solar			Heat/power	
	Wind			Heat/power	
Non-renewable	Fossil fuel	Coal		Lignite	
				Sub-bituminous	
				Anthracite	
		Crude Oil	Refining	Naphtha	Gasoline
				Kerosene	Diesel
				Gas oil	Fuel oil
				Resid	Asphalt
		Natural gas	Refining	Pipeline gas	
				Liquefied petroleum gas	
				Compressed natural gas	
		Oil shale	Retorting	Naphtha	Gasoline
				Kerosene	Diesel
				Gas oil	Fuel oil

Biomass is a *renewable energy* source, unlike the fossil fuel resources (*petroleum, coal,* and *natural gas*) and, like the fossil fuels, biomass is a form of stored solar energy (Table 2.1). The energy of the sun is captured through the process of *photosynthesis* in growing plants. One advantage of biofuel in comparison to most other fuel types is it is biodegradable, and thus relatively harmless to the environment if spilled.

Many different biomass feedstocks can be used to produce liquid fuels. They include crops specifically grown for bioenergy, and various agricultural residues, wood residues and waste streams. Their costs and availability vary widely. Collection and transportation costs are often critical.

Sugarcane, sugar beet, corn, and sweet sorghum are agricultural crops presently grown commercially for both carbohydrate production and animal feeds. Sugarcane, corn and sweet sorghum are efficient at trapping solar energy and use specific biochemical pathways to recycle and trap carbon dioxide that is lost through photorespiration. Sugar beets are efficient because they store their carbohydrate in the ground. Sugarcane was the basis for the World's first renewable biofuel program in Brazil. Corn is the basis for the present renewable ethanol fuel industry in the United States. The sugars produced by these crops are easily fermented by Saccharomyces cerevisiae. The sucrose produced by sugarcane, sugar beet, and sweet sorghum can be fermented directly after squeezing them from the crop. Corn traps its carbohydrate largely in the form of starch which must first be converted into glucose through saccharification with glucoamylase. The residues left over after removing fermentable sugars can also be utilized. In some cases they end up as animal feeds, but many agricultural residues can be converted into additional fermentable sugars through saccharification with cellulase.

By way of clarification, cellulase is any of several enzymes produced chiefly by fungi, bacteria, and protozoans that catalyze cellulolysis which is the decomposition of cellulose and related polysaccharide derivatives. The name is also used for any naturally occurring mixture or complex of various such enzymes, that act serially or synergistically to decompose cellulosic material. Cellulase enzymes break down the cellulose molecule into monosaccharide derivatives (simple sugars.

Bioenergy crops include fast growing trees such as hybrid poplar, black locust, willow, and silver maple in addition to annual crops such as corn, sweet sorghum, and perennial grasses such as switch grass. Switch grass is a thin-stemmed, warm season, perennial grass that has shown high potential as a high yielding crop that can be readily grown in areas that are also suitable for crop production. In fact, there are many perennial crops (grass and tree species) that show high potential for production of cost-competitive cellulosic biomass. Switch grass can be viewed as a surrogate for many perennial energy crops when estimating biomass supply and availability. Many other crops are possible and the optimal crop will vary with growing season and other environmental factors. Most fast-growing woody and annual crops are high in hemicellulose sugars such as xylose.

Waste streams can also be exploited for ethanol production. They are often inexpensive to obtain, and in many instances they have a negative value attributable to current disposal costs. Some principal waste streams currently under consideration include mixed paper from municipal solid waste, cellulosic fiber fines from recycled paper mills, bagasse from sugar manufacture, corn fiber, potato waste, and citrus waste, sulfite waste liquors and hydrolysis streams from fiber board manufacture. Each waste stream has its own unique characteristics, and they generally vary from one source or time to another. Waste streams with lower lignin contents and smaller particle sizes are easier to deal with than those with higher lignin contents and larger particle sizes. Waste paper that has been treated by a chemical pulping process is much more readily converted than is native wood or herbaceous residue.

However, the value of a particular type of biomass depends on the chemical and physical properties of the constituents (usually referred to as biopolymers) that make up the biomass. Knowledge of these chemical structures of these constituents and the chemistry involved in conversion to energy assists the development of process for biomass conversion to fuel especially, in the present context, to biogas.

3.2. Types

Biomass feedstocks are marked by the diversity that exists within the feedstocks, which makes such feedstocks difficult to characterize as a whole. Feedstocks that can be utilized with conversion processes are primarily the organic materials now being landfilled and include forest products wastes, agricultural residues, organic fractions of municipal solid wastes, paper, cardboard, plastic, food waste, green waste, and other waste. Non-biodegradable organic feedstocks, such as most plastics, are not convertible by biochemical processes. Bio-based materials require pre-treatment by chemical, physical, or biological means to open up the structure of biomass.

In the simplest sense, there are three common types of biomass that are considered suitable for conversion to fuel products: (i) wood and wood waste, often referred to as woody fuels, (ii) agricultural residues, which include crop residues and manure, and (iii) urban waste that are frequently referred to, but actually include, municipal solid waste (MSW).

This section presents the three sources of biomass that can be used for conversion to gaseous, liquid, or solid fuels.

3.2.1. Wood and Wood Wastes

Wood fuels are fuels derived from natural forests, natural woodlands and forestry plantations, namely fuelwood and charcoal from these sources. These fuels include sawdust and other residues from forestry and wood processing activities. Over 50% of all wood used in the world is fuelwood. Most of the fuelwood is used in developing countries. In developing countries wood makes up about 80% of all wood used.

Size of the wood waste resource depends upon how much wood is harvested for lumber, pulp and paper. Finally, fuelwood can be grown in plantations like a crop. Fast growing species such as poplar, willow or eucalyptus can be harvested every few years. With short-rotation poplar coppices grown in three 7-year rotations, it is now possible to obtain 10 to 13 tons of dry matter per hectare annually on soil of average or good quality (Demirbas, 2001b). Waste wood from the forest products industry such as bark, sawdust, board ends etc. are widely used for energy production. This industry, in many cases, is now a net exporter of electricity generated by the combustion of wastes.

Overall, wood wastes of all types make excellent biomass fuels and can be used in a wide variety of biomass technologies. Combustion of woody fuels to generate steam or electricity is a proven technology and is the most common biomass-to-energy process. Different types of woody fuels can typically be mixed together as a common fuel, although differing moisture content and chemical makeup can affect the overall conversion rate or efficiency of a biomass project.

There are at least six subgroups of woody fuels: (i) forestry residues which include in-forest woody debris and slash from logging and forest management activities, (ii) mill residues which include byproducts such as sawdust, hog fuel, and wood chips from lumber mills, plywood manufacturing, and other wood processing facilities, (iii) agricultural residues which includes byproducts of agricultural activities including crop wastes, waste from vineyard and orchard pruning, and rejected agricultural products, (iv) urban wood and yard wastes which includes residential organics collected by municipal programs or recycling centers and construction wood wastes, (v) dedicated biomass crops which includes trees, corn, oilseed rape, and other crops grown as dedicated feedstocks for a biomass project, and (vi) chemical recovery fuels – some time known as black liquor) – which includes woody residues recovered out of the chemicals used to separate fiber for the pulp and paper industry. Mill residues are a much more economically attractive fuel than forestry residues, since the in-forest collection and chipping are already included as part of the commercial mill operations. Biomass facilities collocated with and integral to the mill operation have the advantage of eliminating transportation altogether and thus truly achieve a no-cost fuel.

Softwood residues are generally in high demand as feedstocks for paper production, but hardwood timber residues have less demand and fewer competing uses. In the past, as much as 50% of the tree was left on site at the time of harvest. Whole tree harvest systems for pulp chips recover a much larger fraction of the wood. Wood harvests for timber production often generates residues which may be left on the site or recovered for pulp production. Economics of wood recovery depend greatly on accessibility and local demand. Underutilized wood species include Southern red oak, poplar, and various small diameter hardwood species. Unharvested dead and diseased trees can comprise a major resource in some regions. When such timber has accumulated in abundance, it comprises a fire hazard and must be removed. Such low grade wood generally has little value and is often removed by prescribed burns in order to reduce the risk of wildfires.

3.2.2. Agricultural Residues

Agricultural residues are basically biomass materials that are by products of agriculture. This includes materials such as cotton stalks, wheat and rice straw, coconut shells, maize and corn cobs, jute sticks, and rice husks. Many developing countries have a wide variety of agricultural residues in ample quantities. Large quantities of agricultural plant residues are produced annually worldwide and are vastly under-utilized. The most common agricultural residue is the rice husk, which makes up 25% of rice by mass (Demirbas, 2001a).

Corn stalks and wheat straws are the two agricultural residues produced in the largest quantities. However, many other residues such as potato and beet waste may be prevalent in some regions. In addition to quantity it is necessary to consider density and water content (which may restrict the feasibility of transportation) and seasonality which may restrict the ability of the conversion plant to operate on a year-round basis. Facilities designed to use seasonal crops will need adequate storage space and should also be flexible enough to accommodate alternative feedstocks such as wood residues or other wastes in order to operate year-around. Some agricultural residues need to be left in the field in order to increase tilth (the state of aggregation of soil and its condition for supporting plant growth and to reduce erosion) but some residues such as corncobs can be removed and converted without much difficultly.

Agricultural residues can provide a substantial amount of biomass fuel. Similar to the way mill residues provide a significant portion of the overall biomass consumption in areas that are copiously forested, agricultural residues from sugar cane harvesting and processing provide a significant portion of the total biomass consumption in other parts of the world. One significant issue with agricultural residues is the seasonal variation of the supply. Large residue volumes follow harvests, but residues throughout the rest of the year are minimal. Biomass facilities that depend significantly on agricultural residues must either be able to adjust output to follow the seasonal variation, or have the capacity to stockpile a significant amount of fuel.

Animal wastes include manures, renderings, and other wastes from livestock finishing operations. Although animal wastes contain energy, the primary motivation for biomass processing of animal wastes is mitigation of a disposal issue rather than generation of energy. This is especially true for animal manures. Animal manures are typically disposed of through land application to farmlands. Tightening regulations on nutrient management, surface and groundwater contamination, and odor control are beginning to force new manure management and disposal practices.

Dry animal manure, which is typically defined as having a moisture content less than 30% w/w, is produced by feedlots and livestock corrals, where the manure is collected and removed only once or twice a year. Manure that is scraped or flushed on a more frequent schedule can also be separated, stacked, and allowed to dry. Dry manure can be composted or can fuel a biomass-to-energy combustion project. Animal manure does have value to farmers as fertilizer, and a biomass-to-energy project would need to compete for the manure. However, the total volume of manure produced in many livestock operations exceeds the amount of fertilizer required for the farmlands and, in some areas/countries, nutrient management plans are beginning to limit the over-fertilization of farmlands. Therefore, although there are competitive uses for the manure and low-cost disposal options at this time, manure disposal is going to become more costly over time, and the demand for alternative disposal options, including biomass-to-energy, will only increase.

Biomass technologies present attractive options for mitigating many of the environmental challenges of manure wastes. The most common biomass technologies for animal manures are combustion, anaerobic digestion, and composting. Moisture content of the manure and the amount of contaminants, such as bedding, determine which technology is most appropriate.

3.2.3. Urban Wastes

Urban wastes include municipal solid waste generated by household and commercial activities and liquid waste or sewage. Most municipal solid waste is currently disposed of in landfill sites. However, the disposal of this waste is a growing problem worldwide. Much of the waste could be used for energy production though incineration and processes. Japan currently incinerates more than 80% of the municipal solid waste (Demirbas, 2001a). It is also possible to use the methane produced in landfill sites for energy production.

Urban wood and yard wastes are similar in nature to agricultural residues in many regards. A biomass facility will rarely need to purchase urban wood and yard wastes, and most likely can charge a tipping fee to accept the fuel. Many landfills are already sorting waste material by isolating wood waste. This waste could be diverted to a biomass project, and although the volume currently accepted at the landfills would not be enough on its own to fuel a biomass project, it could be an important supplemental fuel and could provide more value to the community in which the landfill resides through a biomass project than it currently does as daily landfill cover.

Municipal solid wastes are produced and collected each year with the vast majority being disposed of in open fields. The biomass resource in municipal solid waste comprises the putrescible materials, paper and plastic and averages 80% of the total municipal solid waste collected. Municipal solid waste can be converted into energy by direct combustion, or by natural anaerobic digestion in the engineered landfill. At the landfill sites the gas produced by the natural decomposition of municipal solid waste (approximately 50% methane and 50% carbon dioxide) is collected from the stored material and scrubbed and cleaned before feeding into internal combustion engines or gas turbines to generate heat and power. The organic fraction of municipal solid waste can be anaerobically stabilized in a high-rate digester to obtain biogas for electricity or steam generation.

4. Preparation for Processing

One extremely important aspect of biomass use as a process feedstock is the preparation of the biomass (also referred to as biomass cleaning or biomass pretreatment) is the removal of any contaminants that could have an adverse effect of the process and on the yields and quality of the products. Thus, feedstock preparation is, essentially, the pretreatment of the biomass feedstock to assist in the efficiency of the conversion process.

Pretreatment of biomass is considered one of the most important steps in the overall processing in a biomass-to-biofuel program and can occur using acidic or alkaline reagents (Table 2.2) as well as using a variety of physical methods (Table 2.3) and the method of choice depends vary much upon the process needs. With the strong advancement in developing lignocellulose biomass-based refinery and algal biomass-based biorefinery, the major focus has been on developing pretreatment methods and technologies that are technically and economically feasible (Pandey et al., 2015).

Table 2.2. Acidic and Alkaline Methods for Biomass Treatment

Method	Conditions	Outcome
Acid based methods	Low pH using an acid (H_2SO_4, H_3PO_4	Hydrolysis of the hemicellulose to monomer sugars
		Minimizes the need for hemicellulases
Neutral conditions	Steam pretreatment and hydrothermolysis	Solubilizes most of the hydrocarbons by conversion to acetic acid
		Does not usually result in total conversion to monomer sugars
		Requires hemicellulases acting on soluble oligomers
Alkaline methods		Leaves a part of the hydrocarbon in the solid fraction
		Requires hemicellulases acting hydrocarbons

Typically, the fundamental step sin the pretreatment of biomass involves processes such as (i) washing/separation of inorganic matter such as stones/pebbles, (ii) size reduction which involves grinding, milling, and crushing, and (iii) separation of soluble matter (Table 2.3). Also, the pretreatment process that is selected is, depending upon the character of the biomass and the process, likely to be different for different raw materials and desired products.

A wide range of biomass feedstocks can be used in pyrolysis processes. The pyrolysis process is very dependent on the moisture content of the feedstock, which should be on the order of 10% w/w. At higher moisture contents, high levels of water are produced and at lower levels there is a risk that the process only produces dust instead of oil. High-moisture waste streams, such as sludge and processing wastes, require drying before subjecting to pyrolysis.

Thus, the efficiency and nature of the pyrolysis process is dependent on the particle size of feedstocks. Most of the pyrolysis technologies can only process small particles to a maximum of 2 mm keeping in view the need for rapid heat transfer through the particle. The demand for small particle size means that the feedstock has to be size-reduced before being used for pyrolysis.

Table 2.3. Summation of the Methods for the Pretreatment of Biomass Feedstocks

Physical methods	Miscellaneous methods
Milling:	Explosion:*
-Ball milling	- Steam, NH_3, CO_2, SO_2 , Acids, Alkali
-Two-roll milling	- NaOH, NH_3, $(NH_4)_2SO_3$
-Hammer milling	Acid:
Irradiation:	- Sulfuric, hydrochloric, and phosphoric acids
-Gamma-ray irradiation	Gas:
-Electron-beam irradiation	- Chlorine dioxide, nitrous oxide, sulfur dioxide
-Microwave irradiation	Oxidation:
Other methods:	- Hydrogen peroxide
- Hydrothermal	- Wet oxidation
- High pressure steaming	- Ozone
- Extrusion	Solvent extraction of lignin:
- Pyrolysis	- Ethanol-water extraction
	- Benzene-water extraction
	- Butanol-water extraction
	Organic solvents Ionic liquids

* The feedstock material is subjected to the action of steam and high-pressure carbon dioxide before being discharged through a nozzle.

Moisture in the biomass is another consideration for feedstock preparation because moisture in the feedstock will simply vaporize during the process and then re-condense with the bio-oil product which has an adverse impact the resulting quality of the bio-oil. It should also be noted that water is formed as part of the thermochemical reactions occurring during pyrolysis. Thus, if dry biomass is subjected to the thermal requirements for fast pyrolysis the resulting bio-oil will still contain water (as much as 12 to 15% w/w). This water is process-originated water that is the result of the dehydration of carbohydrate derivatives in the feedstock s as well as the result of reactions occurring between the hydrogen and oxygen at the high temperature (500°C, 930°F) of the process environment.

Moisture in the feedstock acts as heat sink and competes directly with the heat available for pyrolysis. Ideally it would be desirable to have little or no moisture in the feedstock but practical considerations make this unrealistic. Moisture levels on the order of 5 to 10% w/w are generally considered acceptable for the pyrolysis process technologies currently in use. As with the particle size, the moisture levels in the feedstock biomass are a trade-off between the cost of drying and the heating value penalty paid by having moisture in the feedstock.

If the moisture content in the biomass feedstock is too high, the bio-oil may be produced with high moisture content which eventually reduces its calorific value. Therefore biomass should undergo a pretreatment (drying) process to reduce the water content before pyrolysis is carried out (Dobele et al., 2007). In contrast, high temperature during the drying process could be a critical issue for the possibility of producing thermal-oxidative reactions, causing a cross-linked condensed system of the components and higher thermal stability of the biomass complex.

To achieve high yields of the products (gases, liquids, and solids), it is also necessary to prepare the solid biomass feedstock in such a manner that it can facilitate the required heat transfer rates in the pyrolysis process. There are three primary heat transfer mechanisms available to engineers in designing reaction vessels: (i) convection, (ii) conduction, and (iii) radiation. To adequately exploit one or more of these heat transfer mechanisms as applied to biomass pyrolysis, it is necessary to have a relatively small particle for introduction to the reaction vessel. This ensures a high surface area per unit volume of particle and, as a result of the small particle size the whole particle achieves the desired temperature in a very short residence time. Another reason for the conversion of the feedstock to small particles is the physical transition of biomass as it undergoes pyrolysis when char develops at the surface of the particle. The char can act as an insulator that impedes the transfer of heat into the center of the particle and therefore runs counter to the requirements needed for pyrolysis. The smaller the particle the less of an affect this has on heat transfer (Bridgwater et al., 2001).

One aspect of feedstock preparation in the light of the processes in which the biomass is to be used and converted, is the concept of torrefaction which is used as a pre-treatment step for biomass conversion techniques, such as gasification and cofiring. The thermal treatment not only destructs the fibrous structure and tenacity of biomass, but is also known to increase the calorific value of the biomass (Prins et al., 2006a, 2006b, 2006c).

Typically, torrefaction commence when the temperature reach 200ºC (390ºF) and end when the process is again cooled from the specific temperature to 200ºC (390ºF). During the process the biomass partly devolatilizes leading to a decrease in mass, but the initial energy content of the torrefied biomass is mainly preserved in the solid product so the energy density of the biomass becomes higher than the original biomass which makes it more attractive for i.e., transportation (to a conversion site). After torrefaction at, say, temperatures up to 300ºC (570ºF) the grindability of raw biomass shows an improvement in grindability.

5. The Chemistry of Biomass

Biomass is an organic feedstock that is composed of macromolecular chemical species that have extensive chains of carbon atoms linking the various parts of the macromolecules (Chapter 2). The chemical backbone consists of chemical bonds linking carbon with carbon, or carbon with oxygen, or sometimes linking carbon with other elements such as nitrogen or sulfur. Instead of describing the macromolecules in terms of the chemical structure of the chain, most can be considered to be as assemblages of some larger molecular unit. For example, in the case of cellulose, that unit is the glucan moiety which is a molecule of glucose with one molecule of water missing ($C_6H_{10}O_5$ instead of $C_6H_{12}O_6$) (Figure 4.1). For hemicellulose, the unit is often a 5-carbon sugar (xylose, $C_5H_{10}O_5$)). However, hemicellulose macromolecules are not linear chains (Figure 4.2) as in the cellulose polymer. Some are branched and other monomer units have side chains, with acetyl groups being very common. On the other hand, the lignin macromolecule (Figure 4.3) is composed of phenyl propane subunits linked at various points on the monomer through carbon-carbon (C-C) and carbon-oxygen (C-O) bonds. In addition, there are often side chain feedstocks such as methoxy groups ($-OCH_3$).

Table 2.4. General Analytical Methods to Determine the Processability of Biomass

Method	Outcome
Drying, Grounding, Milling, sieving	Particle size distribution
Proximate Analysis	Chemical content and properties
Ultimate Analysis	Elemental composition (including trace elements)
Bulk composition	Content determination and processability
% w/w Cellulose	
% w/w Hemicellulose	
% w/w Lignin	
% w/w Protein	
% w/w Extractives	
Thermal Analysis	Thermal behavior or kinetic study

Generally the components of biomass raw material are usually studied through ultimate and proximate analysis techniques (Table 2.4, Table 2.5). Proximate analysis involves measuring the values of the moisture content (MC), ash content, volatile matter contents (VM), and fixed carbon contents (FC) of the raw material. On the other hand, ultimate analysis measures carbon, hydrogen, sulfur, nitrogen, and oxygen elements. Since both analysis techniques are able to unveil the main elements that are present in biomass raw material, the corresponding physical and chemical properties of produced pellets from a given raw material may be easily predicted, once the main raw material elements are measured.

Table 2.5. General Categories of Biomass*

Forest residues
Tree branches, tops of trunks, stumps, leaves.

Industrial waste
Citrus peels, sugarcane bagasse, olive husks, milling residues.

Energy crops
Switch grass, miscanthus, bamboo, sweet sorghum, tall fescue, and wheatgrass.

Landfill gas and biogas
Methane can also be produced using energy from agricultural and human wastes. Biogas digesters are airtight containers or pits lined with steel or bricks. Waste placed into the containers is fermented without oxygen to produce a methane-rich gas.

Solid waste
Burning trash turns waste into a usable form of energy. Garbage is not all biomass; perhaps half of its energy content comes from plastics, which are made from petroleum and natural gas. Power plants that burn garbage for energy are called waste-to-energy plants.

Wood and agricultural products
Wood in the form of logs, chips, bark, and sawdust accounts for about 44 percent of biomass energy.
Wood and wood waste are used to generate electricity. Much of the electricity is used by the industries making the waste. Paper mills and saw mills use much of their waste products to generate steam and electricity for their use.

*Listed alphabetically rather than by any preference.

5.1. Chemical Constituents

Biomass is all biologically-produced matter based in carbon, hydrogen and oxygen. Wood remains the largest biomass energy source – examples include forest residues (such as dead trees, branches and tree stumps), yard clippings, wood chips, as well as municipal solid waste (known as MSW). Wood energy is derived by using lignocellulosic biomass (second-generation biofuels) as fuel. Biomass also includes plant or animal matter that can be converted into fibers or other industrial chemicals, including biofuels. Industrial biomass can be grown from numerous types of plants, including miscanthus, switchgrass, hemp, corn sorghum, sugarcane, bamboo and a variety of tree species.

In a chemical context, the chemical types that occur in the various forms of biomass include: (i) carbohydrates – which encompass starch cellulose, hemicellulose, (ii) lignin, (iii) vegetable oils, and (iv) the constituents of plant fibers. Each of these chemical types behave differently in conversion process and, in this respect, a brief refresher of the chemical character of cellulose, hemicellulose, and lignin is warranted in order to understand the type and results of the pretreatment process. Therefore, the chemical aspects of these naturally-occurring compounds are presented in the following sections.

5.1.1. Carbohydrates

A carbohydrate is a molecule consisting of carbon, hydrogen, and oxygen atoms that typically has a hydrogen-oxygen atomic ratio of 2:1 and with the empirical formula $(C_nH_2O)_m$ where $C_m(H_2O)_n$ (where *m* may be different from *n*) – the formula holds true for monosaccharides but some exception do exist. For example, deoxyribose, which is the sugar compete for deoxyribonucleic acid (DNA) has the empirical formula $C_5H_{10}O_4$. The carbohydrates are technically carbon hydrates (as shown by the empirical forma above) but structurally it is more correct to consider the carbohydrates as aldose derivatives and ketone derivatives. Monosaccharides and disaccharides, the smallest (lowest molecular weight) carbohydrates, are commonly referred to as sugars.

Plants capture solar energy as fixed carbon during which carbon dioxide is converted to water and sugars $(CH_2O)_x$:

$$CO_2 + H_2O \rightarrow (CH_2O)_x + O_2.$$

The sugars produced are stored in three types of polymeric macromolecules: (i) starch, (ii) cellulose, and (iii) hemicellulose.

The higher molecular weight carbohydrates (starch, cellulose, sugars): starch readily obtained from wheat and potato, whilst cellulose is obtained from wood pulp. The structures of these polysaccharides can be readily manipulated to produce a range of biodegradable polymers with properties similar to those of conventional plastics such as polystyrene foams and polyethylene film. In addition, these polysaccharides can be hydrolyzed, catalytically or enzymatically to produce sugars, a valuable fermentation feedstock for the production of ethanol, citric acid, lactic acid and dibasic acids such as succinic acid.

In general sugar polymers such as cellulose and starch (Figure 2.1) can be readily broken down to their constituent monomers by hydrolysis, preparatory to conversion to ethanol or other chemicals. Cellulose is a glucose polymer, consisting of linear chains of (1,4)-D-glucopyranose units, in which the units are linked 1-4 in the b-configuration, with an average molecular weight of around 100,000. Hemicellulose is a mixture of polysaccharides, composed almost entirely of sugars such as glucose, mannose, xylose and arabinose and methyl glucuronic acid and galacturonic acid, with a molecular weight that may be up to or on the order of 30,000. In contrast to cellulose, hemicellulose is a heterogeneous branched polysaccharide that binds tightly, but non-covalently, to the surface of each cellulose micro fibril. Hemicellulose differs from cellulose, in consisting primarily of xylose and other five-carbon monosaccharides.

Cellulose (Figure 2.1), which is an abundant component in plants and wood, comes in various forms and a large fraction comes from domestic and industrial wastes (Abella et al., 2007; Balat and Kirtay, 2010). Cellulose is a high molecular weight linear polymer of b-1,4-linked D-glucose units which can appear as a highly crystalline material Demirbaş, 2008a). Glucose anhydride, which is formed via the removal of water from each glucose, is polymerized into long cellulose chains that contain 5000 to 10000 glucose units. The basic repeating unit of the cellulose polymer consists of two glucose anhydride units, called a cellobiose unit (Mohan et al., 2006).

Hemicellulose is a mixture of various polymerized monosaccharides such as glucose, mannose, galactose, xylose, arabinose, 4-O-methyl glucuronic acid and galacturonic acid residues (Mohan et al., 2006). Among the most important sugar of the hemicelluloses component is xylose. In hardwood xylan, the backbone chain consists of xylose units which are linked by b-(1,4)-glycosidic bonds and branched by a-(1,2)-glycosidic bonds with 4-O-methyl glucuronic acid groups (Demirbaş, A. 2009). Hemicelluloses exhibit lower molecular weights than cellulose. The number of repeating saccharide monomers is on the order of 150, compared to the number in cellulose (Mohan et al., 2006).

5.1.2. Lignin

In contrast, lignin is a complex structure containing aromatic groups (Figure 2.3) and is less readily degraded. Although lignocellulose is one of the cheapest and most abundant forms of biomass, it is difficult to convert this relatively unreactive material into sugars. Among other factors, the walls of lignocellulose are composed of lignin, which must be broken down in order to render the cellulose and hemicellulose accessible to acid hydrolysis. For this reason, many programs focused on ethanol production from biomass are based almost entirely on the fermentation of sugars derived from the starch in corn grain.

Lignin is an aromatic polymer that contains phenyl propanoid precursors, other than that generalization, the structure of lignin is largely unknown and remains speculative (Figure 2.3). The hypothesis, with some experimental justification, is that the basic chemical phenyl propane units of lignin (primarily syringyl, guaiacyl and p-hydroxy phenol) are bonded together by a set of linkages to form a very complex matrix. This matrix comprises a variety of functional groups, such as hydroxyl (-OH) groups, methoxyl (-OCH_3) groups, and carbonyl (>C = O) groups, which impart a high polarity to the lignin macromolecule (Feldman et al., 1991).

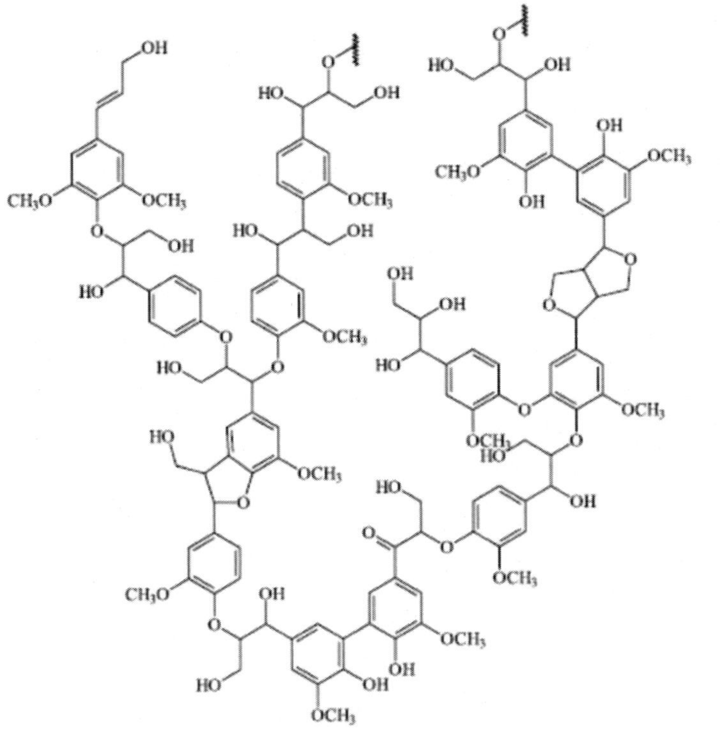

Figure 2.3. Hypothetical structure of lignin to illustrate the complexity of the molecule.

Lignin is a complex chemical compound that is most commonly derived from wood and is an integral part of the cell walls of plants, especially in tracheids, xylem fibres and sclereids. The chemical structure of lignin is unknown and, at best, can only be represented by hypothetical formulas. Lignin can be regarded as a group of amorphous, high molecular-weight, chemically related compounds. The building blocks of lignin are believed to be a three carbon chain attached to rings of six carbon atoms, called phenyl-propane derivatives. These may have zero, one or two methoxyl groups attached to the rings, giving rise to three structures. The proportions of each structure depend on the source of the polymer i.e., structure I is found in plants such as grasses; structure II in the wood of conifers; while structure III is found in deciduous wood.

The term lignin was introduced in 1819 and is derived from the Latin word *lignum* (meaning *wood*). It is one of most abundant organic compounds on earth after cellulose and chitin. By way of clarification, chitin $(C_8H_{13}O_5N)_n$ is a long-chain polymeric polysaccharide of beta-glucose that forms a hard, semitransparent material found throughout the natural world. Chitin is the main component of the cell walls of fungi and is also a major component of the exoskeletons of arthropods, such as the crustaceans (e.g., crab, lobster, and shrimp), and the insects (e.g., ants, beetles, and butterflies), and of the beaks of cephalopods (e.g., squids, and octopuses).

Lignin makes up about one-quarter to one-third of the dry mass of wood and is generally considered to be a large, cross-linked hydrophobic, aromatic macromolecule with molecular mass that is estimated to be in excess of 10,000. Degradation studies indicate that the molecule consists of various types of substructures which appear to repeat in random manner.

The biosynthesis of lignin begins with the synthesis of mono lignol derivatives (e.g., coniferyl alcohol, sinapyl alcohol and paracoumaryl alcohol) starting from an amino acid (phenylalanine). There are a number of other mono-lignol derivatives present in plants but different plants use different mono-lignol derivatives. The mono-lignol derivatives are synthesized as the respective glucosides which are water soluble and allows transportation through the cell membrane to the apoplast where the glucose moiety is removed after which the mono-lignol derivatives form lignin.

Lignin fills the spaces in the cell wall between cellulose, hemicellulose and pectin components and is covalently linked to hemicellulose. Lignin also forms covalent bonds to polysaccharides and thereby crosslinks different plant polysaccharides. It confers mechanical strength to the cell wall (stabilizing the mature cell wall) and therefore the entire plant.

5.1.3. Vegetable Oils

Vegetable oils are obtained from seed oil plants such as Palm, sunflower and soya. The predominant source of vegetable oils in many countries is rapeseed oil. Vegetable oils are a major feedstock for the oleo-chemicals industry (surfactants, dispersants and personal care products) and are now successfully entering new markets such as diesel fuel, lubricants, polyurethane monomers, functional polymer additives and solvents.

Many vegetable oils are consumed directly, or indirectly as ingredients in food – a role that they share with some animal fats, including butter and lard. Oils can be heated and used to cook other foods and such oils suitable for this objective must have a high flash point.

5.1.4. Plant Fibers

Lignocellulosic fibers extracted from plants such as hemp and flax can replace cotton and polyester fibers in textile materials and glass fibers in insulation products. Fiber crops are field crops grown for their fibers, which are traditionally used to make paper, cloth, or rope. Fiber crops are characterized by having a large concentration of cellulose, which is what gives them their strength. Due to cellulose being the main factor of a plant fibers strength, which will allow molecular manipulation to create different types of fibers.

5.2. Decomposition Chemistry

Because of the complexity of biomass, it follows that he complete biological decomposition of organic matter to methane (CH4) and carbon dioxide (CO2) under oxygen-depleted conditions – i.e., anaerobic – is complicated and is an interaction between a number of different bacteria that are each responsible for their part of the task. What may be a waste product from some bacteria could be a substrate (or food) for others, and in this way the bacteria are interdependent.

Compared with the aerobic (oxygen-rich) decomposition of organic matter, the energy yield of the anaerobic process is far smaller. The decomposition of, for example, glucose will under aerobic conditions give a net yield of 38 adenosine triphosphate (ATP) molecules, while anaerobic decomposition will yield only 2 adenosine triphosphate molecules. This means that the growth rate of anaerobic bacteria is considerably lower than that of aerobic bacteria and that the production of biomass (in the form of living bacteria) is less per gram decomposed organic matter. Where aerobic decomposition of 1 g substance results in the production of 0.5 g biomass, the yield under anaerobic conditions is only 1 gm of biomass.

6. Biomass: A Replacement for Petroleum and Natural Gas

Fuels from biomass fuels are organic materials produced in a renewable manner that are being groomed to replace the current, but depletable, source of fossil fuel sources. Two categories of biomass fuels, woody fuels and animal wastes, comprise the vast majority of available biomass fuels. Municipal solid waste (MSW) is also a source of biomass fuel. Biomass fuels have low energy densities compared to fossil fuels. In other words, a significantly larger volume of biomass fuel is required to generate the same energy as a smaller volume of fossil fuel. The low energy density means that the costs of fuel collection and transportation can quickly outweigh the value of the fuel. Biomass fuels tend to have a high moisture content, which adds weight and decreases combustion performance.

There are two primary factors to be considered in the evaluation of biomass fuels: Fuel supply, including the total quantities available, the stability of the supply or of the industry generating the fuel, and competitive uses or markets for the fuel. Cost of biomass fuel collection, processing, and transportation, and who pays these costs. This section discusses three sources of biomass fuel: woody fuels, animal waste, and urban waste, particularly municipal solid waste.

The production of fuels from biomass as a replacement for crude oil and natural gas is in active development, focusing on the use of organic matter (such as cellulose, agricultural waste, and sewage waste) in the efficient production of liquid and gas biofuels. Because the carbon in biofuels was recently extracted from atmospheric carbon dioxide by growing plants, it is considered that burning a biofuel does not result in a net increase of carbon dioxide in the atmosphere of the Earth. As a result, the production and use of biofuels are seen by many observers as a way to reduce the amount of carbon dioxide released into the atmosphere by using them to replace non-renewable sources of energy.

6.1. Gaseous Fuels

In comparison with the recognized fossil fuels, such as coal, tar sand bitumen, and crude oil residua, most biomass materials are relatively easy to gasify because they are relatively reactive with higher ignition stability. This characteristic also makes biomass materials easier to process thermochemically into higher-value fuels such as methanol or hydrogen. The mineral matter content of biomass can be problematic because of the tendency for the minerals to be corrosive as well as catalytic, when catalysis is not required. In fact, some biomass feedstocks stand out for their peculiar properties, such as high content of silicon derivatives and/or alkali metal derivatives which may require special precautions when the biomass is processed and/or for harvesting, processing and combusted. Furthermore, for a specific type of biomass, the content of the mineral matter can vary as a function of soil type and the timing of feedstock harvest. In addition to the bulk constituents of biomass, the nature of the chemical constituents is extremely important which can confer considerably variability within the types of biomass as well as options for processing (Table 2.4, Table 2.5).

The chemical structure and major organic components in biomass are extremely important in the development of processes for producing derived fuels and chemicals (Balat, 2007). The chemical components of lignocellulose can be divided into four major components. They are cellulose, hemicelluloses, lignin and extractives (Yaman, 2004). Generally, the first three components have high molecular weights and contribute much mass, while the last component is of small molecular size, and available in little quantity (Demirbaş, A. 2009). The content of cellulose and hemicellulose tend to be higher in hardwood species than in softwood species but the content of lignin is higher in softwood species than in hardwood species (Balat, 2009).

A number of processes allow biomass to be transformed into gaseous fuels such as methane or hydrogen (Sørensen et al., 2006). One pathway uses algae and bacteria that have been genetically modified to produce hydrogen directly instead of the conventional biological energy carriers. Problems are intermittent production, low efficiency and difficulty in constructing hydrogen collection and transport channels of low cost. A second pathway uses plant material such as agricultural residues in a fermentation process leading to biogas from which the desired fuels can be isolated. This technology is established and in widespread use for waste treatment, but often with the energy produced only for on-site use, which often implies less than maximum energy yields. Finally, high-temperature gasification supplies a crude gas, which may be transformed into hydrogen by a second reaction step. In addition to biogas, there is also the possibility of using the solid by-product as a biofuel.

Hydrogen, the lightest element, is a colorless, odorless, tasteless and nontoxic gas found in the air at concentrations of about 100 ppm (0.01% v/v) and it is the most abundant element in the universe, making up 75% of normal matter by mass and over 90% by number of atoms (Mariolakos et al., 2007). Hydrogen has been recognized as a promising, green and ideal energy carrier of the future due to its high energy yield and clean, efficient, renewable, sustainable and recyclable nature (Mohan et al., 2008). Hydrogen can be used as a transportation fuel, whereas neither nuclear nor solar energy can be used directly. It has good properties as a fuel for internal combustion engines in automobiles.

Hydrogen can be produced from biomass by pyrolysis, gasification, steam gasification, steam-reforming of bio-oils, and enzymatic decomposition of sugars. The yield of hydrogen that can be produced from biomass is relatively low (Demirbas, 2001). In the pyrolysis and gasification processes, water-gas shift is used to convert the reformed gas into hydrogen, and pressure swing adsorption is used to purify the product (Demirbas, 2008b). Gasification coupled with water gas shift is the most widely practiced process route for biomass to hydrogen.

In general, the gasification temperature is higher than that of pyrolysis and the yield of hydrogen from the gasification is higher than that of the pyrolysis (Balat, 2009). Modeling of biomass steam gasification to synthesis gas is a challenge because of the variability (composition, structure, reactivity, physical properties, etc.) of the raw material and because of the severe conditions (temperature, residence time, heating rate, etc.) required (Dupont et al., 2007). Hydrogen can be produced via steam gasification of biomass materials. The yield of H_2 from steam gasification increases with increasing water-to-sample (W/S) ratio (Maschio et al, 1994). The yields of H_2 from steam gasification increase with increasing of temperature.

The technologies for gas production from biomass include: (i) fermentation, (ii) gasification, and (iii) direct biophotolysis.

6.1.1. Fermentation

Traditional fermentation plants producing biogas are in routine use, ranging from farms to large municipal plants. As feedstock they use manure, agricultural residues, urban sewage and waste from households, and the output gas is typically 64% v/v methane. The biomass conversion process is accomplished by a large number of different agents, from the microbes decomposing and hydrolyzing plant material, over the acidophilic bacteria dissolving the biomass in aquatic solution, and to the strictly anaerobic methane bacteria responsible for the gas formation. Operating a biogas plant for a period of some months usually makes the bacterial composition stabilize in a way suitable for obtaining high conversion efficiency (typically above 60%, the theoretical limit being near 100%), and it is found important not to vary the feedstock compositions abruptly, if optimal operation is to be maintained. Operating temperatures for the bacterial processes are only slightly above ambient temperatures, e.g., in the mesophilic region around 30°C (86°F).

A straightforward (but not necessarily economically optimal) route to hydrogen production would be to subject the methane generated to conventional steam reforming. The ensuing biomass-to-hydrogen conversion efficiency would in practice be about 45%.

This scheme could be operated with present technology and thus forms a reference case for assessing proposed alternative hydrogen production routes.

One method is to select bacteria that produce hydrogen directly. Candidates would include *Clostridium* and *Rhodobacter* species. The best reactor operating temperatures are often in the thermophilic interval or slightly above (50 to 80°C, 122 to 176°F). Typical yields are on the order of 2 moles of hydrogen per mole of glucose, which corresponds to conversion efficiency of 17%. The theoretical maximum efficiency for the production of hydrogen is approximately 35%. Acetic acid (CH_3CO_2H) or similar carboxylic acids [$CH_3(CH_2)_nCO_2H$] are not produced in the reaction – these acids, if produced, could be the source of methane and thus additional energy, although not necessarily additional hydrogen.

Operation of this type of gas-producing plant would require pure feedstock biomass (here sugar), because of the specific bacteria needed for hydrogen production, and because contamination can cause decreased yields. Even the hydrogen produced has this negative effect and must therefore be removed continually.

6.1.2. Biophotolysis

The photosynthetic production of gas (e.g., hydrogen) hydrogen employs micro-organisms such as cyanobacteria, which have been genetically modified to produce pure hydrogen rather than the metabolically relevant substances (notably NADPH2). The conversion efficiency from sunlight to hydrogen is very small, usually under 0.1%, indicating the need for very large collection areas.

The current thinking favors ocean locations of the bio-reactors. They have to float on the surface (due to rapidly decreasing solar radiation as function of depth), and they have to be closed entities with a transparent surface (e.g., glass), in order than the hydrogen produced is retained and in order for sunlight to reach the bacteria. Because hydrogen build-up hinders further production, there further has to be a continuous removal of the hydrogen produced, by pipelines to e.g., a shore location, where gas treatment and purification can occur. These requirements make it little likely that equipment cost can be kept so low that the very low efficiency can be tolerated.

A further problem is that if the bacteria are modified to produce maximum hydrogen, their own growth and reproduction is quenched. Presumably, there has to be made a compromise between the requirements of the organism and the amount of hydrogen produced for export, so that replacement of organisms (produced at some central biofactory) does not have to be made at frequent intervals. The implication of this is probably an overall efficiency lower than 0.05%.

In a life-cycle assessment of bio-hydrogen produced by photosynthesis, the impacts from equipment manufacture are likely substantial. To this one should add the risks involved in production of large amounts of genetically modified organisms. In conventional agriculture, it is claimed that such negative impacts can be limited, because of slow spreading of genetically modified organisms to new locations (by wind or by vectors such as insects, birds or other animals).

In the case of ocean bio-hydrogen farming, the unavoidable breaking of some of the glass- or transparent plastic-covered panels will allow the genetically modified organisms to spread over the ocean involved and ultimately the entire biosphere. A quantitative discussion of such risks is difficult, but the negative cost prospects of the bio-hydrogen scheme probably rule out any practical use anyway.

Combustion is a function of the mixture of oxygen with the hydrocarbon fuel. Combustion is a series of complex consecutive homogeneous and heterogeneous reactions that require a fuel source, an oxidizing agent (oxygen but usually air), and a heat sources. In the process, gaseous fuels mix with oxygen more easily than liquid fuels, which in turn mix more easily than solid fuels. Syngas therefore inherently burns more efficiently and cleanly than the solid biomass from which it was made. Biomass gasification can thus improve the efficiency of large-scale power facilities based on biomass as the fuel such as those equipped to for forest industry residues and specialized facilities such as black liquor recovery boilers of the pulp and paper industry, both major sources of biomass power. Like natural gas, syngas can also be burned in gas turbines, a more efficient electrical generation technology than steam boilers to which solid biomass and fossil fuels are limited.

Most electrical generation systems are relatively inefficient, losing half to two-thirds of the energy as waste heat. If that heat can be used for an industrial process, space heating, or another purpose, efficiency can be greatly increased. Small modular power systems are more easily used for such cogeneration than most large-scale electrical generation.

Just as synthesis gas mixes more readily with oxygen for combustion, it also mixes readily with chemical catalysts thereby enhancing the ability of the gas to be converted to other valuable fuels, chemicals and materials. The Fischer-Tropsch process converts syngas to liquid fuels needed for transportation (Chapter 6). The water-gas shift process converts syngas to more concentrated hydrogen for fuel cells. A variety of other catalytic processes can turn syngas into a myriad of chemicals or other potential fuels or products.

6.1.3. Gasification

Gasification occurs through the thermal decomposition of biomass with the help of an oxidant such as pure oxygen or oxygen enriched air to yield a combustible gas such as synthesis gas (syngas) rich in carbon monoxide and hydrogen (Chapter 6). The synthesis gas is post-treated, by steam reforming or partial oxidation, to convert the methane (and any other hydrocarbon derivatives) produced by gasification into hydrogen and carbon monoxide. The carbon monoxide is then put through the shift process to obtain a higher fraction of hydrogen, by carbon dioxide-removal and methanation or by pressure swing adsorption (Speight, 1993; Mokhatab et al., 2006).

$$CH_4 + H_2O \rightleftharpoons \underline{CO} + 3H_2$$
$$CO + H_2O \rightleftharpoons \underline{CO_2} + H_2$$

The gasification of biomass (such as, for example, wood waste or wood scraps) is a well-known process, taking place in pyrolysis reactors in which the oxygen supply less than that required for complete combustion) or fluidized-bed type of reactors. Conditions such as operating temperature determine whether hydrogen is consumed or produced in the process. The hydrogen fraction (in this case typically on the order of 30%) must be separated for most fuel-cell applications, as well as for long-distance pipeline-transmission. In the pyrolysis-type application, gas production is low and most energy is in the oil-type products (often referred to as bio-oil) that must be subsequently reformed in order to produce significant amounts of hydrogen. Alternative concepts use membranes to separate the gases produced (Chapter 7), and many reactor types uses catalysts to help the processes to proceed in the desired direction, notably at a lower temperature on the order of 500°C (930°F).

Environmental concerns include disposal of associated higher-boiling products (often designated as tar even though the products bear no similarity to tar produced from feedstocks such as coal) and ash, particularly for the fluidized bed reactors, where these substances must be separated from the flue gas stream (in contrast to the pyrolysis plants, where most tar and ash deposits at the bottom of the reactor). The ash produced from the mineral matter in the biomass feedstock has the potential to be used as (i) a clarifying agent in water treatment, (ii) a wastewater adsorbent, (iii) a liquid waste adsorbent, (iv) a hazardous waste solidification agent, (v) a lightweight fill aggregate for roadways, parking areas, and structures, (vi) as asphalt mineral filler, or (v) as a mine spoil amendment.

6.2. Liquid Fuels

Among the liquid biomass fuels, biodiesel (vegetable oil ester) is noteworthy for its similarity to petroleum-derived diesel fuel, apart from its negligible sulfur and ash content. Bioethanol has only about 70% the heating value of petroleum distillates such as gasoline, but its sulfur and ash contents are also very low. Both of these liquid fuels have lower vapor pressure and flammability than their petroleum-based competitors – an advantage in some cases (e.g., use in confined spaces such as mines) but a disadvantage in others (e.g., engine starting at cold temperatures).

Currently the production of ethanol by fermentation of corn-derived carbohydrates is the main technology used to produce liquid fuels from biomass resources. Furthermore, amongst different biofuels, suitable for application in transport, bioethanol and biodiesel seem to be the most feasible ones at present. The key advantage of bioethanol and biodiesel is that they can be mixed with conventional gasoline and diesel fuel respectively, which allows using the same handling and distribution infrastructure. Another important strong point of bioethanol and biodiesel is that when they are mixed at low concentrations ($\leq$10% v/v bioethanol in gasoline and $\leq$20% v/v biodiesel in diesel), no engine modifications are necessary.

Alternatively, biomass can be converted into fuels and chemicals indirectly (by gasification to syngas followed by catalytic conversion to liquid fuels) or directly to a liquid product by thermochemical means. Direct thermochemical conversion processes include pyrolysis, liquefaction, and solvolysis (Kavalov and Peteves, 2005). Thermochemical treatment of biomass to produce hydrocarbons, while relatively simple to perform, typically such processes are non-selective, producing a wide range of products.

The production of biodiesel from vegetable oil represents another means of producing liquid fuels from biomass, and one which is growing rapidly in commercial importance. Commercially, biodiesel is produced from vegetable oils, including rapeseed, sunflower and soybean oil, and animal fats (McNeil Technologies Inc. 2005). These oils and fats are typically composed of C_{14} to -C_{20} fatty acid triglycerides (constituting approximately 90 to 95% by weight of the oil). In order to produce a fuel that is suitable for use in diesel engines, these triglycerides are converted to the respective alkyl esters and glycerol by base-catalyzed transesterification with short chain alcohols (generally methanol.

Thus, for every 10 lbs. of biodiesel produced, approximately 1 lb. of glycerol is formed. Glycerol finds application in a wide range of industries (cosmetics, pharmaceuticals, as a plasticizer, etc.), although as biodiesel production grows, new uses will have to be developed to avoid a surplus of glycerol. Biologically produced alcohols, most commonly ethanol and methanol, and less commonly propanol and butanol are produced by the action of microbes and enzymes through fermentation.

Methanol (methyl alcohol, CH_3OH) which is currently produced from natural gas, can also be produced from biomass, although the economic viability must be proven. Ethanol (ethyl alcohol, C_2H_5OH) produced from sugar cane is being used as automotive fuel in some countries and as a gasoline additive in the United States, but direct use as fuel is growing. Cellulosic ethanol (i.e., ethanol produced from cellulose) is being manufactured from straw (an agricultural waste product).

Biologically produced oil (and gas) can be produced from various wastes. For example the thermal depolymerization of organic waste can extract methane and other oils similar to petroleum. Gas-to-liquids (GTL) or biomass-to-liquids (BTL) both produce synthetic fuels out of biomass in the Fischer-Tropsch process (Chapter 6). The synthetic biofuel containing oxygen is used as additive in high quality diesel and gasoline. Furthermore, the diesel fraction produced from biomass is suitable for use in diesel engines.

Feedstocks for such products are (i) vegetable oil, (ii) waste vegetable oil, such as waste cooking oils and greases produced in quantity mostly by commercial kitchens, (iii) the transesterification of animal fats and vegetable oil can yield biodiesel that is directly usable in petroleum diesel engines, (iv) a product known as biologically derived crude oil

– a misnomer and the liquid is more typically known as bio-oil – is produced together with biogas and a carbonaceous solid (char) solid via the thermal depolymerization of complex organic materials including non-crude oil based materials (for example waste products such as old tires, offal, wood and plastic). In addition the bio-oil product (sometimes referred to as pyrolysis oil) may be produced out of biomass, wood waste etc. using heat only in the flash pyrolysis process but the oil has to be treated before using in conventional fuel systems or internal combustion engines (water + pH).

6.3. Solid Fuels

Examples of solid fuels from biofuel feedstocks include wood and wood-derived charcoal and dried manure, particularly cattle manure. This type of product is often referred to as biochar which can provide a feedstock for a gasifier to produce bio-synthesis gas (Chapter 6).

Biochar is a high-carbon, fine-grained carbonaceous residue that today is produced through pyrolysis processes and represents the direct thermal decomposition of biomass in the absence of oxygen (which prevents combustion of the feedstock) and which produces a mixture of (i) gases also called biogas, (ii) liquids also called bio-oil, and (iii) the solid biochar. The specific yield of the products from the pyrolysis of biomass is dependent on process condition, such as temperature, and can be optimized to produce either gases, liquids, or biochar. Temperatures on the order of 400 to 500°C (750 to 930 F) produce higher yields of char, while temperatures above 700°C (1290 F) favor a higher yield of liquid and gaseous products. Pyrolysis occurs more quickly at the higher temperatures, typically requiring seconds instead of hours. By comparison, slow pyrolysis can produce substantially more char (yield: approximately 50% w/w).

One widespread use of biochar is in home cooking and heating. The biofuel may be burned on an open fireplace or in a special stove. The efficiency of this process may vary widely, from 10% for a well-made fire (even less if the fire is not made carefully) up to 40% for a custom designed charcoal stove. Inefficient use of fuel is a cause of deforestation (though this is negligible compared to deliberate destruction to clear land for agricultural use) but more importantly it means that more work has to be put into gathering fuel, thus the quality of cooking stoves has a direct influence on the viability of biofuels. In addition, burning corn in special stoves to reduce their energy bills in parts of North America. Corn-generated heat costs less than a fifth of the current rate for propane and about a third of electrical heat.

In addition to home use, biochar also offers a number of benefits for soil conditioning because of the extremely porous nature of biochar which is effective at retaining both water and water-soluble nutrients. Also, for plants that require high potash (potassium-bearing materials) and elevated pH (alkaline conditions), biochar can be used as a soil amendment to improve yield. Biochar can be used to (i) improve water quality, (ii) reduce soil emissions of greenhouse gases, (iii) reduce nutrient leaching, (iii) reduce soil acidity, (iv) reduce irrigation as well as fertilizer requirements, and above all in the context of this book, (v) be used as a feedstock to produce bi-synthesis gas.

7. Processes

Biomass can be converted into commercial fuels, suitable to substitute for fossil fuels. These can be used for transportation, heating, electricity generation or anything else fossil fuels are used for. The conversion is accomplished through the use of several distinct processes which include both biochemical conversion and thermal conversion to produce gaseous, liquid and solid fuels which have high energy contents, are easily transportable, and are therefore suitable for use as commercial fuels.

Biochemical conversion of biomass is completed through alcoholic fermentation to produce liquid fuels and "anaerobic" digestion or fermentation, resulting in biogas. Alcoholic fermentation of crops such as sugarcane and maize (corn) to produce ethanol for use in internal combustion engines has been practiced for years with the greatest production occurring in Brazil and the U.S., where ethanol has been blended with gasoline for use in automobiles. With slight engine modifications, automobiles can operate on ethanol alone.

Anaerobic digestion of biomass has been practiced for almost a century, and is very popular in many developing countries such as China and India. The organic fraction of almost any form of biomass, including sewage sludge, animal wastes and industrial effluents, can be broken down through anaerobic digestion into methane and carbon dioxide. This "biogas" is a reasonably clean burning fuel which can be captured and put to many different end uses such as cooking, heating or electrical generation.

Wood and many other similar types of biomass which contain lignin and cellulose, (agricultural wastes, cotton gin waste, wood wastes, peanut hulls etc.) can be converted through thermochemical processes into solid, liquid or gaseous fuels. Pyrolysis, used to produce charcoal since the dawn of civilization, is still the most common thermochemical conversion of biomass to commercial fuel.

During pyrolysis, biomass is heated in the absence of air and breaks down into a complex mixture of liquids, gases, and a residual char. If wood is used as the feedstock, the residual char is what is commonly known as charcoal. With more modern technologies, pyrolysis can be carried out under a variety of conditions to capture all the components, and to maximize the output of the desired product be it char, liquid or gas.

There is a consensus amongst scientists that biomass fuels used in a sustainable manner result in no net increase in atmospheric carbon dioxide (CO_2). Some scientists would even go as far as to declare that sustainable use of biomass will result in a net decrease in atmospheric carbon dioxide. This loose conclusion assumes that all the carbon dioxide that is emitted by the use of biomass fuels was recently taken in from the atmosphere by photosynthesis. Increased substitution of fossil fuels with biomass based fuels would therefore help reduce the potential for global warming, caused by increased atmospheric concentrations of carbon dioxide.

Unfortunately, things may not be as simple as has been assumed above. Currently, biomass is being used all over the world in a very unsustainable manner, and the long term effects of biomass energy plantations has not been proven. As well, the natural humus and dead organic matter in the forest soils is a large reservoir of carbon. Conversion of natural ecosystems to managed energy plantations could result in a release of carbon from the soil as a result of the accelerated decay of organic matter.

An ever increasing number of people on this planet are faced with hunger and starvation. It has been argued that the use of land to grow fuel crops will increase this problem. Hunger in developing countries, however, is more complex than just a lack of agricultural land. Many countries in the world today, such as the U.S., have food surpluses. Much fertile agricultural land is also used to grow tobacco, flowers, food for domestic pets and other "luxury" items, rather than staple foods. Similarly, a significant proportion of agricultural land is used to grow feed for animals to support the highly wasteful, meat centered diet of the industrialized world. By feeding grain to livestock we end up with only about 10% of the caloric content of the grain. When looked at in this light, it does not seem to be so unreasonable to use some fertile land to grow fuel. Marginal land and underutilized agricultural land can also be used to grow biomass for fuel.

Acid rain, which can damage lakes and forests, is a by-product of the combustion of fossil fuels, particularly coal and oil. The high sulfur content of these fuels together with hot combustion temperatures result in the formation of sulfur dioxide (SO_2) and nitrous oxides (NO_x), when they are burned to provide energy. The replacement of fossil fuels with biomass can reduce the potential for acid rain. Biomass generally contains less than 0.1% sulfur by weight compared to low sulfur coal with 0.5 to 4% w/w sulfur. Lower combustion temperatures and pollution control devices such as wet scrubbers and electro-static precipitators can also keep emissions of NO_x to a minimum when biomass is burned to produce energy.

The final major environmental impact of biomass energy may be that of loss of biodiversity. Transforming natural ecosystems into energy plantations with a very small number of crops, as few as one, can drastically reduce the biodiversity of a region. Such "monocultures" lack the balance achieved by a diverse ecosystem, and are susceptible to widespread damage by pests or disease.

8. USES

Biomass currently supplies 14% of the world's energy needs, but has the theoretical potential to supply 100%. Most present day production and use of biomass for energy is carried out in a very unsustainable manner with a great many negative environmental consequences. If biomass is to supply a greater proportion of the world's energy needs in the future, the challenge will be to produce biomass and to convert and use it without harming the natural environment. Technologies and processes exist today which, if used properly, make biomass based fuels less harmful to the environment than fossil fuels. Applying these technologies and processes on a site specific basis in order to minimize negative environmental impacts is a prerequisite for sustainable use of biomass energy in the future.

Biodiesel and bioethanol are widely used in automobiles and freight vehicles. For example, in Germany most diesel on sale at gas stations contains a few percent biodiesel, and many gas stations also sell 100% biodiesel. Some supermarket chains in the UK have switched to running their freight fleets on 50% biodiesel, and often include biofuels in the vehicle fuels they sell to consumers, and an increasing number of service stations are selling biodiesel blends (typically with 5% biodiesel).

In Europe, research is being undertaken into the use of biodiesel as a domestic heating oil. A blend of 20% v/v biodiesel with 80% v/v kerosene (known as B20 fuel) has been tested successfully to power modern high efficiency condensing oil boilers. Boilers needed a preheat burner to prevent nozzle blockages and maintain clean combustion. Blends with a higher proportion of biodiesel were found to be less satisfactory, owing to the greater viscosity of biodiesel than conventional fuels when stores in fuel tanks outside the building at typical UK winter temperatures.

Different combustion-engines are being produced for very low prices lately. They allow the private house-owner to utilize low amounts of weak compression of methane to generate electrical and thermal power (almost) sufficient for a well-insulated residential home.

Direct biofuels are biofuels that can be used in existing unmodified petroleum engines. Because engine technology changes all the time, exactly what a direct biofuel is can be hard to define; a fuel that works without problem in one unmodified engine may not work in another engine. In general, newer engines are more sensitive to fuel than older engines, but new engines are also likely to be designed with some amount of biofuel in mind.

Straight vegetable oil can be used in some (older) diesel engines. Only in the warmest climates can it be used without engine modifications, so it is of limited use in colder climates. Most commonly it is turned into biodiesel. No engine manufacturer explicitly allows any use of vegetable oil in their engines.

Biodiesel can be a direct biofuel. In some countries manufacturers cover many of their diesel engines under warranty for 100% biodiesel use. Many people have run thousands of miles on biodiesel without problem, and many studies have been made on 100% biodiesel. In many European countries, 100% biodiesel is widely used and is available at thousands of gas stations.

Ethanol is the most common biofuel, and over the years many engines have been designed to run on it. Many of these could not run on regular gasoline. It is open to debate if ethanol is a direct replacement in these engines though - they cannot run on anything else. In the late 1990s engines started appearing that by design can use either fuel. Ethanol is a direct replacement in these engines, but it is debatable if these engines are unmodified, or factory modified for ethanol.

Butanol is often claimed as a direct replacement for gasoline. It is not in wide spread production at this time, and engine manufacturers have not made statements about its use. While it appears that butanol has sufficiently similar characteristics with gasoline such that it should work without problem in any gasoline engine, no widespread experience exists.

There is some concern about the energy efficiency of biofuel production. Production of biofuels from raw materials requires energy (for farming, transport and conversion to final product), and it is not clear what the overall efficiency of the process is. For some biofuels the energy balance may even be negative. Since vast amounts of raw material are needed for biofuel production, monocultures and intensive farming may become more popular, which may cause environmental damages and undo some of the progress made towards sustainable agriculture.

Finally, biomass will play an important role in the future global energy infrastructure for the generation of power and heat, but also for the production of chemicals and fuels. The dominant biomass conversion technology will be gasification, as the gases from biomass gasification are intermediates in the high-efficient power production or the synthesis from chemicals and fuels. In the discussion on the utilization of gases from biomass gasification it is important to understand that the composition of the gasification gas is very dependent on the type of gasification process and especially the gasification temperature.

In the past several definitions for gasification gases were used without making this distinction. However, clear understanding of the relation between gasification technologies, the generated gas, its typical applications, and the corresponding specifications is crucial in today's decision-making processes. Incorrect perceptions resulting from confusing terminologies may result in delayed developments, too high expectations, disappointments, and in the worst-case into loss of support because promises are not made true.

REFERENCES

Abella, L., Nanbu, S., and Fukuda, K. 2007. A Theoretical Study on Levoglucosan Pyrolysis Reactions Yielding Aldehydes and a Ketone in Biomass. In: *Memoirs of Faculty of Engineering*, Kyushu University. 67: 67-74.

Balat, M. 2007. Hydrogen in Fueled Systems and the Significance of Hydrogen in Vehicular Transportation. *Energy Sources Part B,* 2: 49-61.

Balat M. 2009. Gasification of Biomass to Produce Gaseous Products. *Energy Sources Part A*, 31: 516-526.

Balat, H., and Kirtay, E. 2010. Hydrogen from Biomass – Present Scenario and Future Prospects. *International Journal of Hydrogen Energy*, 35: 7416-7426.

Boerrigter H., Van Der Drift A. 2004. *Biosyngas: Description of R&D trajectory necessary to reach large-scale implementation of renewable syngas from biomass.* Energy Research Center of the Netherlands, Petten, The Netherlands.

Bridgwater, A. V.; Czernik, S.; Piskorz, J. 2001. An Overview of Fast Pyrolysis. *Prog. Thermochem. Biomass Convers*, 2: 977-997.

Cobb, J. T. Jr. 2007. Production of Synthesis Gas by Biomass Gasification. Proceedings. Spring National Meeting AIChE. Houston, Texas, April 22-26, 2007.

Crocker, M., and Crofcheck, C. 2006. Biomass Conversion to Liquid Fuels and Chemicals. *Energeia*, 17: 1-3.

Demirbas, A. 2001. Yields of Hydrogen of Gaseous Products via Pyrolysis from Selected Biomass Samples. *Fuel,* 80: 1885-1891.

Demirbaş, A. 2008a. Products from Lignocellulosic Materials via Degradation Processes. *Energy Sources Part A*, 30: 27-37.

Demirbas, A. 2008b. Biohydrogen generation from organic waste. *Energy Sources Part A,* 30:475-482.

Demirbaş, A. 2009. Pyrolysis of Biomass for Fuels and Chemicals. *Energy Sources Part A,* 31: 1028-1037.

Dobele, G., Urbanovich, I., Volpert, A., Kampars, V., and Samulis, E. 2007. Fast pyrolysis – Effect of wood drying on the yield and properties of bio-oil. *Bioresources,* 2: 699-706.

Dupont, C., Boissonnet, G., Seiler, J. M., Gauthier, P., and Schweich, D. 2007. Study about the Kinetic Processes of Biomass Steam Gasification. *Fuel,* 86: 32-40.

Feldman, D., Banu, D., Natansohn, A., and Wang, J. 1991. Structure-Properties Relations of Thermally Cured Epoxy-Lignin Polyblends. *J. App. Polym. Sci.*, 42: 1537-1550.

Kavalov, B., and Peteves, S. D. 2005. Status and Perspectives of Biomass-to-Liquid Fuels in the European Union. European Commission. Directorate General Joint Research Centre (DG JRC). Institute for Energy, Petten, The Netherlands.

Mariolakos, I., Kranioti, A., Markatselis, E., and Papageorgiou, M. 2007. Water, Mythology and Environmental Education. *Desalination* 213:141-146.

Maschio, G., Lucchesi, A., and Stoppato, G. 1994. Production of Syngas from Biomass. *Biores. Technol.*, 48: 119-126.

McKendry, P. 2002a. Energy Production from Biomass (Part 1): Overview of Biomass. *Bioresource Technology,* 83: 37-46.

McKendry, P. 2002b. Energy Production from Biomass (Part 2): Conversion Technologies. *Bioresource Technology,* 83: 47-56.

McNeil Technologies Inc. 2005. Colorado Agriculture IOF: Technology Assessments Liquid Fuels. Prepared Under State of Colorado Purchase Order # 01-336. Governor's Office of Energy Conservation and Management, Denver, Colorado.

Mohan, D., Pittman, C. U., and Steele, P. H. 2006. Pyrolysis of Wood/Biomass for Bio-oil: a critical review. *Energy Fuels*, 20: 848-889.

Mohan, S. V., Mohanakrishna, G., and Sarma, P. N. 2008. Integration of acidogenic and methanogenic processes for simultaneous production of biohydrogen and methane from wastewater treatment. *Int J Hydrogen Energy*, 33:2156-2166.

NREL. 2003. *Dollars from Sense*. National Renewable Energy Laboratory, Golden, Colorado. http://www.nrel.gov/docs/legosti/fy97/20505.pdf.

Pandey, A., Negi, S., Binod, P., and Larroche, C. (Editors) 2015. *Pretreatment of Biomass: Processes and Technologies*. Elsevier BV, Amsterdam, Netherlands.

Prins, M. J., Ptasinski, K. J., and Janssen, F. J. J. G. 2006a. More Efficient Biomass Gasification via Torrefaction. *Energy,* 31(15): 3458-3470.

Prins, M. J., Ptasinski, K. J., and Janssen, F. J. J. G. 2006b. Torrefaction of Wood: Part 1. Weight Loss Kinetics. *Journal of Analytical and Applied Pyrolysis*, 77(1): 28-34.

Prins, M. J., Ptasinski, K. J., and Janssen, F. J. J. G. 2006c. Torrefaction of Wood: Part 2. Analysis of Products. *Journal of Analytical and Applied Pyrolysis*, 77(1): 35-40.

Ruth, M. 2004. *Development of a Biorefinery Optimization Model.* Renewable Energy Modeling Series Forecasting the Growth of Wind and Biomass. National Bioenergy Centre, National Renewable Energy Laboratory, Golden, Colorado. (http://www.epa.gov/cleanenergy/pdf/ruth2_apr20.pdf).

Sørensen, B. E., Njakou, S., and Blumberga, D. 2006. Gaseous Fuels Biomass. *Proceedings. World Renewable Energy Congress IX*. WREN, London.

Speight, J. G. 2008. *Synthetic Fuels Handbook: Properties, Processes, and Performance*. McGraw-Hill, New York.

Speight, J. G. (Editor). 2011a. *The Biofuels Handbook.* The Royal Society of Chemistry, London, United Kingdom.

Speight, J. G. 2011b. *An Introduction to Petroleum Technology, Economics, and Politics.* Scrivener Publishing, Beverly, Massachusetts.

Speight, J. G. 2014. *The Chemistry and Technology of Petroleum* 5th Edition. CRC Press, Taylor & Francis Group, Boca Raton, Florida.

Speight, J. G., and Islam, M. R. 2016. *Peak Energy – Myth or Reality*. Scrivener Publishing, Beverly.

Speight, J. G. 2017. *Handbook of Petroleum Refining*. CRC Press, Taylor & Francis Group, Boca Raton, Florida.

Speight, J. G. 2019a. *Handbook of Petrochemical Processes*. CRC Press, Taylor & Francis Group, Boca Raton, Florida.

Speight, J. G. 2019b. *Heavy Oil Recovery and Upgrading*. Gulf Professional publishing Company, Elsevier, Oxford, United Kingdom.

Van Der Drift A., Van Ree R., Boerrigter H., Hemmes K. 2004. Bio-Syngas: Key Intermediate for Large Scale Production of Green Fuels and Chemicals," *Proceedings. 2nd World Conference and Technology Exhibition "Biomass for Energy, Industry and Climate Protection,"* Rome. http://www.conference-biomass.com/Biomass2004/conference_Welcome.asp.

Wright, L., Boundy, R, Perlack, R., Davis, S., and Saulsbury. B. 2006. Biomass Energy Data Book: Edition 1. Office of Planning, Budget and Analysis, Energy Efficiency and Renewable Energy, United States Department of Energy. Contract No. DE-AC05-00OR22725. Oak Ridge National Laboratory, Oak Ridge, Tennessee.

Yaman, S. 2004. Pyrolysis of Biomass to Produce Fuels and Chemical Feedstocks. *Energy Convers. Manage,* 45: 651-671.

PART 2: BIOGAS PRODUCTION AND PROPERTIES

Chapter 3

BIOGAS BY ANAEROBIC DIGESTION

1. INTRODUCTION

The conversion of organic materials into biogas is a relatively simple technology that is, however, dependent on the type of biomass (the feedstock) type as well as on the operational parameters (Molino et al., 2013a, 2013b, 2018). The anaerobic digestion process is actually is a collection of processes by which microorganisms break down biodegradable organic material in the absence of oxygen (Braun, 2007; Bharathiraja, 2018). The process occurs naturally in some soils as well as in lake sediments and in ocean basin sediments and typically begins with the bacterial hydrolysis of the organic feedstock. The production of biogas provides a versatile carrier of renewable energy, as methane can be used for replacement of fossil fuels in both heat and power generation and as a vehicle fuel (Chapter 8). In this context, biogas from wastes, residues, and energy crops will play a vital role in future. Biogas is a versatile renewable energy source, which can be used for replacement of fossil fuels in power and heat production, and it can be used also as gaseous vehicle fuel. Methane-rich biogas (biomethane) can replace also natural gas as a feedstock for producing chemicals and materials.

In the anaerobic digestion process, insoluble organic high molecular weight organic materials (sometimes referred to as high molecular weight polymeric materials), such as carbohydrate derivatives, are broken down to produce simpler materials (often soluble derivatives) that become available for other bacteria. Acidogenic bacteria then convert the products (such as sugar derivatives and amino acid derivatives) into carbon dioxide (CO_2), hydrogen (H_2), ammonia (NH_3), and organic acids (RCO_2H). Other bacteria (acetogenic bacteria) convert these resulting organic acids into acetic acid (CH_3CO_2H) as well as additional ammonia, hydrogen, and carbon dioxide. In the final stages of the process, methanogenic bacteria (often referred to as methanogens) convert these products to methane (CH_4) and carbon dioxide. In the current environmentally-concerned world, the anaerobic digestion of energy crops, residues, and wastes is of increasing interest in order to reduce the greenhouse gas emissions and to facilitate a sustainable development of energy supply.

As previously mentioned (Chapter 1), methane is a greenhouse gas that contributes significantly to the enhanced greenhouse effect – part of the methane currently emitted into the atmosphere originates from agricultural activities where the largest sources are farm animal digestion and handling of animal manure – and caution is advised to ensure the leakage of the methane in the atmosphere does not occur.

To be able to compare methane (CH_4) and nitrous oxide (N_2O) to carbon dioxide, methane and nitrous oxide are usually converted to carbon dioxide-equivalents (CO_2-equivalents) when assessing the global warming potential of these gases. For example:

1 kg CH_4 ≡ approximately 22 kg CO_2
1 kg N_2O ≡ approximately 310 kg CO_2

Thus, nitrous oxide thus has a global warming potential that is 310 times larger than that of carbon dioxide and very little additional nitrous oxide is needed for the greenhouse effect to be affected.

As a note of caution in this respect, nitrous oxide, which can also be a constituents of biogas is formed during the conversion (denitrification) of nitrogenous fertilizers on farmland under anaerobic conditions and in the presence of easily degradable organic matter. The anaerobic digestion of the feedstock slurry (including manure) in biogas plants reduces the content of easily degradable matter due to the production of biogas. During subsequent storage, less methane will therefore be released from the storage facility tank and less nitrous oxide will be released in the field when the slurry is later used as a fertilizer.

For biogas production by the anaerobic digestion process, various process types are applied which can be classified in wet and dry fermentation systems. Thus, in general terms, if the solid concentration is lower than 10% w/w of the feedstock, the digestion process is considered to be a wet process (wet fermentation). On the other hand, the dry process (dry fermentation) is the process when the concentration of solids in the feedstocks is on the order of 15 to 35% w/w. The wet digestion process is dominant among biogas production plants since it allows the application of completely stirred tank digesters plus a continuous feeding process. The dry digestion process requires batch feeding to the tank and it is more challenging to apply proper mixing, which can increase a chance of swimming layers and sediments (Weiland, 2009). The term swimming layer (sometime called the *floating layer*) refers to feedstock addition which results in floating material. This floating layer can dry out and, with insufficient mixing, will form a crust that can take up digester capacity thereby reducing digester efficiency.

At this point, it is necessary to differentiate between wet digester and dry digester. From a general aspect, the subdivision of the digestion processes into wet processes and dry processes (sometime referred to as solid-state processes) can be is confusing since the microorganisms involved in the digestion process always require a liquid medium in which to survive and grow. Confusion also arises when defining the dry matter content of the fresh mass that is to be digested, since it is common practice to use several different feedstocks, each with a different dry matter content.

In this connection it must be made clear that it is not the dry matter content of the individual feedstocks that determines the classification of the process but the dry matter content of the feedstock mixture that is introduced into the digester. Therefore, the classification of the process into wet digestion or dry digestion therefore is arbitrary and is based on the dry matter content of what is contained *in the digester* and in both cases the microorganisms require sufficient water for their activity in the environment of the digester. Thus, although there is no precise definition of the dividing line between wet and dry digestion, in practice it has become customary to refer to a wet digestion process when using energy crops with a dry matter content of up to approximately 12% w/w in the digester, because the digester contents are generally still pumpable with this water content. If the dry matter content in the digester rises to 15 to 16% w/w or more, the material is usually no longer pumpable and the process is referred to as dry digestion.

Most systems are wet digester systems using vertical stirred tank digester with different stirrer types dependent on the origin of the feedstock. The digestate from anaerobic fermentation (the material remaining after the anaerobic digestion of a biodegradable feedstock) contains most of the nutrients originating from the feedstock is a valuable fertilizer due to the increased availability of nitrogen and the better short-term fertilization effect. It is, in fact, the solid remnants of the original input material to the digesters that the microbes use for sustenance. It also consists of the mineralized remains of the dead bacteria from within the digesters. However, depending on the feedstock, the digestate has different properties including the potential for pathogens (microorganisms, such as a viruses bacteria, or fungi) to be present in the digestate. However, in some cases, anaerobic treatment tends to have an adverse effect on the survival of pathogens thereby reducing any risks when the digested residue to be used as fertilizer.

The production of biogas through anaerobic digestion offers significant advantages over other forms of bioenergy production. It has been evaluated as one of the most energy-efficient and environmentally beneficial technology for bioenergy production (Fehrenbach et al., 2008). The digestion process is widely used as part of the process to treat biodegradable waste and sewage sludge and, as part of an integrated waste management system, anaerobic digestion reduces the emission of landfill gas into the atmosphere. However, anaerobic digestion is widely used as a source biogas, consisting predominantly of methane and carbon dioxide, but there are traces of other contaminant gases. This biogas can be used directly as lean fuel (low-quality methane) or upgraded to contaminant-free methane (often referred to, in this context, as biomethane).

Many microorganisms affect anaerobic digestion, including acetic acid-forming bacteria (acetogens) and methane-forming archaea (methanogens) which promote a number of chemical processes in the conversion of biomass to biogas. Digestate can come in three forms: (i) a fibrous product, (ii) a liquid product, sometimes called *liquor* or *digester liquor*, or (iii) a sludge-based combination of the first two fractions. In two-stage digester systems, different forms of digestate come from different digestion tanks whereas in single-stage digestion systems, the two fractions will be combined and, if desired, separated by further processing (Bertin et al., 2013).

The second byproduct (acidogenic digestate) is a stable, organic material consisting largely of lignin and cellulose, but also of a variety of mineral components in a matrix of dead bacterial cells; some plastic may be present. The material often resembles domestic compost and can be used as such or to make low-grade building products, such as fiberboard. The solid digestate can also be used as an organic feedstock for ethanol production. The third byproduct is a liquid (methanogenic digestate) rich in nutrients, which can be used as a fertilizer, depending on the quality of the digester feedstock.

In any situation where the digestate is to be used for further production, the levels of potentially toxic elements (PTEs) should be chemically assessed. Though small quantities of mineral ions like sodium, potassium stimulates the growth of bacteria, the high concentration of heavy metals and detergents have negative impact in gas production rate (often referred to as *inhibition*). Detergents such as soap, antibiotics, and organic solvents are toxic to the growth of microbes inside the digester. Addition of these substances along with the feedstock should be avoided.

In the case of most clean and source-separated biodegradable waste streams, the levels of potentially toxic elements will be (or should be) low but caution should prevail. . In the case of wastes originating from an industrial process, the levels of potentially toxic elements may be higher than from non-industrial processes (but this is not always the case) and the potential of the presence of toxic elements should be given serious consideration before deciding upon a suitable end use for the material.

On the organic side, the digestate may (typically) contain, for example, lignin which cannot be broken down by the anaerobic microorganisms. Also, the digestate may contain ammonia that is phytotoxic (a toxic effect by a compound on plant growth) insofar as it can hamper the growth of plants if it is used as a soil-improving material. For these two reasons, a maturation or composting stage may be necessary after digestion. However, lignin and other complex constituents of a feedstock may be amenable to degradation by aerobic microorganisms, such as fungi, helping reduce the overall volume of the material for transport. During this process, the ammonia will be oxidized into nitrate derivatives ($-NO_3^-$) thereby improving the material and making it more suitable as a fertilizer and/or a soil improver.

Gaseous oxygen is excluded from the reactions in the digester by physical containment since the anaerobic species utilize electron acceptors (molecular species that act as oxidizing agents in chemical reactions) from sources other than oxygen gas (O_2). The acceptors can be the organic material itself or may be derived from any inorganic oxide derivatives that occur in the feedstock. When the oxygen source in an anaerobic system is derived from the organic material itself, the intermediate products are primarily alcohol derivatives (ROH), aldehyde derivatives (RCHO), and organic acid derivatives (RCO_2H) as well as carbon dioxide. In the presence of specialized methanogens (microorganisms that produce methane as a metabolic byproduct in hypoxic conditions), the intermediate products undergo further conversion to methane, carbon dioxide, and trace levels of hydrogen sulfide (H_2S). In an anaerobic system, thc majority of the chemical energy contained within the starting material is released by methanogenic bacteria as methane. However, anaerobic microorganisms typically take a significant period of time to be fully effective and it is common practice is to introduce anaerobic microorganisms from materials with existing populations (seeding the digesters) which is typically accomplished with the addition of a promoter, such as sewage sludge or slurry of cattle manure.

The final output from anaerobic digestion systems is water, which originates both from the moisture content of the original waste that was treated and water produced during the microbial reactions in the digester. This water may be released from the dewatering of the digestate or may be implicitly separate from the digestate. If the wastewater is exiting the anaerobic digestion facility, it will typically have elevated levels of biochemical oxygen demand (BOD) and chemical oxygen demand (COD). These data are indications of the reactivity of the effluent indicate an ability of the effluent to pollute land system. Some of this material cannot be accessed by the anaerobic bacteria for conversion into biogas and if this effluent was directly introduced into (or allowed to enter) into watercourses, it would negatively affect the watercourse by causing eutrophication (the situation when a body of water becomes overly enriched with minerals and nutrients that induce excessive growth of plants and algae). As such, further treatment of the wastewater is often required and will typically be an oxidation stage wherein air is passed through the water in a sequencing batch reactors or reverse osmosis unit.

2. History

The history of anaerobic digestion is a long one, beginning as early as tenth century BC in Assyria where biogas was used to heat bath water. Reported scientific interest in the manufacture of gas produced by the natural decomposition of organic matter dates from the 17th Century, when Robert Boyle (1627-1691) and Stephen Hales (1677-1761) reported that when the sediments of streams and lakes are disturbed there is (potentially) released of a flammable gas and, in 1778, the Italian physicist Alessandro Volta (1745-1827), identified the gas as methane. In 1808, Sir Humphry Davy (1778-1829) proved that methane also occurred in the gases emitted from the manure of cattle. Following from this, the first attempts to deliberately produce methane from waste organic materials resulted in the construction of an anaerobic digester in 1859 at Mumbai (formerly known as Bombay, India). In 1895, the technology was further developed in the City of Exeter (England), where a septic tank was used to generate gas for the sewer gas destructor lamp, a type of gas lighting.

By the early 20th Century, anaerobic digestion systems began to resemble the current modern technology. For example, in 1904 the first dual-purpose tank for both sedimentation and sludge treatment was installed in Hampton (London, England). In 1906, an early form of anaerobic digester (the Imhoff tank) was developed and, after 1920, closed tank systems began to replace the previously common use of anaerobic lagoons-covered earthen basins that were commonly used as digesters up to that time. This led to the design of balanced digester systems in which the stages the rates of degradation were equal in size. If the first degradation step runs too fast, the acid concentration rises, and the pH drops below 7.0 there may be an inhibition effect that interferes with the action of the bacteria. Inhibition occurs when a substance (such as a product of one of the stages) has a negative effect on bacteria without directly killing them. The delicately-balanced digestion process can be inhibited in many ways and the ways are often divided into endogenous and exogenous causes. Endogenous inhibition is due to conditions or material created during the process itself that under certain circumstances may inhibit the process. On the other hand, exogenous inhibition is due to influence of condition that are external to the process. However, inhibition is not always automatics where there is a change in the pH of the digester system. Whether the process is single-stage or multi-stage, the pH value is established automatically within the system by the alkaline and the acidic products of the metabolic processes formed in the course of anaerobic de- composition.

3. THE PROCESS

In principle, all organic materials can ferment or be digested. However, only homogenous and liquid feedstocks can be considered for simple biogas plants. When the plant is filled, the feedstock has to be diluted with about the same quantity of liquid, if possible, the urine should be used. Waste and wastewater from food-processing industries are only suitable for simple plants if they are homogenous and in liquid form. The maximum of gas production from a given amount of raw material depends on the type of feedstock (Pavlostathis and Gossett, 1988; Pavlostathis and Giraldo-Gomez, 1991). However there are several parameters that are critical to the successful operation of a digester and these are (i) acid-base effects, (ii) the chemistry of the process, and (iii) the effects of the temperature.

3.1. Acid-Base Effects

The level of acidity-alkalinity is an important factor in the operation of an anaerobic digester. To maintain a constant supply of gas, it is necessary to maintain a suitable pH range in the digester. The digester process occurs in a defined narrow pH interval ranging from approximately 6 to 8.5 and the preferred level is pH = 7.2 – i.e., mildly acid to mildly basis conditions – since the methanogenic microbes that use using organic acids for some of their food intake are unable to operate (even unable to survive) in an acidic environment. When the process is in balance, the acidity in the reactor will be within this range and as the buffer capacity in the reactor is very large.

Table 3.1. Examples of Mono-carboxylic Acid Derivatives

Volatile fatty acids (VFAs)	Formic acid $HCOOH$
	Acetic acid CH_3COOH
	Propionic acid C_2H_5COOH
	Butyric acid C_3H_7COOH
Long chain fatty acids (LCFAs)	Lauric acid $C_{11}H_{23}COOH$
	Palmitic acid $C_{15}H_{31}COOH$
	Stearic acid $C_{17}H_{35}COOH$
	Oleic acid $C_{17}H_{33}COOH$ (one double bond)
	Linoleic acid $C_{17}H_{31}COOH$ (two double bonds)
	Linolenic acid $C_{17}H_{29}COOH$ (three double bonds)

By way of explanation, a buffer solution is used in certain systems (such as an anaerobic digester) to retain almost constant pH when small amount of acid/base is added – the quantitative measure of this resistance to a change in the pH is the buffer capacity. The buffer capacity quantifies the ability of a solution to resist changes in pH by either absorbing or desorbing H+ and OH- ions. When an acid or base is added to a buffer system – such as a digester – the effect on pH change can be large or small, depending on both the initial pH and the capacity of the buffer to resist change in pH. The buffer capacity (β) is as the number of moles of an acid or base that is necessary to change the pH of a solution by 1, divided by the pH change and the volume of buffer in liters. A buffer resists changes in pH due to the addition of an acid or base though consumption of the buffer. As long as the buffer has not been completely reacted, the pH will not change drastically. The pH change will increase (or decrease) more drastically as the buffer is depleted: it becomes less resistant to change.

In case the pH of the reactor exceeds these limits then the process is deteriorated resulting in a dramatic decrease in methane production. However, changes in the pH value of the digester system can be correlated with other operational parameters: (i) an accumulation of organic acids – acidification – will typically lower the pH, (ii) an increase in the concentration of ammonia or the removal of carbon dioxide will lead to a variation in the pH of the system. Furthermore, the decrease in the pH value (into acid territory) due to the accumulation of volatile fatty acid derivatives (Table 3.1) is dependent on the feedstock to the digester. Some organic residues, as for example cattle manure, have high buffer capacity and, as a result, are able to maintain a balanced pH in the system. A decrease in the pH of the system will occur only in cases that the concentration of volatile fatty acid derivatives is remarkably high by which time it is likely that the process has been severely influenced. As a result, the accumulation of volatile fatty acid derivatives can be considered to be a result of an already inhibited process and may not be the actual cause of the inhibition.

In general, hydroxides and carbonates of calcium, magnesium, sodium, potassium and ammonium produce alkalinity in the wastewater. Alkalinity plays an important role in anaerobic digestion process as it control the pH by buffering the acidity created in the acidogenesis process. The alkalinity of the digester in general is proportional to the solids feed concentration of the digester and the growth of anaerobic process microorganism significantly depends on the pH value of the system. Most methanogens prefer a narrow pH range and the optimal is reported to be 7 to 8. Acidogens usually have a lower value of optimum pH.

3.2. Process Chemistry

The anaerobic digestion process is generally considered to consist of four steps in the following order of reaction: (i) hydrolysis, (ii) acidogenesis, (iii) acetogenesis, and (iv) methanogenesis. In some considerations of the process, a pre-hydrolysis step – disintegration – is also noted. Disintegration is the step in which the complex biomass is disintegrated into organic polymers such as carbohydrates, proteins, and lipids. Disintegration includes several steps such as lysis, non-enzymatic decay, phase separation, and physical breakdown (Kiran et al., 2016).

3.2.1. Hydrolysis

Typically, biomass is made up of naturally-occurring high molecular weight organic polymers (Chapter 2) and, if the bacteria in anaerobic digesters are to be successful, the organic feedstock must first be broken down into the lower molecular weight constituents. These constituents (monomers, such as sugar derivatives, amino acid derivatives, and fatty acid derivatives) are then readily available to other bacteria. Thus, the process of breaking these chains and dissolving the smaller molecules into solution (hydrolysis) which is the necessary first step in anaerobic digestion.

In this first step (hydrolysis), the organic matter is attacked externally by extracellular enzymes (cellulase, amylase, protease and lipase) of microorganisms. Bacteria decompose the long chains of the complex carbohydrates, proteins and lipids into shorter parts. During this step, long-chain molecules, such as protein, carbohydrate and fat polymers, are broken down to monomers (small molecules). Different specialized bacteria produce a number of specific enzymes that catalysed the decomposition, and the process is extracellular – i.e., it takes place outside the bacterial cell in the surrounding liquid.

Proteins, simple sugars (such as monosaccharide and disaccharides), and starch hydrolyze easily under anaerobic conditions. Other polymeric carbon compounds somewhat more slowly, while lignin, which is an important plant component, cannot be decomposed under anaerobic conditions at all. Cellulose – a polymer composed of a number of glucose molecules – and hemicellulose – composed of a number of other sugars – are complex polysaccharides that, are easily hydrolyzed by specialized bacteria. In plant tissue both cellulose and hemicellulose are tightly packed in lignin and are therefore difficult for bacteria to get at. This is why only approximately 40% of the cellulose and hemicellulose in pig slurry is decomposed in the biogas process. Normally the decomposition of organic matter to methane and carbon dioxide is not complete and is frequently on the order of 30 to 60% for animal manure and other feedstocks that have a high concentration of complex molecular constituents.

Acid derivatives (typically referred to as acetates) and the hydrogen produced in the first stage can be used directly by methanogens. Other products, such as volatile fatty acids (VFAs) with a chain length greater than that of acetic acid must first be catabolized into compounds that can be directly used by methanogens.

By way of clarification, catabolism is the set of metabolic reactions that break down feedstock constituents into lower molecular weight products that are then either (i) oxidized to release energy or (ii) used in other anabolic reactions. Anabolic reactions (also referred to as *anabolism*) use energy to build more complex molecules from relatively simple raw materials. Thus, examples of catabolic reactions are:

Fats → fatty acids
Lipids → fatty acids, glycerol
Proteins → amino acids
Polysaccharides → monosaccharides

If the stages flowing the hydrolysis stage are too rapid, methane production is limited by the hydrolysis stage and thus, the rate-limiting step depends on the compounds of the feedstock which is used for biogas production. Undissolved feedstock constituents such as cellulose, proteins, or fats are converted slowly into the related monomers within several days whereas the hydrolysis of soluble carbohydrate derivatives can occur within hours.

Therefore, the process design must be well adapted to the feedstock properties for achieving a complete degradation without process failure.

3.2.2. Acidogenesis

Acidogenesis (sometimes referred to as fermentation) is the biological process that results in further breakdown of the remaining components by acidogenic (fermentative) bacteria. In this process, volatile fatty acids are produced, along with ammonia, carbon dioxide, and hydrogen sulfide, as well as other byproducts.

In a balanced bacterial process approximately 50% of the monomers (glucose, xylose, amino acids) and long-chain fatty acids (LCFA) are broken down to acetic acid (CH_3COOH). Twenty percent is converted to carbon dioxide (CO_2) and hydrogen (H_2), while the remaining 30% is broken down into short-chain volatile fatty acids (VFAs). Fatty acids are monocarboxylic acids that are found in fats and have fewer than six carbon atoms whereas long-chain fatty acids. If there is an imbalance in the digester process, the relative level of volatile fatty acids will increase with the risk of accumulation and, since the bacteria that degrade the volatile fatty acids have a slow growth rate and cause an imbalance between the process stages. A steady degradation of the volatile chain fatty acids is therefore crucial and often a limiting factor for the biogas process.

Hydrolysis of simple fats results in 1 mol glycerol and 3 mol long-chain fatty acids and, therefore, high proportions of fat in the digester feedstock will result in large amounts of long-chain fatty acids, while large amounts of protein, which contain nitrogen in amino groups ($-NH_2$), will produce large amounts of ammonium/ammonia (NH^{4+}/NH_3). In both cases this can lead to inhibition of the subsequent decomposition phase, particularly if the composition of the biomass feedstock varies.

3.2.3. Acetogenesis

Acetogenesis is the third stage of anaerobic digestion in which simple molecules created through the acidogenesis phase are further digested by acetogens to produce largely acetic acid, as well as carbon dioxide and hydrogen. The acid-producing bacteria, involved in the second step, convert the intermediates of fermenting bacteria into acetic acid (CH_3COOH), hydrogen (H_2) and carbon dioxide (CO_2). These bacteria are anaerobic and can grow under acid conditions. To produce acetic acid, the bacteria need oxygen and carbon. For this, they use the oxygen solved in the solution or bounded-oxygen whereby the acid-producing bacteria create an anaerobic condition which is essential for the methane producing microorganisms. Moreover, these bacteria reduce the compounds with a low molecular weight into alcohols, organic acids, amino acids, carbon dioxide, hydrogen sulfide and traces of methane. From a chemical standpoint, this process is partially endergonic (i.e., only possible with energy input), since bacteria alone are not capable of sustaining that type of reaction. An endergonic reaction is a chemical reaction in which total amount of energy is a loss – it takes more energy to initiate the reaction than the energy produced by the reaction and, thus, the total energy is a negative net result.

An acetogen is a microorganism that generates acetate (CH_3COO^-) as an end product of anaerobic respiration (fermentation) and can produce, in most cases, acetate as the end product) from two molecules of carbon dioxide (CO_2) and four molecules of molecular hydrogen (H_2). This process (acetogenesis) is different from acetate fermentation, although both occur in the absence of molecular oxygen (O_2) and produce acetate. Acetogens are found in a variety of habitats, generally those that are anaerobic (lack oxygen). Thus, acetogenesis is a process through which acetate is produced from carbon dioxide and an electron source (such as hydrogen and carbon monoxide) by anaerobic bacteria. In this reaction, carbon dioxide is reduced to carbon monoxide and formic acid (HCO_2H) or directly into a formyl group, the formyl group is reduced to a methyl group and then combined with the carbon monoxide and Coenzyme A produce acetyl-CoA. Two specific enzymes participate on the carbon monoxide side of the pathway: (i) CO-dehydrogenase, which catalyzes the reduction of the carbon dioxide and (ii) acetyl CoA synthase, which combines the resulting carbon monoxide with a methyl group to give acetyl-CoA.

The key aspects of the acetogenic pathway are several reactions that include the reduction of carbon dioxide to carbon monoxide and the attachment of the carbon monoxide to a methyl group. The first process is catalyzed by specific enzymes (carbon monoxide dehydrogenase enzymes) and the coupling of the methyl group (provided by methylcobalamin) and the carbon monoxide is catalyzed by acetyl CoA synthetase.

$$2CO_2 + 4H_2 \rightarrow CH_3COOH + 2H_2O$$

The accumulation of hydrogen can inhibit the metabolism of the acetogenic bacteria and present knowledge suggests that hydrogen may be a limiting feedstock for methanogens (Bagi et al., 2007). This assumption is based on the fact that addition of hydrogen-producing bacteria to the natural biogas-producing consortium increases the daily biogas production. At the end of the degradation chain, two groups of methanogenic bacteria produce methane from acetate or hydrogen and carbon dioxide. These bacteria are strict anaerobes and require a lower redox potential for growth than most other anaerobic bacteria (Schink 1997).

3.2.4. Methanogenesis

Methanogenesis is the terminal stage of anaerobic digestion in which methanogens use the intermediate products of the preceding stages and convert them to them into methane, carbon dioxide, and water. These components make up the majority of the biogas emitted from the system. Methanogenesis is sensitive to both high pH (alkaline) and low pH (acidic) and occurs between pH 6.5 and pH 8, in near neutral conditions. The remaining, indigestible material the microbes cannot use and any dead bacterial remains constitute the digestate.

Methane-producing bacteria decompose compounds with a low molecular weight. For example, they utilize hydrogen, carbon dioxide and acetic acid to form methane and carbon dioxide. Under natural conditions, methane producing microorganisms occur to the extent that anaerobic conditions are provided, e.g., under water (for example in marine sediments), in ruminant stomachs and in marshes. They are obligatory anaerobic and very sensitive to environmental changes. In contrast to the acidogenic and acetogenic bacteria, the methanogenic bacteria belong to the archaebacteria genus, i.e., to a group of bacteria with a very heterogeneous morphology and a number of common biochemical and molecular-biological properties that distinguish them from all other bacterial general. The main difference lies in the makeup of the bacteria's cell walls.

The last step in the production of methane is undertaken by the so-called methanogenic bacteria or methanogens. Thus:

Acetic acid → methane + carbon dioxide
Hydrogen + carbon dioxide → methane + water

Two different groups of bacteria are responsible for the methane production. One group degrades acetic acid to methane and the other produces methane from carbon dioxide and hydrogen. Under stable conditions, the majority (approximately 70% v/v) of the methane production comes from the degradation of acetic acid, while the remaining methane production (approximately 30% v/v) comes from carbon dioxide and hydrogen. The two processes are finely balanced and inhibition of one will also lead to inhibition of the other. The methanogens have the slowest growth rate of the bacteria involved in the process, they also become the limiting factor for how quickly the process can proceed and how much material can be digested.

In the overall process, the acid-producing bacteria and the methane-producing bacteria act in a symbiotic way. On the one hand, acid-producing bacteria create an atmosphere with ideal parameters for methane-producing bacteria (anaerobic conditions, compounds with a low molecular weight feedstock). On the other hand, methane-producing microorganisms use the intermediates of the acid-producing bacteria. Without removal (consumption) of these intermediate products from the system by the methane-producing microorganisms, toxic conditions for the acid-producing microorganisms would develop leading to an inhibition effect and a loss of process efficiency.

Essentially, the four phases of anaerobic degradation take place simultaneously in a single-stage process. However, as the bacteria involved in the various phases of degradation have different requirements in terms of habitat (regarding pH value and temperature, for example), a compromise has to be found in the process technology. Since the methanogenic microorganisms are the weakest link in the four stages because of the low rate of growth of the methanogens and they are the most sensitive to respond to disturbances, the digester conditions have to be adapted to the requirements of the methane-forming bacteria. However, any attempt to physically separate hydrolysis and acidogenesis from methanogenesis by implementing two distinct process stages (two-phase process management) will succeed to only a limited extent because, despite the low pH value in the hydrolysis stage ($pH < 6.5$), some methane will still be formed. The resulting hydrolysis gas therefore also contains methane in addition to carbon dioxide and hydrogen. In multi-stage processes, different environments can become established in the individual digester stages depending on thc dcsign of the biogas plant and its operating regime, as well as on the nature and concentration of the fresh mass used as substrate. In turn, the ambient conditions affect the composition and activity of the microbial biocoenosis and thus have a direct influence on the resulting metabolic products.

As a recap (Chapter 3), Phase 1 of the process is the hydrolysis reaction in which high-molecular weight naturally- occurring compounds (cellulose, hemicellulose and lignin) are converted to lower molecular weight products. Typically sugars and oxygenated aromatic derives. During acidogenesis, soluble monomers are converted into small organic compounds, such as short chain (volatile) acids (propionic, formic, lactic, butyric, succinic acids), ketones (glycerol, acetone), and alcohols (ethanol, methanol). Generally, the chemistry of Phase 2 (acidogenesis) and Phase 3 (acetogenesis) can be represented by a series of simple equations:

Acidogenesis:

$$C_6H_{12}O_6 + 2H_2 \rightarrow 2CH_3CH_2COOH + 2H_2O$$
$$C_6H_{12}O_6 \rightarrow 2CH_3CH_2OH + 2CO_2$$

Acetogenesis:

$$CH_3CH_2COO^- + 3H_2O \rightarrow CH_3COO^- + H^+ + HCO_3^- + 3H_2$$
$$C_6H_{12}O_6 + 2H_2O \rightarrow 2CH_3COOH + 2CO_2 + 4H_2$$
$$CH_3CH_2OH + 2H_2O \rightarrow CH_3COO^- + 2H_2 + H^+$$
$$2HCO_3^- + 4H_2 + H^+ \rightarrow CH_3COO^- + 4H_2O$$

Phase IV: is the phase of methane formation (methanogenesis) in which the concentrations of the intermediate acids are usually small in proportion to their production and degradation rates, and they are quickly transformed to methanogenic substrates, including acetate, methanol, and formate. Thus, the last phase of anaerobic digestion is the methanogenesis phase. Several reactions take place using the intermediate products from the other phases, with the main product being methane. The reactions are commonly represented as:

$2CH_3CH_2OH + CO_2 \rightarrow 2CH_3COOH + CH_4$

$CH_3COOH \rightarrow CH_4 + CO_2$

$CH_3OH \rightarrow CH_4 + H_2O$

$CO_2 + 4H_2 \rightarrow CH_4 + 2H_2O$

$CH_3COO^- + SO_4^{2-} + H^+ \rightarrow 2HCO_3 + H_2S$

$CH_3COO^- + NO^- + H_2O + H^+ \rightarrow 2HCO_3 + NH_4^+$

Several bacterial contribute to methanogenesis, including: *Methanobacterium, methanobacillus, methanococcus,* and *methanosarcina.*

Any kind of organic matter can be fed to an anaerobic digester, including manure and litter, food wastes, green wastes, plant biomass, and wastewater sludge. The materials that compose these feedstocks include polysaccharides, proteins, and fats/oils. Some of the organic materials degrade at a slow rate; hydrolysis of cellulose and hemicellulose is rate limiting. (Amani et al., 2010). There are some organic materials that do not biodegrade: lignin, peptidoglycan (a polymer consisting of sugar derivatives and amino acid derivatives that forms a mesh-like layer outside the plasma membrane of most bacteria, forming the cell wall), and membrane-associated proteins. The organic residues contain water and biomass composed of volatile solids and fixed solids (minerals or ash after combustion) and it is the volatile solids that can be non-biodegradable and biodegradable. However, unlike the chemistry of the anaerobic digester, the organic waste in the landfill is not available for separate pretreatment unless there is a form of pretreatment before he material is discharged into the landfill.

3.3. Effects of Temperature

In the anaerobic digestion process, temperature is not only important for microbial metabolic activities but also for the overall digestion rate, specifically the rates of hydrolysis and methane formation. In general, anaerobic digestion process can occur within a wide range of temperatures. This temperature range has been divided into three groups: (i) psychrophilic: less than 20°C, 68°F, (ii) mesophilic: 30 to 42°C, 86 to 108°F, and (iii) thermophilic- 43 to 55°C, 109 to 131°F. In practice, most of the anaerobic digestion system are designed to operate at mesophilic range, between 30 to 38°C (86 to 100°F), and some of them are designed for thermophilic temperature range of 50 to 57°C (122 to 135°F).

However, in actual practice, the boundaries between the temperature ranges presented above (and elsewhere in this text) are variable but it is the rapid changes in temperature that cause harm to the microorganisms. For example, if the temperature changes slowly the methanogenic microorganisms are able to adjust to different temperature levels. It is therefore not so much the absolute temperature that is crucial for stable management of the process, but constancy at a certain temperature level.

More generally, the anaerobic digestion process occurs in anaerobic reactors that operate under mesophilic temperature conditions [30 to 40°C (86 to 104°F), but more typically at 35 to 37°C (95 to 99°F) or thermophilic temperature conditions 50 to 60°C (122 to 140°F), but more typically at mainly 52 to 55°C (126 to 131°F). The selection of the operating temperature and the temperature control is extremely important as this parameter has a major influence on the development of the microbial colony in the digester. Furthermore, fluctuations in the temperature can (will) cause imbalance of the process that is associated with accumulation of volatile fatty acids (VFAs) as well as a decrease in biogas production (Angelidaki and Batstone, 2010; Jansen, 2010).

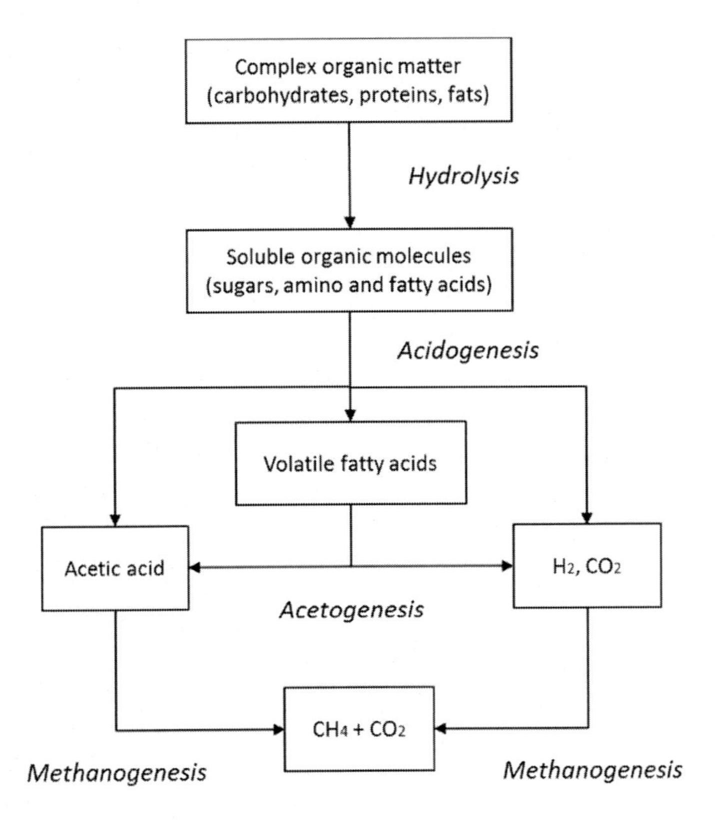

Figure 3.1. The Four Stages of the Anaerobic Digestion Process.

Assuming that the temperature is controlled to the point of digester efficiency, there are four key stages in the anaerobic digestion process: (i) hydrolysis, (ii) acidogenesis, (iii) acetogenesis, and (iv) methanogenesis (Figure 3.1). In the simplest terms, the overall anaerobic process can be described by a chemical reaction using glucose as the example, where organic material such as glucose ($C_6H_{12}O_6$) which is biochemically converted into carbon dioxide (CO_2) and methane (CH_4):

$$C_6H_{12}O_6 \rightarrow 3CO_2 + 3CH_4$$

The individual degradation steps are carried out by different consortia of microorganisms, which partly stand in syntrophic interrelation and place different requirements on the environment (Angelidaki et al., 1993). Hydrolyzing and fermenting microorganisms are responsible for the initial attack on polymers and monomers and produce mainly acetate and hydrogen and varying amounts of volatile fatty acids such as propionate and butyrate. Hydrolytic microorganisms excrete hydrolytic enzymes, e.g., cellulase, cellobiase, xylanase, amylase, lipase, and protease. A complex consortium of microorganisms participates in the hydrolysis and fermentation of organic material. Most of the bacteria are strict anaerobes such as bacteriocides, clostridia, and bifidobacteria. Furthermore, some facultative anaerobes such as Streptococci and Enterobacteriaceae take part. The higher volatile fatty acids are converted into acetate and hydrogen by obligate hydrogen-producing acetogenic bacteria. The hydrogen-producing acetogenic bacteria are not well characterized. Typical homoacetogenic bacteria are Acetobacterium woodii and Clostridium aceticum.

The rate of biochemical processes generally increases with temperature – a general rule of thumb, the rate is doubled for every 10-degree rise (10°C or 18°F) in temperature within certain limits. This is also the case with the biogas process and although temperature ranges for optimal operation have been presented (above), there are several types or strains of bacteria involved that have adapted to different temperature ranges: (i) psychrophiles 0 to 20°C, 32 to 68°F, (ii) mesophiles 15 to 45°C, 59 to 113°F, and (iii) thermophiles 40 to 65°C, 104 to 149°F. One common to these bacteria is that they sensitive to changes in temperature and the sensitivity increases with temperature.

Thus, it is important to keep a constant temperature during the digestion process, as temperature changes or fluctuations will influence the biogas production negatively. In most instances, methanogenic diversity is lower in plants operating at thermophilic temperatures (Karakashev et al., 2005; Leven et al., 2007). Therefore, thermophilic processes are more sensitive to temperature fluctuations and require longer time to adapt to a new temperature. Mesophilic bacteria tolerate temperature fluctuations of ±3°C (±5.4°F) without significant reductions in methane production. The growth rate of methanogenic bacteria is higher at thermophilic process temperatures making the process faster and more efficient. Therefore, a well-functioning thermophilic digester can be loaded to a higher degree or operated at a lower hydraulic retention time (HRT, the time a given biomass stays in the reactor before it is pumped out again) than at mesophilic conditions. But the thermophilic process temperature results in a larger degree of imbalance and a higher risk for ammonia inhibition. Ammonia toxicity increases with increasing temperature, and washout of microbial population can occur (Kroiss 1985; Dornack 2009).

Finally, the phenomenon of self-heating is frequently observed in practice, and should be mentioned in this connection. This effect occurs when substrates consisting largely of carbohydrates are used in combination with an absence of liquid input materials and well insulated containers. Self-heating is attributable to the production of heat by individual groups of microorganisms during the decomposition of carbohydrates. The consequence can be that in a system originally operating under mesophilic conditions the temperature rises to the region of 43 to 48°C (109 to 118°F). Given adequate monitoring of the process, the temperature change can be managed with small reductions in gas production for short periods. However, without necessary interventions in the process (such as reduction of the input quantities), the microorganisms are unable to adapt to the change in temperature and, in the worst case, a temperature-inhibition effect occurs and gas production ceases.

3.4. Process Inhibitors

In the current context, a reaction inhibitor is a substance that decreases the rate of, or prevents, a chemical reaction in contrast to a catalyst which is a substance that increases the rate of a chemical reaction. In this same sense, and in this context, a bacterial (microbial) inhibitor is a molecule that binds to the bacterium (microbe) and decreases the activity of the bacterium (microbe) (Rozzi and Remigi, 2004; Chen et al., 2008).

There are various reasons for the inhibition of gas production and the chemicals (or physical events) that can not only inhibit the process but also slow down the process. In addition, there is a distinction must be drawn between inhibitors that enter the digester through the addition of substrate and those that are formed as intermediate products from the individual stages of decomposition.

Thus, during anaerobic digestion, there are some compounds that, if the concentration of these compounds exceeds compound-specific certain limits, can reduce the biogas production or in worst conditions can cause fatal deterioration of the process. These compounds are either toxic substances or intermediate metabolic products. In general, methanogens are considered to be more sensitive to a potential exposure to toxicants compared to bacteria (Kougias and Angelidaki, 2018).

Furthermore, due to the complex composition of biomass, it is difficult to reach a steady state during anaerobic digestion. Methanogenesis occurs in the pH range of between 6 and 8.5 with an optimum pH range of 7.0 to 8.0 and the methanogenesis stage is dramatically inhibited when the pH decreases below 6.0 or increases above 8.5. The imbalances in the pH occur (especially) during the bioconversion of proteins and lipids, ie, by the accumulation of free ammonia and volatile fatty acids. This, in turn, affects the functions of the extracellular enzymes and the hydrolysis rate. Nutrient-rich biomasses can be easily acidified into volatile fatty acids by fermentative microorganisms. Volatile fatty acids accumulation results in a decrease in pH, and so might inhibit the methanogenic system.

One of the most common inhibitors of anaerobic digestion process is the increased ammonia concentration. Ammonia is present in a wide variety of organic residues, as for example certain types of manure and high proteinaceous sludge (Kougias and Angelidaki,

2018). Moreover, ammonia can also be formed during protein degradation or can originate from other compounds, such as urea.

One of the most significant endogenous inhibitors is ammonia (NH_3) which is created during the bacterial degradation of nitrogen-containing substances such as proteins. Nitrogen is essential for bacterial growth and ammonia is an important source of nitrogen. But ammonia at high concentrations is highly toxic to the bacteria. In an aqueous solution ammonia is always found in an equilibrium with ammonium (NH_4^+). This equilibrium is determined by the acidity, pH and temperature of the environment and, as ammonium is not as toxic as ammonia, this equilibrium is important:

$$NH_4^+ \rightarrow NH_3 + H^+$$

At a high pH (alkaline conditions), the equilibrium is moved to the right-hand side of the equation and the environment becomes more toxic to bacteria. Higher temperatures will also shift this equilibrium to the right and, as a result, a thermophilic biogas process – all other things being equal – is more sensitive than a mesophilic process to ammonia inhibition. There will be a certain inhibition of the bacteria already at relatively low ammonia concentrations which is a benefit because the biomass typically used in biogas production, such as slurry, usually have a relatively high concentration of ammonia. However, the bacteria do have difficulties with a sudden increase in the concentration of ammonia and a consistent and even input of biomass is important for the process, especially at higher temperatures. When the process is inhibited by ammonia, an increase in the concentration of volatile fatty acids (VFA) will lead to a decrease in pH which will partly counteract the effect of ammonia (Nielsen and Angelidaki 2008).

However, the absolute concentration value above which ammonia leads to process inhibition is difficult to be quoted as this is additionally dependent on other factors, such as temperature, pH or inoculum source. More specifically, free ammonia (NH_3, rather than bound ammonia, NH_4^+) is in equilibrium with ammonium ion and its concentration depends on the pH value.

Similarly, the equilibrium is affected by the operational temperature; higher temperature leads to higher concentration of free ammonia, resulting in more intense toxicity phenomena. Ammonia inhibition causes also the accumulation of volatile fatty acid derivatives, which will in turn decrease the pH of the reactor. The lowering of the pH will partially alleviate the toxicity effect of ammonia as the concentration of free ammonia will be decreased. However, this homeostatic mechanism (i.e., the tendency of a system to maintain internal stability) will maintain the operation of the reactor in a relapsed phase (the *inhibited steady state* condition).

Excess of free ammonia can inhibit methane-synthesizing enzymes which can create a proton (H^+) imbalances and potassium deficiency, causing cell lysis i.e., breakdown of the cell membrane and loss of cell efficiency). The inhibitory ammonia concentration to methanogenesis is still ambiguous and, although a high concentration of ammonia can result in operational difficulties in an anaerobic digester, microorganisms can adapt to a higher free ammonia concentration with time which makes it difficult to predict the precise concentration of ammonia at which process failure or instability may occur. In order to avoid such a problem, it is necessary to remove the excess ammonia – many different strategies such as ammonia stripping, biological nitrogen elimination processes, and electrochemical conversion can be employed.

Other important endogenous process inhibitors are the organic acids formed during the process. If these are not removed as soon as they are formed – which can happen during an overload – this can lead to an increase in the acidity (acidification) of the process. Among the exogenous causes, antibiotics and disinfection agents are inhibitors of the process, because both – by definition – are toxic because they are used to kill microorganisms. Both substances are used in livestock production to treat sick animals and to keep animal houses and milking parlous clean and can therefore also be found in the slurry, but apparently only at concentrations so low that they do not have a negative impact on the biogas plant. A slow adaptation to these substances can also take place if the supply is fed in continuously.

A series of compounds that is associated with toxicity effects of biogas production process is the long chain fatty acid derivatives (Kougias and Angelidaki, 2018). Various agro-industrial residues, as for example slaughterhouse wastes, food wastes and olive mill wastewater, contain high concentrations of long chain fatty acid derivatives (Table 3.1). The inhibition caused by long chain fatty acid derivatives is attributed to the accumulation of oxidation products which cannot be further oxidized as the required reactions are thermodynamically unfavorable. Therefore, long chain fatty acid derivatives can negatively affect the activity of hydrolytic, acidogenic, acetogenic bacteria and methanogenic archaea. However, methanogens are more tolerant to the inhibitory effect of long chain fatty acid derivatives compared to the bacterial community. Moreover, hydrogenotrophic methanogens (organisms that are able to metabolize molecular hydrogen as a source of energy) are more resilient to the toxicity of long chain fatty acid derivatives than acetoclastic methanogens. However, inhibition by long chain fatty acid derivatives does not necessarily lead to a fatal deterioration of the process, but is rather a reversed phenomenon – a biogas microbial community, which has been exposed to long chain fatty acid derivatives, recovers faster from the inhibitory shock compared to a non- adapted microbial consortium and also that the process is less deteriorated (Kougias and Angelidaki, 2018).

Another problem of the anaerobic digester is related to foaming incidents, which are caused by operational problems (such as poor mixing and feedstock overload) or by specific biosurfactants produced during anaerobic digestion process. Foaming is not due to the accumulation of long chain fatty acid derivatives or acidification of the process liquor. The process imbalance is attributed to the thick layer that is formed on the reactors surface, entrapping the produced biogas, reducing the reactor active volume and thus creating dead zones. Several strategies, mainly based on the addition of chemical agents, have been employed to counteract foaming incidents either by preventing the formation of foam or by destructing it once it is created.

In summary, the presence of inhibitory compounds (often referred to as toxic compounds in the contest of the anaerobic digestion process) can adversely affect the anaerobic process microorganism. A wide variety of inorganic and organic toxic and inhibitory substances can cause failure of the anaerobic digester process. Thus, the commonly present toxic substance in anaerobic digesters include, but are not limited to: (i) oxygen and light – a high amount of either could inhibit the activity of methane producing bacteria, (ii) disinfectants such as herbicides, heavy metals or antibiotics found in poultry/chicken manure can also disturb the process if present in high concentration, (iii) hydrogen sulfide, which is a product of the digestion process but can be found in the organic material as well, along with sulfuric acid is highly corrosive and could seriously affect the components of digester, and (iv) a high ammonia concentration, which is a cellular poison, could be caused by high nitrogen and ammonium (NH_4^+) concentrations under certain circumstances and a feedstock with a high nitrogen concentration – such as chicken manure and pig slurry – should be diluted or mixed with a nitrogen-poor feedstock in order to dilute the effect of the nitrogen constituents.

Finally, the rate of introduction (or the rate of loading) of the feedstock can also produce an inhibitory effect. For example, if the rate of introduction of the feedstock into the digester it too high in any given period of time or if methanogenesis is inhibited for some other reason, the acid metabolic products of acidogenesis will accumulate. Typically, the pH value is in the neutral range by the carbonate and ammonia buffer but if the buffer capacity is exhausted, i.e., if the concentration of the organic acids is too high, the pH value drops. This, in turn, increases the inhibitory effect of hydrogen sulfide (H_2S) and propionic acid ($CH_3CH_2CO_2H$) and the digestion process is halted. On the other hand, the pH value will rise if the ammonia is released as a result of the decomposition organic nitrogen compounds the feedstock and the ammonium ion (NH_4^+) is formed, inhibition of the process occurs.

4. The Anaerobic Digester

The digester, which is the heart of the process, is typically an enclosed and insulated steel tank – alternatively it can also be a concrete tank covered by an airtight seal. The tank is usually fitted with heating coils that (if necessary) warm the digesting biomass or the heat supply can be external heat exchange system. The tank is also equipped with a stirrer that can keep the entire volume fully agitated to prevent the formation of a surface crust. Other equipment include an overflow outlet, temperature and pressure gauges, as well as any other necessary ancillary equipment (Weiland, 2009).

Energy crops digestion requires prolonged hydraulic retention times of several weeks to month to achieve complete fermentation with high gas yields and minimized residual gas potential of the digestate. For dry fermentation, several batch processes with percolation and without mechanical mixing are applied mainly for some energy crops in which the solid feedstock is loaded into a batch gas-tight fermenter box and mixed with inoculum from a previous batch digestion. The requisite necessary share of solid inoculum has to be determined individually for each feedstock – for example, yard manure from cows requires only small ratios of solid inoculum but as much as 70% w/w of the input is the necessary amount of inoculum for energy crops (Kusch et al., 2005).

Another approach is to apply a continuous dry fermentation process for feedstocks that contain more than 25% w/w dry matter. The type of digester that is suited for such a process is a horizontal mechanically mixed digester or a vertical plug flow digester. The latter digester (the vertical digester) needs no mixing and the feedstock flows from the top to the bottom by gravity only but before the feedstock is fed to the top of the digester, it must be mixed with digestate coming from the bottom of the digester. In terms of process control, methane production is normally the only continuously measured parameter at agricultural biogas plants, but this parameter cannot reflect a process imbalance, if the biogas plant treats feedstocks with changing composition.

4.1. Digester Configuration

An anaerobic digester can be designed and engineered to operate using a number of different configurations and can be categorized into (i) a batch mode or a continuous process mode, (ii) mesophilic temperature conditions or thermophilic temperature conditions, (iii) high-solid feedstock or low-solid feedstock, and (iv) a single stage process or a multistage process. Typically, a larger volume batch process digester is needed to handle the same amount of waste as a continuous process digester. Also, higher heat energy is demanded in a thermophilic system compared to a mesophilic system and has a larger gas output capacity and higher methane gas content. In terms of the solids content of the feedstock, above the level of 15% w/w/ solids, the feedstock is considered to be a high-solids feedstock and the process can also be known as dry digestion.

In the *fixed-bed anaerobic digester*, the wastewater is introduced from the bottom or the top of a column which is filled with inert material (rocks, cinder, plastic, or gravel). The filling material provides the surface upon which microorganisms are attached forming a biofilm. The microorganisms can also be retained through entrapment in the microporous structure of the filling material. However, clogging is a typical problem with this type of digester and the organic load of the wastewater must be low to medium. Wastewater containing significant amounts of suspended solids or constituents that cause precipitation of organic and inorganic compounds are not suitable for this type of bioreactor.

The *expanded and fluidized bed anaerobic digester* allows a more effective mass transfer from the liquid phase to the membrane, because fine filling material is used. The upflow velocity must be high enough (through recirculation) to maintain the expansion of the bed between 15% and 30% of the non-fluidized volume, while if the expansion rises up to 300%, the bed is characterized as fluidized (Hall, 1992).

The *upflow anaerobic sludge bed digester* was designed as an alternative to wastewater treatment without the operating problems of bioreactors with filling materials but incorporating the concept of biomass immobilization. In this type of digester, the microorganisms are agglomerated to form a dense granule-type) conformation with excellent settling properties and strength under adverse conditions. The granular sludge blanket remains in the bottom of the bioreactor. The feed is introduced from the bottom and the motion of the flow is upwards – thus, the upflow velocity is very important since it influences the formation of the granules. The biogas produced is often entrapped in the granules making them lighter and buoyant with their potential wash out. An effective three-phase separator on the top of the bioreactor results in the retention of the granule and their return to the sludge blanket (Chong et al., 2012). Also, configurations of the upflow anaerobic sludge bed digester type digesters with other bioreactors (such as, for example, a septic tank or a sequential batch reactor) in series have proved to increase the efficiency of the upflow anaerobic sludge bed digester (Chong et al., 2012).

Many variations of the upflow anaerobic sludge bed digester system have been developed. The *expanded granular sludge blanket* is a high upflow velocity that is maintained though the high recycle ratio applied and its long and narrow geometry. As a result the contact between the granules and the wastewater is intensified (Lim and Kim, 2014). Hybrid systems have been developed to combine the characteristics of an upflow anaerobic sludge bed digester and an anaerobic filter, expanded or fluidized bed digester (Banu and Kaliappan, 2008; Ramakrishnan and Gupta, 2006; Sandhya and Swaminathan, 2006; Sunil Kumar et al., 2007).

The *anaerobic baffled digester* is a rectangular tank with baffles in which the wastewater flows above and below the baffles and coming successively into contact with the biomass which is accumulated in the bottom of the digester. This is an efficient system at low retention times and its operation is stable under sudden changes in the organic loading rate. A modification of this bioreactor type led to the development of the *periodic anaerobic baffled digester* which is based on the periodic feeding mode to all compartments and the ability of vary the switching frequency of the feedstock allows flexibility in the operation of the digester. In addition, the periodic anaerobic baffled digester can be operated as a simple *anaerobic baffled reactor* if the switching frequency is set to zero, and, in the extreme case of very high switching frequency, as a single-compartment upflow bioreactor (Stamatelatou et al., 2009).

The *anaerobic membrane digester* has evolved from the *aerobic membrane digester* by retaining the same configuration concept under anaerobic conditions. The anaerobic membrane digester can be used for municipal and industrial wastewater treatment and are improved with respect to the problems of membrane fouling and operational cost (Liao et al., 2004, 2006; Lin et al., 2013). The main feature of the membrane digester is the membrane which can be located immersed inside or outside the digester bioreactor. The membrane separates the microorganisms from the permeate, allowing for high solid retention times and produced a high-quality effluent (Skouteris et al., 2012; Stuckey, 2012). The size of pores ranges from ultrafiltration to fine microfiltration level. The basic disadvantage for the wide application of the anaerobic membrane digester is the tendency for fouling of the membrane fouling, which is caused by the presence of different types of matter (inorganic precipitates, soluble microbial products, colloids) in combination with the membrane type, the shear forces and fluxes prevailing, the hydraulic retention time, and the biological and chemical system developed in the bioreactor (Berube et al., 2006; Liao et al. , 2004, 2006; Lin et al., 2013; Meng et al., 2009; Skouteris et al., 2012; Stuckey, 2012).

The *plug flow digester* is a long narrow insulated and heated tank in which the feedstock flows from one end of the tank to the other – the tank bioreactor can be placed horizontally or vertically. It is used in the case of solid feedstocks and, in order to provide mixing, various practices are applied (De Baere, 2008). For example, in the Dranco process (dry anaerobic composting process) in which the digester is a vertical, downflow plug-flow unit, the fresh feedstock is mixed with a portion of the digested material and is introduced from the top of the digester. The same concept can be applied while the plug-flow reactor is placed horizontally (Kompogas process) and slowly rotating impellers inside the reactor can aid the horizontal movement of the mixture, also serving for mixing, degassing and suspension of the heavier particles. In another plug-flow type configuration (the Valorga process), the horizontal flow is circular and biogas injection at intervals under pressure through a network of nozzles provides mixing.

In the *leach-bed digester*, the feedstock is loaded in a vertical bioreactor to form a bed through which a liquid stream percolates as a leachate and is recirculated to the top of the same reactor where it is produced (the Biocel process) (ten Brummeler, 1999). The *complete mixed anaerobic digester* usually consists of a round insulated tank, above or below ground. Heating is provided through coils with hot water inside the tank or an external heat exchanger and mixing is achieved through a motor-driven mixer, recirculation of the mixed liquor or biogas. In the case of low solid content feedstocks and in order to enhance the biomass concentration in the digester, a modification of the complete mixed anaerobic digester led to the anaerobic contact process. In this configuration, the digester is followed by a settling tank (or inclined parallel plates or membranes) to separate the sludge from the supernatant. The sludge is recycled to the digester increasing the biomass concentration. The *covered anaerobic lagoon* is a large earthen impoundment, lined with appropriate geomembranes and covered with a flexible or floating gas-tight cover. They are used mostly for manure treatment and heat or mixing is not provided – thus, the ambient temperature is prevailed making this type of digester unsuitable in cold climatic conditions.

The *floating drum digester* (constant pressure digester) is made of bricks and is circular in shape. The digester is constructed typically underground to lessen the heat loss. The partition wall is constructed (dividing the digester into two parts) for higher size capacity plants to avoid the short-circuiting of digested slurry with the fresh feed. The separate gasholder is fabricated and fixed to store the gas produced during digestion besides acting as an anaerobic seal for the process. As the volume of gas production increases drum starts to rise and if the stored gas is withdrawn the level of drum drops to lower level. A central guide frame is provided to hold the gasholder and to allow it to move vertically during gas production. Salient features of this type of digester include weight of drum which assists in the discharge of the gas produced at constant pressure. In the fixed dome digester (constant volume digester), the dome act as gasholder in place of a high cost drum. The digester is completely underground to maintain a perfect environment for anaerobic fermentation to take place besides avoiding cracking of dome due to difference in temperature and moisture.

Other digester configurations include (i) the *Janata digester* in which the slurry feedstock is allowed to undergo anaerobic fermentation and the gas produced as a result rises up and is collected in the dome, and (ii) the *Deenbandhu digester* which is an improved version of Janata digester with design modifications that include the elimination of the loss of biogas through inlet chamber and maximum utilization of digester volume to make the operational lower hydraulic retention time close to the designed lower hydraulic retention time.

4.2. Process Configuration

In a single stage process, one reactor houses the four anaerobic digestion steps whereas a multistage process utilizes two or more reactors for digestion to separate the methanogenesis phase and the hydrolysis phase. Also, in a single-stage digestion system, all of the biological reactions occur within a single, sealed reactor or holding tank. Using a single stage reduces construction costs, but results in less control of the reactions occurring within the system. Acidogenic bacteria, through the production of acids, reduce the pH of the tank. Methanogenic bacteria, as outlined earlier, operate in a strictly defined pH range. Therefore, the biological reactions of the different species in a single-stage reactor can be in direct competition with each other. Another one-stage reaction system is an anaerobic lagoon. These lagoons are pond-like, earthen basins used for the treatment and long-term storage of manures. Here the anaerobic reactions are contained within the natural anaerobic sludge contained in the pool.

In a two-stage digestion system (multistage), the different digestion vessels are optimized to allow maximum control over the bacterial communities living within the digesters. For example, the acidogenic bacteria produce organic acids and more quickly grow and reproduce than methanogenic bacteria while the methanogenic bacteria require stable pH and temperature to optimize their performance.

In the typical digester, the hydrolysis, acetogenesis, and acidogenesis processes occur within the first reaction vessel. The organic product is then heated to the required operational temperature (either mesophilic or thermophilic conditions) prior to being pumped into the methanogenic reactor. The initial hydrolysis or acidogenesis tanks prior to the methanogenic reactor can provide a buffer to the rate at which feedstock is added. In some digesters, there may be the need for a degree of elevated heat treatment to kill harmful bacteria in the input waste. In this instance, there may be a pasteurization or sterilization stage prior to digestion or between the two digestion tanks.

In some digesters, plants are sometimes used, such as those using an upflow anaerobic sludge blanket (UASB) which can treat biomass with a low dry matter content (Chong et al., 2012). The advantage of this type of digester is that the hydraulic retention time is relatively short, typically in the range of a few hours to 48 hours, and that the reactor tank therefore does not need to be quite so large.

In terms of the process configuration, one issue that often arises is the issue of batch digestion or continuous digestion. In the batch digester system, the biomass feedstock is added to the digester at the start of the process after which the digester is sealed for the duration of the process. In the simplest form of the batch process, there is the need for inoculation of the fresh feedstock with feedstock that has already been processed (sometimes referred to as *aged feedstock*) to commence the anaerobic digestion. In a typical operation, the biogas product will be formed with a normal distribution pattern over time which can be used to determine when they believe the process of digestion of the organic matter has completed. However, there can be severe odor issues if a batch reactor is opened and emptied before the process has been completed.

A more advanced type of batch approach has limited the odor issues by integrating anaerobic digestion with in-vessel composting. In this approach inoculation takes place through the use of recirculated degasified percolate. After anaerobic digestion has completed, the biomass is retained in the digester which is then used for in-vessel composting before the digester is opened. Since the batch digestion is a simple system which and requires less equipment and lower levels of design work, it is typically a cheaper form of digestion. Using more than one batch reactor at a plant can ensure constant production of biogas. In the continuous digestion process, the organic matter is constantly added (continuous completely mixed) or added in stages to the reactor (continuous plug flow; first in-first out). In this system, the end products are constantly or periodically removed, resulting in constant production of biogas. A single digester or multiple digesters in sequence may be used.

Finally, in actual practice, the digestion process is a continuous process and, thus, on the one hand, there will be no adaptation period needed for the bacteria if the biomass composition is constant but on the other hand, the daily in- and outflow of biomass from the reactor means that in a fully stirred reactor some of the biomass pumped out will have spent a relatively short time in the reactor.

4.3. Digester Parameters

Most wet digesters are operated at mesophilic temperatures with optimum temperatures between 38 and 42°C (100 and 108°F). At higher temperatures, the degradation rate is faster and, thus, a shorter hydraulic retention time and a smaller reactor volumes are required but the ultimate methane yield from organic matter is not influenced. Decreasing the temperature to 50°C (122°F) or below reduces the toxicity of ammonia, but the growth rate of the thermophilic microorganisms will drop drastically, and a risk of washout of the microbial population can occur, due to a growth rate lower than the actual hydraulic retention time. The increased energy requirement for maintaining the reactor at thermophilic temperatures is not an important factor at present because surplus heat from the combined heat and power station (CHP) is often wasted today. But with increasing selling of heat to local residential houses and industry or by injection of upgraded biogas into the natural gas grid, thermophilic processes become lower efficient. The parameters of major importance are (alphabetically) (i) nutrient supply, (ii) residence time of the feedstock in the digester, and (iii) temperature.

4.3.1. Nutrient Supply

The microorganisms involved in anaerobic degradation have species-specific needs in terms of macronutrients, micronutrients and vitamins. The concentration and availability of these components affect the rate of growth and the activity of the various populations. There are species-specific minimum and maximum concentrations, which are difficult to define because of the variety of different cultures and their respective adaptability. In order to obtain as much biogas (i.e., methane) as possible from the feedstock, an optimum supply of nutrients to the microorganisms must be ensured. The amount of biogas that can ultimately be obtained from the substrates will depend on the proportions of proteins, fats and carbohydrates they contain. These factors likewise influence the specific nutrient requirements.

Thus, the anaerobic digestion process is dependent on the growth of microorganisms and, thus, there is a necessity to supply nutrients in sufficient amounts and in the correct proportions to sustain an optimal growth of the bacteria and archaea to obtain an efficient biogas production from a given feedstock. The carbohydrates and lipids of an organic feedstock mostly provides carbon, oxygen and hydrogen, while nitrogen and sulfur are supplied via proteins and phosphorus from, for example, nucleic acids phospholipids. Together with these elements, most organic feedstocks provide potassium, sodium, magnesium, calcium and iron and to some extent trace metals (micro-nutrients). The trace elements are often found in the active sites of enzymes essential for digestion.

The proportions and availability of the nutrients in a given feedstock will determine the growth rate of the constituents of the microbial community in the digester and the nutrient balance will govern the degree of degradation of the feedstock and therefore the biogas production efficiency at a given hydraulic retention time and organic loading rate. In fact, the limitation of any of the nutrients will lead to a limitation of the production efficiency of the biogas. Furthermore, several of the enzyme systems needed for the anaerobic utilization of organic matter for biogas production have demands for specific trace elements, e.g., the methanogens need nickel and cobalt and tungsten and selenium are required by the latter e.g., during fat and long fatty acid degradation (c.f. Gustavsson et al., 2013a, 2013b). Thus, optimization of biogas production from any feedstock should be based on an analysis of the nutrient composition to investigate the need for complements.

The concentration and availability of macronutrients, micronutrients and vitamins affect the rate of growth and the activity of the various populations. There are species-specific minimum and maximum concentrations, which are difficult to define because of the variety of different cultures and their adaptability, which is often considerable. In order to obtain the maximum amount of gas (methane) from the feedstock, an optimum supply of nutrients to the microorganisms must be ensured. The amount of methane that can ultimately be obtained from the feedstocks will depend on the proportions of proteins, fats and carbohydrates in the feedstock as well as the specific nutrient requirements. A balanced ratio between macronutrients and micronutrients is needed to ensure stable management of the process.

As well as macronutrients, an adequate supply of certain trace elements is vital for the survival of the microorganisms. The demand for micronutrients is generally satisfied in most agricultural biogas plants, particularly when the plant is fed with animal manure since a deficiency in trace elements is common in the energy crops. The elements that methanogenic archaea require are cobalt (Co), nickel (Ni), molybdenum (Mo) and selenium (Se), and sometimes also tungsten (W). Nickel, cobalt, and molybdenum are needed in cofactors for essential reactions in their metabolism. Magnesium (Mg), iron (Fe) and manganese (Mn) are also important micronutrients that are required for electron transport and the function of certain enzymes.

The concentration of trace elements in the reactor is therefore a crucial reference variable. A comparison of various sources in the literature on this topic reveals a strikingly large range of variation (sometimes by a factor of as much as 100) in the concentrations of trace elements considered essential. An analysis of the concentrations of trace elements in the feedstock can therefore provide no reliable information about the availability of trace elements, as it merely determines the total concentration. Consequently, larger quantities of trace elements have to be added to the process than would be needed solely to compensate for a deficient concentration. When determining requirements it is always necessary to take account of the trace element concentrations of all feedstocks.

It is well known from analyses of the trace element concentrations of various animal feeds that they are subject to considerable fluctuation which can make it difficult to optimize the dosing of trace elements in situations where there is a deficiency. Nevertheless, in order to prevent overdosing of trace elements, the concentration of micronutrients in the digester should be determined before trace elements are added. Overdosing can result in the concentration of heavy metals in the digestate (fermentation residue) exceeding the permissible limit for agricultural use, in which case the digestate cannot be used as organic fertilizer.

In summary, nutrients such as carbon, nitrogen, phosphorus and sulfur are very important for the survival and growth of anaerobic digestion process organism. Different micronutrients/microelements (trace elements) such as iron, nickel, cobalt, selenium, molybdenum or tungsten are also essential for the efficient activity of the microorganisms in the anaerobic digestion process. Insufficient amount of these nutrients and trace elements can cause inhibition and instability in anaerobic digestion process. The ideal atomic carbon to nitrogen (C: N) ratio for anaerobic digestion ranges from approximately 20:1 to 30:1. The optimal nutrient ratio for the atomic carbon, nitrogen, phosphor, and sulfur (C: N: P: S) is considered to be 600:15:5:1 and, in order to maintain optimum methanogenic activity, a desirable liquid phase concentration of nitrogen, phosphorus and sulfur should be on the order of 50, 10 and 5 mg/liter.

4.3.2. Residence Time

The residence time (retention time) is the time that particular volume of feedstock remains in the digester. The residence time is the length of time that the feedstock is subjected to the anaerobic reaction and is calculated as the volume of digester divided by the feedstock added per day and it is expressed in days. Under anaerobic condition, the decomposition of the organic substances is slow and hence the need to remain in the digester for prolonged periods in order to complete the digestion.

The residence time in a digester varies with the amount and type of feed material, and with the configuration of the digestion system. In a typical two-stage mesophilic digestion, residence time varies between 15 and 40 days, while for a single-stage thermophilic digestion, residence times is normally faster and takes approximately 14 days. The plug-flow nature of some of these systems will mean the full degradation of the material may not have been realized in this timescale. In this event, digestate exiting the system will be darker in color and will typically have more odor.

In the case of upflow anaerobic sludge blanket digestion (UASB), hydraulic residence times can be as short as 1 hour to 1 day, and solid retention times can be up to 90 days. In this manner, an upflow anaerobic sludge blanket digestion system is able to separate solids and hydraulic retention times with the use of a sludge blanket. Continuous digesters have mechanical or hydraulic devices, depending on the level of solids in the material, to mix the contents, enabling the bacteria and the food to be in contact. They also allow excess material to be continuously extracted to maintain a reasonably constant volume within the digestion tanks.

In summary, the anaerobic reactions and anaerobic reactor size are directly related to the solids residence time and the hydraulic residence time. Each of the anaerobic digestion reactions requires a minimum solids residence time to be completed and if the designed solids residence time is less than the minimum residence time, the digestion process will fail. In a completely mixed reactor with no recycle, solids and hydraulic retention times are the same.

Related to the residence time, the loading rate is defined as the amount of raw material fed to the digester per day per unit volume. If the reactor is overloaded (high residence time), acid accumulation will be more obviously affecting daily gas production. On the other hand, under loading of digester have negative impact in designed gas production.

Furthermore, the organic loading rate is the amount of organic dry matter that can be fed into the digester per unit volume of its capacity per day and is calculated based on the mass of volatile solids added per day per unit volume of digester capacity. Alternatively, the loading rate is the amount of volatile solids added to the digester each day per mass of volatile solids in the digester; although the first approach is favorable. Whichever method is used, consistency of the use of the method is necessary.

The loading rate is an important operational factor for digester because if it is too high, valuable methane former can washout from the system. In addition, toxic materials like ammonia can accumulate and upset the process. On the other hand, if the lading rate is too low, it can result in lower organic solids destruction and lower biogas production. Moreover, larger uneconomical digester will require higher heats. For these reason, the optimum loading rate should be a compromise between the highest possible biogas generation and having a justifiable plant economy.

Also, the stability of the anaerobic digestion process also depends on concentration of some products produced during the organic break-down process, such as the volatile fatty acids. During the acidogenesis process, different fatty acids like acetate, propionate, butyrate, lactate etc. are produced. Excessive accumulation of these acids can drop the pH value inside the reactor when the buffering capacity of the digester is exhausted. The buffering capacity of the digesters and how they will react to a maximum concentration of the volatile fatty acids (feedstock and reactor dependent) which will vary from digester to digester based on the microbial population in the digester.

4.3.3. Temperature

Temperature is one of the key parameters affecting the anaerobic digestion because it influences the microbial growth, enzymes activities, feedstock characteristics, and consequently the yield of biogas. Moreover, temperature affects the rate of reaction happening inside the digester. Increase in the ambient temperature increases the rate of reaction thus increasing the biogas production as well. Although methane-generating bacteria operate more efficiently at work at a temperature on the order of 30 to 38°C (86 to 100°F) – more specifically at 35 to 38°C (95 to 100°F) – the decrease in the rate of gas production starts at 20°C (68°F) and is essentially zero at 10°C (50°F).

Thermophilic anaerobic digestion has the advantages of higher specific growth rate, a faster metabolism, higher load-bearing capacity, and consequently results in a higher methane yield. However, it is necessary to recognize that the anaerobic digestion process can be affected negatively if the temperature is not constant. In fact, when compared to mesophilic anaerobic digestion, thermophilic anaerobic digestion is more sensitive to temperature change and it takes longer to adapt to a new temperature, as thermophilic methanogens have lower methanogenic diversity than mesophilic methanogens. Also, thermophilic anaerobic digestion has a higher risk of acidification, particularly when the biomass is rich in protein, which may consequently inhibit biogas production.

Generally, mesophilic species outnumber thermophiles, and they are also more tolerant to changes in environmental conditions than thermophiles and mesophilic systems are, therefore, considered to be more stable than thermophilic digestion systems. In contrast, while thermophilic digestion systems are considered less stable, their energy input is higher, with more biogas being removed from the organic matter in an equal amount of time. The increased temperatures facilitate faster reaction rates, and thus faster gas yields. Operation at higher temperatures facilitates greater pathogen reduction of the digestate.

Additional pre-treatment can be used to reduce the necessary retention time to produce biogas. For example, certain processes shred the feedstocks to increase the surface area or use a thermal pretreatment stage (such as pasteurization) to significantly enhance the biogas output. The pasteurization process can also be used to reduce the pathogenic concentration in the digestate leaving the anaerobic digester. Pasteurization may be achieved by heat treatment combined with maceration of the solids.

4.3.4. In Situ Methane Enrichment

In situ methane enrichment is achieved using desorption process by using modest recirculation rates. It is predicted that system is capable of giving biogas containing 94% v/v methane and it could be modified to achieve carbon dioxide removal efficiency greater than 60 (Hayes et al., 1990; Kashyap et al., 2003). A simple in situ technique was developed to separate carbon dioxide and methane from biogas by using solubility. The methane purity was found more than 98% v/v however leachate recycle rates and alkalinity affects the resulting off gas methane contents [219]. On the other hand high recycle rates takes the pH of digester to above 8 which results in volatile fatty acid accumulation and methane production rates (Jewell et al., 1993; Richards et al., 1994; Lindberg and Rasmuson, 2006, 2007).

5. FEEDSTOCKS

All types of biomass can be used as feedstocks for biogas production as long as they contain carbohydrates, proteins, fats, cellulose, and hemicelluloses as main components. When the feedstock is woody or contains high proportions of lignin, digestion becomes difficult – to obtain as efficient digestion, these feedstocks are combined in proportions that allow the digestion process to proceed. Pre-digestion and finely chopping will be helpful in the case of some materials – for example animal wastes are predigested and plant wastes do not need pre-digestion but excessive amounts of plant material in the feedstock may cause digester inefficiency.

The composition of biogas and the methane yield depends on (i) the feedstock type, (ii) the digestion system, and (iii) the retention time. The theoretical gas yield varies with the content of carbohydrate derivatives, protein derivatives, and fat derivatives. Only strong lignified organic substances, such as wood, are not suitable due to the slow rate of the anaerobic decomposition of lignin.

The biogas yield from the individual feedstocks varies considerably dependent on their origin, content of organic substance, and feedstock composition. Fat derivatives provide the highest biogas yield but require a long retention time due to their poor bioavailability. Carbohydrate derivatives and protein derivatives have much higher rates of conversion than fats but the yield of the biogas is lower. All feedstocks should be free of pathogens and other organisms; otherwise, pasteurization at 70°C (158°F) or sterilization at 130°C (266°F) is necessary prior fermentation.

The most important co-feedstocks are energy crops which have the highest potential for biogas production and the most important parameter for choosing energy crops is their net energy yield per hectare. Many conventional forage crops produce large amounts of easily degradable biomass which is necessary for high biogas yields. Forage crops have the advantage of being suitable for harvesting and storing with existing machinery and methods. The biogas (methane) yield is affected by the chemical composition of the crop which changes as the plant matures. Harvesting time and frequency of harvest are, thus, important for the feedstock quality and biogas yield.

Semi-continuously feeding of high amounts of silage (fermented high-moisture stored fodder which can be fed to cattle and sheep or used as a feedstock for anaerobic digesters) results in a sudden change of the gas quality because carbon dioxide is stripped out due to the local reduction of the pH in the fermenter. A pretreatment by mechanical, thermal, chemical, or enzymatic processes can be applied to increase the rate of degradation of feedstocks. The decomposition process is faster with decreasing particle size but does not necessarily increase the methane yield (Mshandete et al., 2006) and as a result, feedstock crushing is usually directly connected to the feeding system by application of an extruder or by ultrasonic treatment of a side stream of the fermenter (Kim et al., 2003).

Treatment by a combination of a thermal pressure hydrolysis reaction (230°C, 445°F, 300 to 450 psi) results in degradation of organic polymers by hydrolysis into short chain, biologically good available compounds which increases the biogas yield while the retention time in the digester can be reduced drastically (Mladenovska et al., 2006). The addition of hydrolytic enzymes can improve the decomposition of structural polysaccharides resulting in an increased biogas yield of up to 20%. The addition of enzymes reduces the viscosity of the feedstock mixture in the digester significantly and avoids the formation of floating layers. But the effect of enzymes can be strongly reduced if the protease enzymes of anaerobic microorganisms degrade the added enzymes (Morgavi et al., 2001). Protease enzymes are those enzymes that assist proteolysis – protein catabolism (breakdown of protein molecules into simpler ones) by hydrolysis of peptide bonds.

Thus, the most important initial issue when considering the application of anaerobic digestion systems is the feedstock to the process. However, in the present context where the production of biogas is the goal, the level of putrescibility (the liability of an organic waste to become putrid or the liability of the waste to decay) is the key factor in its successful application. The more putrescible (digestible) the material, the higher the gas yields possible from the system.

Feedstocks can include biodegradable waste materials, such as waste paper, grass clippings, leftover food, sewage, and animal waste. Woody wastes are the exception, because they are largely unaffected by digestion, as most anaerobes are unable to degrade lignin. Xylophalgeous anaerobes (lignin consumers) or using high temperature pretreatment, such as pyrolysis, can be used to break lignin down. Anaerobic digesters can also be fed with specially grown energy crops, such as silage, for dedicated biogas production. A co-digestion or co-fermentation plant is typically an agricultural anaerobic digester that accepts two or more input materials for simultaneous digestion.

In fact, co-digestion of a mixture of two or more feedstocks is amenable to geographic areas where there is a coexistence of different types of residues in the same geographic area. In fact, due to the different characteristics of waste streams treated together, co-digestion may enhance the performance of the anaerobic digestion process owing to a positive synergism established in the digestion medium by providing a balanced nutrient supply and sometimes by suitably increasing the moisture content required in the digester (Horváth et al., 2016).

Finally, in terms of the moisture content of the feedstock, a strict subdivision of the processes into wet and solid-state (dry) digestion can be misleading, since the microorganisms involved in the digestion process always require a liquid medium in which to survive and grow. Misunderstandings also repeatedly arise when defining the dry matter content of the fresh mass that is to be digested, since it is common practice to use several different feedstocks (feedstocks), each with a different dry matter content. In this connection it must be clear to the operator that it is not the dry matter content of the individual feedstocks that determines the classification of the process but the dry matter content of the feedstock mixture fed into the digester.

Biogas produced by the anaerobic digestion process from any one (or more) of the above-mentioned resources has been reported to be much more competitive than alcohol-based liquid biofuels on the basis of net energy generation (Edelmann et al., 2000; Chynoweth et al., 2001). In particular, biomass that contains carbohydrate derivatives, protein derivatives, fat derivatives, cellulose, and hemicellulose are suitable feedstocks for the production of biogas. Furthermore, the addition of co-feedstocks increases the organic content is advantageous to the process insofar as higher yields of biogas are achieved higher gas yield. The co-feedstocks include components such as organic waste materials from the agriculture-related industries as well as food waste and/or collected municipal biowaste from households. Chemically, feedstocks containing high proportions of carbohydrate derivatives and protein derivatives tend to exhibit faster conversion rates than feedstocks containing fats but tend to provide a higher yield of biogas.

Furthermore, the feedstock should be free of pathogens (infectious microorganisms) and the atomic nitrogen-to-carbon (N/C) ratio should not be excessive in order to avoid the production of excess ammonia in the biogas (Zubr, 1986). The atomic nitrogen-to-carbon (N/C) ratio in the feedstock is a crucial factor in maintaining perfect environment for digestion. Carbon is used for energy and nitrogen for building the cell structure and the bacteria use up carbon about 20 to 30 times faster than they use up nitrogen. When there is too much carbon in the raw wastes, nitrogen will be used up first and carbon left over which will cause digester inefficiency and the digestion process will eventually cease. On the other hand if there is too much nitrogen, the carbon soon becomes exhausted and fermentation stops and any nitrogen remaining will combine with hydrogen to form ammonia which can inhibit the growth of bacteria specially the methane producers.

Animal manure with the presence of diverse microbial flora has 75 to 92% w/w of moisture content and 72 to 93% w/w of volatile solids (VS) of the total solids (TS) present, good buffering capacity, and eliminate the step of inoculating the digester is a decided benefit which make the manure an ideal feedstock for anaerobic digestion (Muller et al., 2004; Fujino et al., 2005). But, the anaerobic digestion of animal manure tends to be a slow process and, as cautioned above, the production of high concentrations of ammonia in the product gas has a poisoning effect on the methanogenic bacteria (Kayhanian, 1999).

Large quantities of lignocellulosic waste are collected from agricultural, municipal, and other activities. Generally, the composition of lignocellulose is highly variable when it is selected from different sources – the composition is dependent on various conditions such as the origin and seasonal factors (Prassad et al., 2007; Bertero et al., 2012; Iqbal et al., 2013). Cellulose is a linear polymer and is linked by several b-1,4 glycoside bonds.

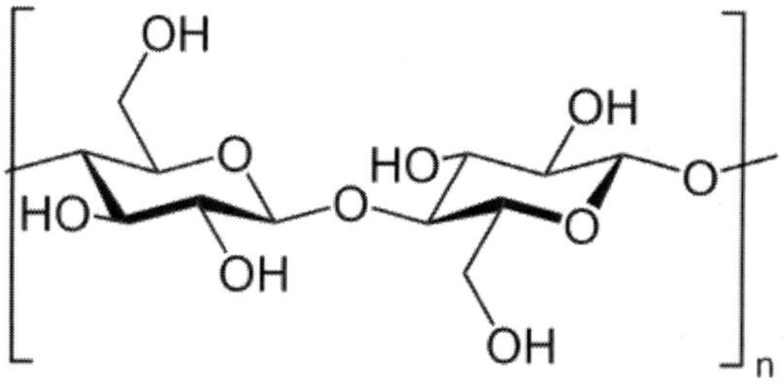

The structure contains parts with a crystalline structure and parts with an amorphous arrangement. The crystalline structure based on hydrogen linkages which results in higher toughness and solidity to the molecule. Moreover, crystalline cellulose can be converted to a non-organized structure-based cellulose by applying a temperature of 320°C (610°F) and a pressure of 3500 psi (Deguchi et al., 2006).

On the other hand, hemicellulose is a complex and changeable structure consisting of different polymers such as pentose derivatives (xylose, arabinose), hexose derivatives (mannose, glucose, galactose), and uronic acid derivatives (glucuronic acid, galacturonic acid, and methyl galacturonic acid) (Figure 3.2).

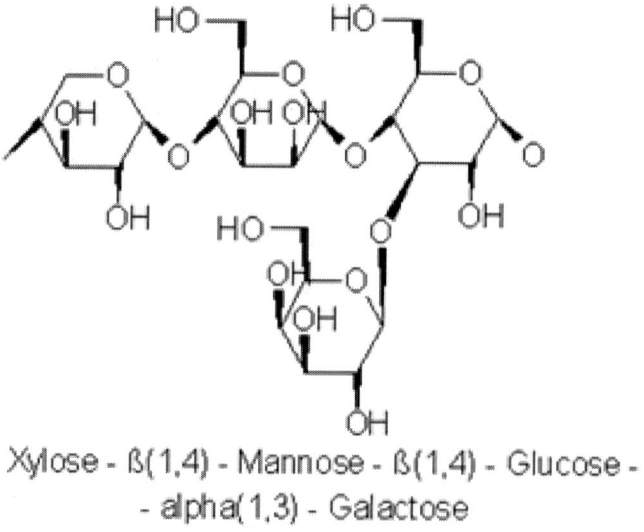

- Xylose - ß(1,4) - Mannose - ß(1,4) - Glucose -
- alpha(1,3) - Galactose

Figure 3.2. Schematic (and Hypothetical) Representation of the Structures in Hemicellulose.

Uronic acids are a class of sugar-related acid derivatives with both carbonyl and carboxylic acid functional groups. They are sugars in which the hydroxyl group on the terminal carbon atom has been oxidized to a carboxylic acid. The names of uronic acids are generally based on their parent sugars, for example, the uronic acid analog of glucose is glucuronic acid. Uronic acids derived from hexose derivatives (six-carbon sugars) are known as hexuronic acids and uronic acids derived from pentose derived derivatives (five-carbon sugars) are known as penturonic acids. Thus:

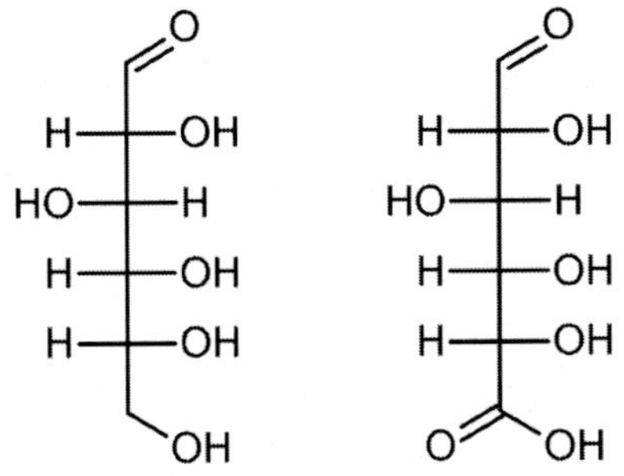

Glucose Glucuronic acid

The dominant compound in the hemicellulose is xylan (up to 90%) which is a group of hemicellulose derivatives that represents an abundant naturally-occurring biopolymer but which can vary according to the origin of the feedstock. Recent studies refer that hemicellulose requires a wide variety of enzymes in order to be fully hydrolyzed into free monomers (Saha, 2003; Ebringerova et al., 2005; Laureano-Perez et al., 2005; Girio et al., 2010).

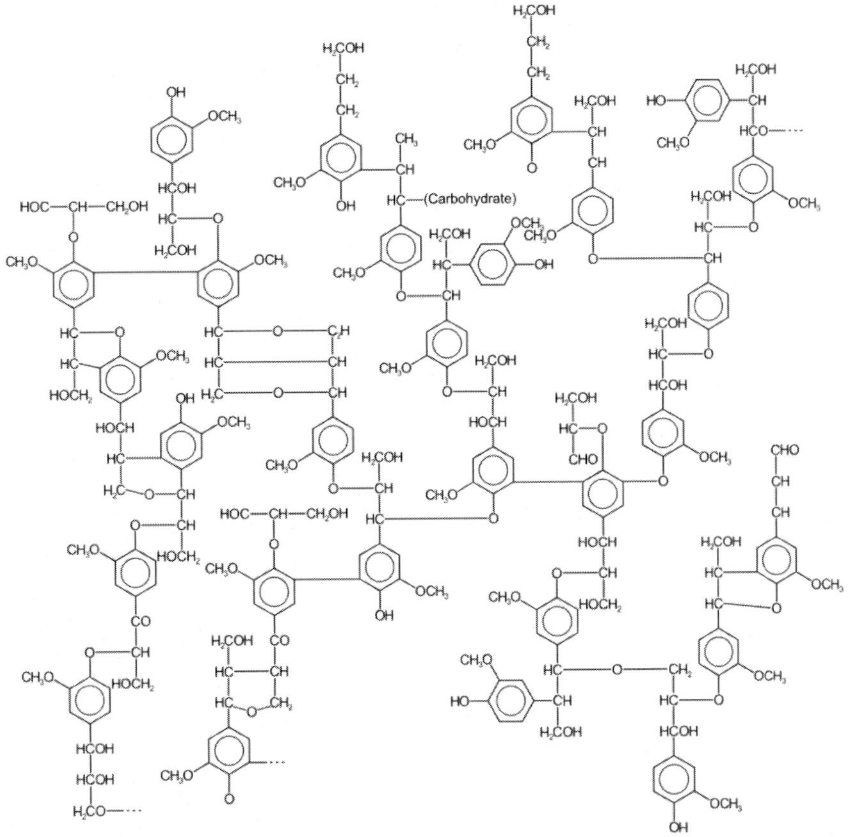

Figure 3.3. Hypothetical Structure for Lignin – Used Here to Indicate the Potential Complexity of the Molecule.

Lignin is a natural heteropolymer that occurs in cellular. The structure is complex and largely unknown although hypothetical structures have been proposed (Figure 3.3). These structures contain covalent bonds and consists of three phenyl propane-based units (p-coumaryl, coniferyl, and sinapyl alcohol units) that are held together by linkages (Mielenz, 2001). However, the structural compactness of lignin provides resistance in microbial attack and the non-water solubility of lignin makes for difficult degradation and the characteristics of lignin, such as composition and structure, can positively affect the hydrolysis process in order to increase the biogas production efficiency and pretreatment is necessary (Grabber, 2005).

The application of pretreatment methods enhances the degradation of feedstocks and the process efficiency and chemical, thermal, mechanical or enzymatic processes can be applied in order to fast the decomposition process. However the pretreatment process does not necessarily mean there will be an increase in the yield of biogas. Pretreatment involving particle size reduction and co-digestion with other biomass such as municipal sludge or animal manures substantially increases methane production from municipal solid waste (Del Borghi et al., 1999; Gomez-Lahoz, et al., 2007). The usual form of feedstock pretreatment is to remove the inorganic fraction (e.g., metals and glass) prior to anaerobic digestion in order to assure a more proficient operation of anaerobic digesters by increasing the digestibility of the organic feedstock constituents and prevention of equipment failure involved in material handling.

Algae possess a number of potential advantages compared to higher plants as alternative feedstocks for biogas production because the relatively rapid growth as well as the higher production rate compared to the terrestrial biomass and the possibility of cultivation on non-cultivable land areas or in lakes or the ocean, therefore abating food and feed competition (Rittmann, 2008; Stephens et al., 2010; Kroger and Muller-Langer, 2012). Moreover, microalgae can be grown using nutrient rich wastewater or reject water which would conceivably provide all of the vital nutrients needed for growth (Pittman et al, 2011; Rusten and Sahu, A. K. 2011). During the growth throughout the photosynthesis process, microalgae can reduce emissions of carbon dioxide by carbon uptake and remove nutrients from wastewater. These distinct advantages support the use of microalgae as an alternative and promising feedstocks for energy production. Additionally, some microalgae like the green microalga *Chlamydomonas reinhardtii* have the remarkable ability to produce hydrogen via hydrolysis of water during illumination. This represents an environmentally friendly gaseous fuel (Nguyen et al., 2008).

Hydrogen generation is a two-phase process with an aerobic stage and an anaerobic stage, during which the cells undergo major physiological changes. After hydrogen production, algal biomass residues serve as a waste product and, in the context of bioenergy production using microalgae, there is the suggestion that residues from the use of algal biomass should be converted into biogas via anaerobic fermentation. In fact, both, raw microalgae biomass and residues from other biofuels production could be utilized as feedstocks for the anaerobic digestion process. Using residues would allow algal biomass (due to the impact of algae cell wall structure on the volume of produced biomethane) to assist in the reduction of the amounts of algal waste and additional requirements, such as landfilling.

As might be anticipated, and in accordance with the dependency on the type of feedstock, the yield of methane (biogas) depends on (i) the species of microalgae, (ii) the pretreatment of the feedstock, and (iii) on the presence or absence of the inhibitors of methanogenesis (Mussgnug et al., 2010; Jones and Mayfield, 2012). The main benefit of using microalgae in an anaerobic digester is thc much higher energy efficiency in comparison to biofuels. It is mainly due to the fact that oil and lipids extraction is that the main product – methane – is captured in the gaseous phase. During the methanogenic fermentation all of the macromolecules (proteins, lipids and sugars – all parts of the microalgae structure) are utilized and in addition, the nutrients such as organic nitrogen or phosphorus could be mineralized and later recycled for algae cultivation.

The biomethane so produced could be burned in a combined heat and power unit to produce heat and electricity or it could be upgraded and injected into natural gas grid or used as a car fuel (Chapter 8). It is also worthy of note that the residues after the anaerobic digestion of microalgae could be further utilized as natural fertilizer (Harun et al., 2010). Therefore the drawbacks are conditioned by considering the feedstock concentration pH, temperature, cell wall characteristic, feedstock pretreatment and co-fermentation which will in turn decrease the atomic nitrogen-to-carbon (N/C) ratio, e.g., addition of biomass with a high concentration of organic carbon to a feedstock mixture (Saharan et al., 2013).

5.1. Composition

Feedstock composition is a major factor in determining the methane yield and methane production rates from the digestion of biomass. The length of time required for anaerobic digestion depends on the chemical complexity of the material. Material rich in easily digestible sugars breaks down quickly, whereas intact lignocellulosic material rich in cellulose and hemicellulose polymers can take much longer to break down. Anaerobic microorganisms are generally unable to break down lignin, the recalcitrant aromatic component of biomass.

Techniques to determine the compositional characteristics of the feedstock are available, while parameters such as solids, elemental, and organic analyses are important for digester design and operation. Methane yield can be estimated from the elemental composition of feedstock along with an estimate of its degradability (the fraction of the feedstock that is converted to biogas in a reactor). In order to predict biogas composition (the relative fractions of methane and carbon dioxide) it is necessary to estimate carbon dioxide partitioning between the aqueous and gas phases, which requires additional information (such as reactor temperature, pH, and feedstock composition) and a chemical speciation model.

Anaerobic microorganisms can metabolize different organic compounds (carbohydrates, proteins, lipids, etc.) (Kiran et al., 2016). The methane content of the biogas mixture depends on the biomass, particularly the oxidative state of carbon found in the feedstocks. Highly-reduced carbon leads to a higher the content of methane in the biogas and, thus, the atomic carbon/nitrogen ratio of the feedstock should be well-balanced (Kiran et al., 2016).

The lipid content of biomass also affects anaerobic digestion performance. High lipid content can also create problems during the digestion. Although the biomethane potential of the lipids (1014 L/kg volatile solids is much higher than carbohydrates (370 L/kg volatile solids, long-chain fatty acids (LCFA) are inhibitory to the anaerobic digestion process and cause system failures at high concentrations. Long-chain fatty acids are absorbed on the cell surface and spoil cell transport mechanism and, furthermore, they are also adsorbed on the biomass and cause microbial flocs. The inhibition of the process by long-chain fatty acids can be mitigated defeated by the addition of inoculum or by co-digestion with lipid-poor feedstocks, such as cattle manure and sludge.

Macronutrients, such as phosphorus, potassium, magnesium, and sulfur are required for the activation and functioning of anaerobic microorganisms. Apart from macro-elements, other trace elements are also crucial for cell growth and enzymatic activities. As examples, nickel is crucial for coenzyme F430 synthesis, iron is an essential constituent of electron carriers, calcium stabilizes the cellular wall and is required for the thermal stability of the endospores, cobalt is the component of the vitamin B12, and zinc is required for the synthesis of several enzymes. Still, they are toxic at elevated concentrations. To improve the anaerobic digestion performance, insufficient trace elements might be supplemented, while excess trace elements might be diluted by co-digestion to prevent trace element inhibition (Kiran et al., 2016).

5.2. Contamination

The level of contamination of the feedstock material is a key consideration. If the feedstock to the digesters has significant levels of physical contaminants, such as plastic components, glass components, or metal components, pre-processing the feedstock to remove the contaminants will be required before the feedstock is added to the digester. If such contaminants are not removed, the digesters can be blocked and will not function efficiently. It is with this understanding that mechanical biological treatment plants are designed. The higher the level of pretreatment a feedstock requires, the more processing machinery will be required, and, hence, the project will have higher capital costs.

The anaerobic digestion process can be inhibited by several compounds, affecting one or more of the bacterial groups responsible for the different organic matter degradation steps. The degree of the inhibition depends, among other factors, on the concentration of the inhibitor in the digester. Potential inhibitors are ammonia, sulfide, light metal ions (Na^+, K^+, Mg^{2+}, Ca^{2+}, Al^{3+}), heavy metals (i.e. metals with relatively high density, atomic weight, or atomic number), some organics (such as chlorophenol derivatives, halogenated aliphatic derivatives, N-substituted aromatic derivatives, and long chain fatty acid derivatives).

One of the most significant endogenous inhibitors is ammonia (NH_3) which is created during the bacterial degradation of nitrogen-containing substances such as proteins. Nitrogen is essential for bacterial growth and ammonia is an important source of nitrogen. But ammonia at high concentrations is highly toxic to the bacteria. In an aqueous solution ammonia is always found in an equilibrium with ammonium (NH4+). This equilibrium is determined by the acidity, pH and temperature of the environment and, as ammonium is not as toxic as ammonia, this equilibrium is important:

$$NH_4^+ \rightarrow NH_3 + H^+$$

At a high pH, the equilibrium is shifted to the right, and the environment becomes more toxic to bacteria. Higher temperatures will also shift this equilibrium to the right. This is why a thermophilic biogas process – all other aspects of the process being in order – is more sensitive than a mesophilic process to ammonia inhibition. There will be a certain inhibition of the bacteria already at relatively low ammonia concentrations. But with a longer adaptation period, bacteria are able to adapt to a higher concentration. This is fortunate, because the biomasses typically used in biogas production, such as slurry, usually have an ammonia concentration at the higher end of the scale. What the bacteria will have difficulties with is a sudden increase in the concentration, and a consistent and even input of biomass is therefore important for the process and even more so at higher temperatures.

After sorting or screening to remove any physical contaminants from the feedstock, the material is often shredded, minced, and mechanically or hydraulically pulped to increase the surface area available to microbes in the digesters and, hence, increase the speed of digestion. The maceration of solids can be achieved by using a chopper pump to transfer the feedstock material into the airtight digester, where anaerobic treatment takes place.

5.3. Moisture content

A second consideration related to the feedstock is moisture content. Drier, stackable feedstocks, such as food and yard waste, are suitable for digestion in tunnel-like chambers.

These systems typically have near-zero wastewater discharge, as well, so this style of system has advantages where the discharge of digester liquids are a liability. The wetter the material, the more suitable it will be to transportation using standard pumps instead of energy-intensive concrete pumps and physical means of movement. Also, the wetter the material, the more volume and area it takes up relative to the levels of biogas produced. The moisture content of the target feedstock will also affect what type of system is applied to its treatment. To use a high-solids anaerobic digester for dilute feedstocks, bulking agents, such as compost, should be applied to increase the solids content of the input material. Another key consideration is the carbon/nitrogen ratio of the input material. This ratio is the balance of food a microbe requires to grow; the optimal carbon-nitrogen atomic ratio is (20 to 30):1. Excess nitrogen lead to the inhibition (by ammonia) of the digestion process.

5.4. Solids Content

In a typical scenario, three different operational parameters are associated with the solids content of the feedstock to the digesters: (i) high solids – dry, stackable feedstock, (ii) high solids – wet, pumpable feedstock, and (iii) low solids – wet, pumpable feedstock.

High solids (dry) digesters are designed to process materials with a solids content between 25 and 40% w/w. Unlike the wet digesters, which process pumpable slurries, high solids (dry – stackable feedstock) digesters are designed to process solid feedstocks without the addition of water. The primary styles of dry digesters are continuous vertical plug flow and batch tunnel horizontal digesters. Continuous vertical plug flow digesters are upright, cylindrical tanks where feedstock is continuously fed into the top of the digester, and flows downward by gravity during digestion. In batch tunnel digesters, the feedstock is deposited in tunnel-like chambers with a gas-tight door. Neither approach has mixing inside the digester.

The amount of pretreatment, such as contaminant removal, depends both upon the nature of the waste streams being processed and the desired quality of the digestate. Size reduction (grinding) is beneficial in continuous vertical systems, as it accelerates digestion, while batch systems avoid grinding and instead require structure (e.g., yard waste) to reduce compaction of the stacked pile. Continuous vertical dry digesters have a smaller footprint due to the shorter effective retention time and vertical design. Wet digesters can be designed to operate in either a high-solids content, with a total suspended solids (TSS) concentration greater than approximately 20% w/w, or a low-solids concentration less than 15% w/w.

High solids (wet) digesters process a thick slurry that requires more energy input to move and process the feedstock. The thickness of the material may also lead to associated problems with abrasion. High solids digesters will typically have a lower land requirement due to the lower volumes associated with the moisture. High solids digesters also require correction of conventional performance calculations (e.g., gas production, retention time, kinetics, etc.) originally based on very dilute sewage digestion concepts, since larger fractions of the feedstock mass are potentially convertible to biogas.

Low solids (wet) digesters can transport material through the system using standard pumps that require significantly lower energy input. Low solids digesters require a larger amount of land than high solids due to the increased volumes associated with the increased liquid-to-feedstock ratio of the digesters. There are benefits associated with operation in a liquid environment, as it enables more thorough circulation of materials and contact between the bacteria and their food, which enables the bacteria to more readily access the substances on which they are feeding, and increases the rate of gas production.

6. Feedstock Pretreatment

All organic matter, with the exception of lignin, can be decomposed anaerobically to produce biogas, although the time taken to do so differs greatly. How long it takes depends on the composition of the biomass – the more complicated the molecules, the longer it takes a microorganism to break it down. Sometimes a biomass feedstock consists of a number of different substances that have an impact on the gas yield (Table 3.2). The ratios between carbohydrate, protein and fat – has an effect on how much methane the biomass contains and therefore on its calorific value. The methane content from a biogas plant where the input is mainly animal manure is typically on the order of 65%, which is because there is some carbon dioxide in solution in the alkaline liquid.

Landfills can be considered as large anaerobic plants with the difference that the decomposition process is discontinuous and depends on the age of the landfill site. Recovery of landfill gas is not only essential for environmental protection and reduction of emissions of methane and other landfill gases but it is also a cheap source of energy, generating benefits through faster stabilization of the landfill site and revenues from the gas utilization. Landfill gas recovery can be optimized through the management of the site such as shredding the waste, re-circulating the organic fraction and treating the landfill as a bioreactor.

The hydrolysis of feedstock into soluble organics is the rate-limiting step for anaerobic digestion (Zhang et al., 2014) and, as a consequence, anaerobic digestion often suffers from long solid residence time and low efficiency for the process (Quiroga et al., 2014). Therefore, the effects of different pretreatment methods such as (i) mechanical, (ii) thermal, (iii) chemical, and (iv) enzymatic pretreatments are available for improving the digestion (hydrolysis) of biomass feedstocks. Also, it has been established that the overall biogas yield is directly linked with the type of interaction within different waste streams that interfere with digestibility of wastes in the anaerobic digestion process (Mahanty et al., 2014). This has, in turn, resulted in the finding that the co-digestion on sewage sludge, the organic fraction of municipal solid waste, agricultural crops, lignocellulosic wastes and algal biomass give the best biomethane yield in terms of quality and quantity. However, the most used basic feedstock in agriculture is animal (pig or cow) manure in co-fermentation with biogas crops (Rodriguez et al., 2017).

Table 3.2. Examples of the Chemistry of the Conversion of Biogas Constituents

Constituent	Chemistry
Cellulose	$(C_6H_{10}O_5)_n$ + nH_2O + $3nCH_4$ + $3nCO_2$
Protein	$2C_5H_7NO_2$ + $8H_2O$ + $5CH_4$ + $3CO_2$ + $2(NH_4HCO_3)$
Fat	$C_{57}H_{104}O_6$ + $28H_2O$ + $40CH_4$ + $17CO_2$

Note: The fat is represented by glycerol trioleic acid.

Besides co-digestion of feedstocks, studies have also delineated that the appropriate pretreatment technique can (i) improve the rate of anaerobic digestion process, (ii) increase the biogas yield, (iii) provide a wide range of new and/or locally available feedstocks for use, (iv) increase the biogas production rate, as well as (v) reduce the production of volatile solids (Carlsson et al., 2012). The various process performance enhancers are (alphabetically rather than by any preference) are: (i) anaerobic co-digestion, (ii) biological pretreatment, (iii) chemical pretreatment, (iv) mechanical pretreatment, (v) thermal pretreatment.

6.1. Anaerobic Co-Digestion

Anaerobic digestion has great potential for energy recovery from and stabilization of the waste biomass (Mata-Alvarez et al., 2014; Zhang et al., 2014). The process parameters such as composition of the feedstocks and their mixing ratio are very important for the performance of the co-digestion process. The ratio of feedstocks is generally optimized based on atomic nitrogen-to-carbon ration but other parameters such as macronutrients, micronutrients, acidity-alkalinity (pH) and free ammonia and inhibitory compounds should be considered (Dai et al., 2013; Koch et al., 2015; Zhang et al., 2011).

Thermophilic anaerobic digestion usually provides a faster metabolic rate and higher system performance than mesophilic anaerobic digestion (Zhang et al., 2014). The biomethane yield can be increased significantly by co-digestion of organic wastes. However, the transportation of waste influences this effect negatively. In order to achieve a sustainable waste treatment strategy, the transportation of the wastes should be minimized.

6.2. Biological Pretreatment

Biological pretreatment using enzymes may improve the solubility of the biomass without producing any inhibitory compounds (Moon and Song, 2011; Parawira, 2012). Biological pretreatments by the addition of microorganisms have also been found to improve biogas production from cattle manure and agricultural residues (Angelidaki and Ahring, 2000; Chen et al., 2010; Zhong et al., 2011). The bacteria and fungi basically degrade lignin and hemicellulose and increase the accessibility of cellulose in an environmentally friendly way. Another biological pretreatment method is ensiling, which is particularly applied for energy crops and is applied using starter cultures or enzymes to convert soluble sugars to organic acids such as lactic and acetic acids in order to inhibit the growth of undesirable microorganisms during the storage (Weiland, 2010).

6.3. Chemical Pretreatment

Chemical pretreatment with strong acid, strong alkali, or an oxidant are applied to digester feedstock to enhance the hydrolysis rate and subsequent biogas production and, as expected, the effect of the pretreatment depends on the type of method applied and the characteristics of the feedstock. For example, pretreatment using concentrated acid is not a suitable option for easily degradable feedstocks which have high amount of carbohydrates, due to the accumulation of volatile fatty acids and the degradation of soluble sugars into inhibitory compounds such as furfural and hydroxymethyl furfural (Vavouraki et al., 2012).

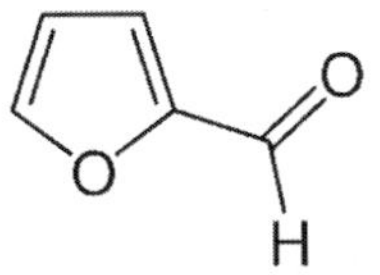

Furfural

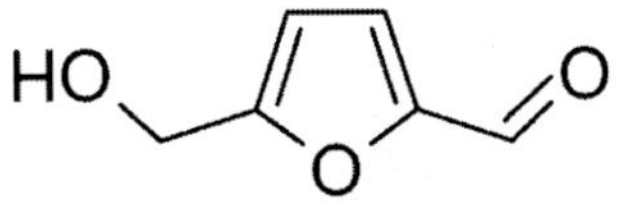

Hydroxymethyl furfural

Dilute acid pretreatment is generally used and coupled with thermal pretreatment for enhancing the hydrolysis and biogas production from such biomasses, ie, food waste (Ariunbaatar et al., 2014). On the other hand, these methods facilitate hemicellulose hydrolysis and modify the complex structure of lignin, improving the accessibility of the cellulose to microbial attacks in lignocellulosic biomasses. Therefore, the hydrolysis of and biogas production from lignocellulosic biomasses significantly is improved by chemical/thermochemical pretreatments (Zheng et al., 2014). However, the cations released from the salts and hydroxides, such as sodium (Na^+), potassium (K^+), and calcium (Ca^{2+}), and sodium, can exert an inhibitory effect on the anaerobic digestion process if the concentration of the cations is allowed to build up. In addition, the pH of the pretreated biomass should be neutralized before anaerobic digestion. Therefore, the cost of the chemicals and the addition of neutralization agents should be considered in large-scale applications.

Controlled doses of alkali solutions in anaerobic digestion systems can enhance the yield of biogas and at the same time reducing cellulose production especially when using plant material as feedstock (Taherdanak and Zilouei, 2014). However, alkali solutions often lead to saponification reactions in continuous plants. These reactions tend to yield generate compounds leading to tremendous a reduction in the acetate degradation rate and in the glucose degradation rate (Mouneimne et al., 2003). The *addition of metals* such as cadmium (Cd^{2+}), nickel (Ni^{2+}) and zinc (Zn^{2+}) has been observed to improve the anaerobic digestion of a combination of cattle manure and potato waste and, as a result, the yield of biogas was enhanced greatly(Kumar et al., 2006). Also, *preheating the feedstock* before anaerobic digestion can improve the production of methane as well reduce the production of volatile solids. Furthermore, it has also been demonstrated that preheating the feedstock that has previously been treated with chemical additives (thermo-chemical) enhanced the yield of methane even further and cause further reduction in the yield of volatile solids thereby enhancing the biodegradability of the feedstock (Ardic and Taner, 2005; Carrere et al., 2009).

Oxidation using ozone (O_3, also called ozonation or ozonization) is another chemical pretreatment option for enhancing the biogas production. The process has some advantageous over alkaline or acid pretreatment because it does not release any inhibitory compounds like furfural and hydroxymethyl furfural or discharge any cations. It is also useful for the disinfection of the pathogens. Therefore, the method is often used for sludge treatment and wastewater treatment to improve the hydrolysis and, hence, the anaerobic digestion process.

6.4. Mechanical Pretreatment

Mechanical pretreatment (also often referred to as *milling*) is usually required, before any other kind of pretreatment, for feedstocks that have high content of solids and is the process by which the particle size reduction of the feedstock is accomplished. The process causes the disintegration of the solid constituents and/or decreases the particle size of the biomass thereby increasing the surface area which results in increased feedstock availability for the microorganisms. Consequently, microbial growth and anaerobic digestion process are improved (Zhu et al., 2009; Izumi et al., 2010; De la Rubia et al., 2011; Subramani and Ponkumar, 2012; Ariunbaatar et al., 2014; Zheng et al., 2014).

Microwave pretreatment has been claimed to enhance the yield of biogas (Shahriari et al., 2012) and ultrasonic pretreatment, which is commonly used in sewage sludge treatment, is claimed to improve biogas production from anaerobic digestion. This technique introduces ultrasonic cavitation into the system that in-turn builds up mechanical shear forces that ultimately aid the sludge dis-integration as well as the collapse of cavitation bubbles which improve the feedstock's physical properties (Tichm et al., 2001).

An additional option is the use of an ultrasonic technique that can improve the hydrolysis and anaerobic digestion of sludge. The process disintegrates sludge by disrupting the microbial cell walls in sludge, resulting in the release of organic substances of sludge into the liquid phase leading to an increase in the biogas generation (Pilli et al., 2011). As might be expected, the efficiency of the process depends on the sludge characteristics and ultrasonication parameters, therefore, not all studies confirm the enhancement of hydrolysis or improved biogas production by ultrasonication (Sandino et al., 2005; Marin et al., 2010; Quiroga et al., 2014).

6.5. Thermal Pretreatment

Thermal pretreatment is another pretreatment method, which is applied for the enhancement of anaerobic digestion processes from various kinds of biomass in large scale. In the process, the cell membranes are disintegrated thereby enhancing the solubilization of the feedstock constituents leading to improvement in the anaerobic digestion process and shorten the retention time in the digester. Moreover, the process helps to remove the pathogens and reduces the viscosity of the biomass. However, the effects of thermal pretreatment depend on the feedstock type, pretreatment temperature, and duration and is not suitable for all feedstocks (Skiadas et al., 2005; Wang et al., 2006; Appels et al., 2010; Rafique et al., 2010; Chamchoi et al., 2011; Vavouraki et al., 2012; Gonzalez-Fernandez et al., 2012; Tampio et al., 2014).

On the other hand, steam explosion is one of the most promising pretreatments for lignocellulosic feedstocks. The process is operated at high temperatures (150 to 250°C, 300 to 480°F) for a few seconds to minutes, which is followed by a sudden pressure drop (Weiland, 2010). It helps to reduce the crystallinity, release soluble compounds and improve the biogas yield (Shafiei et al., 2013).

6.6. Other Forms of Pretreatment

Seeding is a way of initiating a newly commissioned biogas plant by feeding it with previously digested material from another established set up. The method is, essentially, the introduction of an inoculum into the system to reduce the plant start-up time. Alternatively, feedstocks such as animal manure or municipal sewerage are often used to seed a newly commissioned biogas digester. Particle size reduction is often practiced because the particle sizes of the feedstock directly affect digestion as it has direct indication on the available surface area for hydrolyzing enzymes especially when plant fiber is part pf the feedstock. Methane yield and fiber degradation have been found to improve with decreasing particle sizes within the feedstock from, for example, from 100 mm to 2 mm.

Liquid hot water treatment will (i) assist in the solubilization of hemicellulose and lignin, (ii) reduce the risk of inhibitors production such as furfural, and (iii) increase enzyme accessibility. However, the process demand high heat and this type of stimulus may only be effective up to a certain temperature (Laser et al., 2002). *Acid pretreatment* will assist in the solubilization of hemicellulose derivatives and any methanogens present are capable of adapting to inhibiting compounds. However, the additional cost of the acid and the of forming inhibiting compounds as well as introducing corrosion problems may persist and offer disadvantages to this method (Xiao and Clarkson, 1997).

REFERENCES

Amani, T., Nosrati, M., and Sreekrishnan, T. R. 2010. Anaerobic Digestion from the Viewpoint of Microbiological, Chemical, and Operational Aspects – A Review. *Environ. Rev.,* 18: 255-278.

Angelidaki, I., Ellegaard, L., and Ahring, B. K. 1993. A Mathematical Model for Dynamic Simulation of Anaerobic Digestion of Complex Feedstocks: Focusing on Ammonia Inhibition. *Biotechnol Bioeng*., 42: 159-166.

Angelidaki, I., and Ahring, B. K., 2000. Methods for Increasing the Biogas Potential from the Recalcitrant Organic Matter Contained in Manure. *Water Science and Technology*, 41(3): 189-194.

Angelidaki, I., and Batstone, D. J. 2010. Anaerobic Digestion: Process. In: *Solid Waste Technology & Management Volume 2*. T. Christensen. (Editor). John Wiley & Sons Inc., Hoboken, New Jersey. Chapter 9.4. Page 583-585.

Appels, L., Degreve, J., Bruggen, B. V., Impe, J. V., and Dewil, R. 2010. Influence of Low Temperature Thermal Pre-Treatment on Sludge Solubilization, Heavy Metal Release and Anaerobic Digestion. *Bioresource Technology*, 101: 5743-5748.

Ardic, I., and Taner, F. 2005. Effects of Thermal, Chemical and Thermochemical Pretreatment to Increase Biogas Production Yield of Chicken Manure. *Fresenius Environ, Bull,* 14: 373-380.

Ariunbaatar, J., Panico, A., Frunzo, L., Esposito, G., Lens, P. N. L., and Pirozzi, F. 2014. Enhanced Anaerobic Digestion of Food Waste by Thermal and Ozonation Pretreatment Methods. *Journal of Environmental Management*, 146: 142-149.

Bagi, Z., Acs, N., Balint, B., Horvath, L., Dobo, K., Perei, K. R., Rakhely, G., and Kovacs, K. L. 2007. Biotechnological Intensification of Biogas Production. *Appl. Microbiol. Biotechnol.,* 76: 473-482.

Bertero, M., de la Puente, G., and Sedran, U. 2012. Fuels from Bio-oils: Bio-Oil Production from Different Residual Sources, Characterization and Thermal Conditioning. *Fuel,* 95: 263-271.

Bertin, L., Grilli, S., Spagni, A., and Fava, F. 2013. Innovative Two-Stage Anaerobic Process for Effective Codigestion of Cheese Whey and Cattle Manure. *Bioresource Technology,* 128; 779-783.

Bharathiraja, B., Sudharsana, T., Jayamuthunagai, J., Praveenkumar, R., Chozhavendhan, S., and Iyyappan, J. 2018. Biogas Production – A Review on Composition, Fuel Properties, Feedstock and Principles of Anaerobic Digestion. *Renewable and Sustainable Energy Reviews*, 90: 570-582.

Braun R. 2007. Anaerobic Digestion: A Multi-Faceted Process for Energy, Environmental Management and Rural Development: *Improvement of Crop Plants for Industrial End Uses.* 1: 335-415.

Carlsson, M., Lagerkvist, A., and Morgan-Sagastume, F. 2012. The Effects of Feedstock Pretreatment on Anaerobic Digestion Systems: A Review. *Waste Manag,* 32: 1634-1650.

Carrere, H., Sialve, B., and Bernet, N. 2009. Improving Pig Manure Conversion into Biogas by Thermal and Thermo-Chemical Pretreatment. *Bioresour. Technol,* 00: 3690-3694.

Chamchoi, N., Garcia, H., and Angelidaki, I., 2011. Methane Potential of Household Waste: Batch Assays Determination. *Journal of Environmental Resource*, 33: 13-26.

Chen, G., Zheng, Z., Yang, S., Fang, C., Zou, X., and Luo, Y. 2010. Experimental Co-Digestion of Corn Stalk and Vermicompost to Improve Biogas Production. *Waste Management,* 30: 1834-1840.

Chen, Y., Cheng, J. J., and Creamer, K. S. 2008. Inhibition of Anaerobic Digestion Process: A Review. *Bioresource Technology*, 99(10): 4044-4064.

Chong, S., Sen, T. K., Kayaalp, A., and Ang, H. M. 2012. The Performance Enhancements of Upflow Anaerobic Sludge Blanket (UASB) Reactors for Domestic Sludge Treatment: A State-of-the Art Review. *Water Research*, 46(11): 3434-3470.

Chynoweth, D. P., Owens, J. M., and Legrand, R. 2001. Renewable Methane from Anaerobic Digestion of Biomass. *Renew Energy,* 22: 1-8.

Dai, C., Duan, N., Dong, B., and Dai, L. 2013. High-solids Anaerobic Co-Digestion of Sewage Sludge and Food Waste in Comparison With Mono-digestion: Stability and Performance. *Waste Management*, 33: 308-316.

Deguchi, S., Mukai, S. A., Tsudome, M., and Horikoshi, K. 2006. Facile Generation of Fullerene Nanoparticles by Hand-Grinding. *Adv. Mater.,* 18: 729-32.

De la Rubia, M. A., Fernandez-Cegri, V., Raposo, F., and Borja, R. 2011. Influence of Particle Size and Chemical Composition on the Performance and Kinetics of Anaerobic Digestion Process of Sunflower Oil Cake in Batch Mode. *Biochemical Engineering Journal,* 58/59: 162-167.

Del Borghi, A., Converti, A., Palazzi, E., and Del Borghi, M. 1999. Hydrolysis and Thermophilic Anaerobic Digestion of Sewage Sludge and the Organic Fraction of Municipal Solid Waste. *Bioprocess. Eng.*, 20: 553-560.

Driehuis, F., Elferink, S. J., and Spoelstra, S. F. 1999. Anaerobic Lactic Acid Degradation during Ensilage of Whole Crop Maize Inoculated With Lactobacillus Buchneri Inhibits Yeast Growth and Improves Aerobic Stability. *J. Appl. Microbiol.,* 87: 583-594.

Ebringerova, A., Hromadkova, Z., and Heinze, T. 2005. Hemicellulose. *Adv. Polym. Sci.,* 186: 1-67.

Edelmann, W., Schleiss, K., and Joss, A. 2000. Ecological, Energetic and Economic Comparison of Anaerobic Digestion with Different Competing Technologies to Treat Biogenic Wastes. *Water Sci. Technol.,* 41: 263-73.

Fujino, J., Morita, A., Matsuoka, Y., and Sawayama S. 2005. Vision for Utilization of Livestock Residue as a Bioenergy Resource in Japan. *Biomass Bioenergy*, 29: 367-74.

Girio, F. M., Fonseca, C., Carvalheiro, F., Duartem, L. C., Marquesm, S., and Bogel-Lukasic, R. 2010. Hemicelluloses for Fuel Ethanol: A Review. *Bioresour. Technol.,* 101: 4775-4800.

Gomez-Lahoz, C., Fernandez-Gimenez, B., Garcia-Herruzo, F., Rodriguez-Maroto, J. M., and Vereda-Alonso, C. 2007. Biomethanization of Mixtures of Fruits and Vegetables Solid Wastes and Sludge from a Municipal Waste Water Treatment Plant. *J. Environ. Sci. Health*, 42: 481-487.

Gonzalez-Fernandez, C., Sialve, B., Bernet, N., and Steyer, J. P. 2012. Thermal Pretreatment to Improve Methane Production of Scenedesmus Biomass. *Biomass Bioenergy,* 40: 105-111.

Grabber, J. H. 2005. How Do Lignin Composition, Structure, and Cross-Linking Affect Degradability? A Review of Cell Wall Model Studies. *Crop. Sci.,* 45: 820-831.

Gustavsson, J., Shakeri, Y. S., Sundberg, C., Karlsson, A., Ejlertsson, J., Skyllberg, U., Svensson, B. H. 2013a. Bioavailability of Cobalt and Nickel during Anaerobic Digestion of Sulfur-Rich Stillage for Biogas Formation. *Appl. Energy,* 112: 473-477.

Gustavsson, J., Yekta, S. S., Karlsson, A., Skyllberg, U., and Svensson, B. H. 2013b. Potential Bioavailability and Chemical Forms of Co and Ni in the Biogas Process – An

Evaluation Based on Sequential and Acid Volatile Sulfide Extractions. *Eng. Life Sci.,* 13: 572-579.

Harun, R., Singh, M., Forde, G. M., and Danquah, M. K. 2010. Bioprocess engineering of microalgae to produce a variety of consumer products. *Renew, Sustain, Energy, Rev.,* 14: 1037-1047.

Hayes, T. D., Issacson, H. R., Pfeffer, J. T., and Liu, Y. M. 1990. In Situ Methane Enrichment in Anaerobic Digestion. *Biotechnol. Bioeng*, 35(1): 73-86.

Horváth, I. S., Tabatabaei, M., Karimi, K., and Kumar, R. 2016. Recent Updates on Biogas Production - A Review. *Biofuel Research Journal*, 10: 394-402.

Iqbal, H. H. G., Kyazze, G., and Keshavarz, T. 2013. Advances in Valorization of Lignocellulosic Materials by Bio-Technology: An Overview. *Bioresources,* 8(2): 3157-3176.

Izumi, K., Okishio, Y., Nagao, N., Niwa, C., Yamamoto, S., and Toda, T., 2010. Effects of Particle Size on Anaerobic Digestion of Food Waste. *International Biodeterioration & Biodegradation,* 64 (7): 601-608.

Jansen, J. C. Anaerobic Digestion: Technology. In: *Solid Waste Technology & Management Volume 2.* T. Christensen (Editor). John Wiley & Sons Inc., Hoboken, New Jersey. Chapter, 9.4, page 607-612.

Jewell, W. J., Cummings, R. J., and Richards, B. K. 1993. Methane Fermentation of Energy Crops: Maximum Conversion Kinetics and in Situ Biogas Purification. *Biomass Bioenergy,* 5(3-4): 261-278.

Jones, C. S., and Mayfield, S. P. 2012. Algae Biofuels: Versatility for the Future of Bioenergy. *Curr. Opin. Biotechnol*, 23: 346-351.

Karakashev, D., Bastone, D., and Angelidaki, I. 2005. Influence of Environmental Conditions on Methanogenic Compositions in Anaerobic Biogas Reactors. *Appl Environ Microbiol* 71: 331-338.

Kashyap, D. R., Dadhich, K. S., and Sharma, S. K. 2003. Biomethanation under Psychrophilic Conditions: A Review. *Bioresource Technology,* 87: 147-153.

Kayhanian, M. 1999. Ammonia Inhibition in High-Solids Biogasification: An Overview and Practical Solutions. *Environ. Technol.,* 20: 355-365.

Kim, J., Park, C., Kim, T. H. L. M., Kim, S., and Lee, S. W. 2003. Effects of Various Pretreatments for Enhanced Anaerobic Digestion With Waste Activated Sludge. *J. Biosci. Bioeng,* 95: 271-275.

Kiran, E. U., Stimulator, K., Antonopoulou, G., and Lyberatos. G. 2016. Production of Biogas via Anaerobic Digestion. *Handbook of Biofuels Production* 2nd Edition. R. Luque, Lin, C., Wilson, K., and Clark, J. (Editors). Woodhead Publishing, Elsevier, B.V., Amsterdam, Netherlands. Chapter 10. Page 259-301.

Koch, K., Helmreich, B., and Drewes, J. E. 2015. Co-digestion of Food Waste in Municipal Wastewater Treatment Plants: Effect of Different Mixtures on Methane Yield and Hydrolysis Rate Constant. *Applied Energy,* 137: 250-255.

Kougias, P. G. and Angelidaki, I. 2018. Biogas and Its Opportunities – A Review. *Front. Environ. Sci. Eng,* 12(3): 14-25.

Kroger, M., and Muller-Langer, F. 2012. Review on Possible Algal-Biofuel Production Processes. *Biofuels,* 3(3): 333-349.

Kumar, A., Miglani, P., Gupta, R. K., Bhattacharya, T. K. 2006. Impact of Ni(II), Zn(II) and Cd (II) on the Biogasification of Potato Waste. *J. Environ. Biol,* 27: 61-66.

Laser, M., Schulman, D., Allen, S. G., Lichwa, J., Antal, M. J. Jr., and Lynd, L. R. 2002. A Comparison of Liquid Hot Water and Steam Pretreatments of Sugar Cane Bagasse for Bioconversion to Ethanol. *Bioresour. Technol,* 81(1): 33-44.

Laureano-Perez, L., Teymouri, F., Alizadeh, H., and Dale, B. E. 2005. Understanding Factors That Limit Enzymatic Hydrolysis Of Biomass. *Appl. Biochem. Biotechnol,* 1081-1099.

Leven, L., Eriksson, A. R. B., and Schnürer, A. 2007. Effect of Process Temperature on Bacterial and Archael Communities in Two Methanogenic Bioreactors Treating Organic Household Waste. *FEMS Microbiol. Ecol,* 59: 683-693.

Liao, B. Q., Bagley, D. M., Kraemer, H. E., Leppard, G. G., and Liss, S. N. 2004. A Review of Biofouling and Its Control in Membrane Separation Bioreactors. *Water Environment Research,* 76(5): 425-436.

Liao, B. Q., Kraemer, J. T., and Bagley, D. M. 2006. Anaerobic Membrane Bioreactors: Applications and Research Directions. *Critical Reviews in Environmental Science and Technology,* 36(6): 489-530.

Lin, H., Peng, W., Zhang, M., Che, N. J., Hong, H., and Zhang, Y. 2013. A Review on Anaerobic Membrane Bioreactors: Applications, Membrane Fouling and Future Perspectives. *Desalination,* 314: 169-188.

Lindberg, A., and Rasmuson, A. C. 2006. Selective Desorption of Carbon Dioxide from Sewage Sludge for In Situ Methane Enrichment Part I: Pilot Plant Experiments. *Biotechnol. Bioeng,* 95(5): 794-803.

Lindberg, A., and Rasmuson, A. C. 2007. Selective Desorption of Carbon Dioxide from Sewage Sludge for In Situ Methane Enrichment Part II: Modeling and Evaluation of Experiments. *Biotechnol Bioeng,* 97(5): 1039-1052.

Mahanty, B., Zafar, M., Han, M. J., and Park, H. S. 2014. Optimization of Co-Digestion of Various Industrial Sludges for Biogas Production and Sludge Treatment: Methane Production Potential Experiments and Modeling. *Waste Manag,* 34: 1018-1024.

Marin, J., Kennedy, K. J., and Eskicioglu, C. 2010. Effect of Microwave Irradiation on Anaerobic Degradability of Model Kitchen Waste. *Waste Management,* 30: 1772-1779.

Mata-Alvarez, J., Dosta, J., Romero-Guiza, M. S., Fonoll, X., Peces, M., and Astals, S. 2014. A Critical Review on Anaerobic Co-Digestion Achievements between 2010 and 2013. *Renewable and Sustainable Energy Reviews,* 36: 412-427.

Mielenz, J. R. 2001. Ethanol Production from Biomass: Technology and Commercialization Status. *Curr. Opin. Microbiol*, 4: 324-325.

Mladenovska, Z., Hartmann, H., Kvist, T., Sales-Cruz, M., Gani, R., and Ahring, B. K. 2006. Thermal Pretreatment of the Solid Fraction of Manure: Impact of the Biogas Reactor Performance and Microbial Community. *Water Sci. Technol.,* 53: 59-67.

Molino, A., Nanna, F., Ding, Y., Bikson, B., and Braccio, G. 2013a. Biomethane Production by Anaerobic Digestion of Organic Waste. *Fuel*, 103: 1003-1009.

Molino, A., Migliori, M., Ding, Y., Bikson, B., Giordano, G., and Braccio, G. 2013b. Biogas Upgrading Via Membrane Process: Modelling of Pilot Plant Scale and the End Uses for the Grid Injection. *Fuel,* 107: 585-592.

Moon, H. C., and Song, I. S. 2011. Enzymatic Hydrolysis of Food Waste and Methane Production Using UASB Bioreactor. *International Journal of Green Energy*, 8(3): 361-371.

Molino, A., Larocca, V., Chianese, S., and Musmarra, D. 2018. Biofuels Production by Biomass Gasification: A Review. *Energies,* 11: 811-843. https://www.researchgate.net/publication/324179284_Biofuels_Production_by_Biomass_Gasification_A_Review; accessed December 1, 2018.

Morgavi, D. P., Beauchemin, K. A., Nsereko, L. M. 2001. Resistance of Feed Enzymes to Proteolytic Inactivation by Rumen Microorganisms and Gastrointestinal Proteases. *J. Anim. Sci.*, 79: 1621-1630.

Mouneimne, A. H., Carrere, H., Bernet, N., and Delgenes, J. P. 2003. Effect of Saponification on the Anaerobic Digestion of Solid Fatty Residues. *Bioresour. Technol,* 90: 89-94.

Mshandete, A., Bjornsson, L., Kivaisi, A. K., Rubindamayugi, M. S. T., and Matthiasson B. 2006. Effect of Particle Size on Biogas Yield from Sisal Fiber Waste. *Renew. Energy,* 31: 2385-2392.

Muller, J. A., Winter, A., and Strunkmann G. 2004. Investigation and Assessment of Sludge Pretreatment Processes. *Water Sci. Technol*, 49: 97-104.

Mussgnug, J. H., Klassen, V., Schluter, A., and Kruse O. 2010. Microalgae as a Feedstock for Fermentative Biogas Production in a Combined Biorefinery Concept. *J. Biotechnol*, 150(1): 51-56.

Nguyen, A. V., Thomas-Hall, S. R., Malnoe, A., Timmins, M., Mussgnug, J. H., and Rupprecht, J. 2008. Transcriptome for Photobiological Hydrogen Production Induced by Sulfur Deprivation in the Green Alga Chlamydomonas Reinhardtii. Eukaryot. *Cell,* 7:1965-1979.

Parawira, W., 2012. Enzyme Research and Applications in Biotechnological Intensification of Biogas Production. *Critical Reviews in Biotechnology,* 32(2): 172-186.

Pavlostathis, S. G., and Giraldo-Gomez, E. 1991. Kinetics of Anaerobic Treatment: A Critical Review. *Critical Reviews in Environmental Control*, 21(5/6): 411-490.

Pavlostathis, S. G., Gossett, J. M., 1988. Preliminary Conversion Mechanisms in Anaerobic Digestion of Biological Sludges. *Journal of Environmental Engineering and Science,* 114: 575-592.

Pilli, S., Bhunia, P., Yan, S., Le Blanc, R. J., Tyagi, R. D., and Surampalli, R. Y. 2011. Ultrasonic Pretreatment of Sludge: A Review. *Ultrasonics Sonochemistry*, 18(1): 1-18.

Pittman, J. K., Dean, A. P., and Osundeko O. 2011. The Potential of Sustainable Algal Biofuel Production Using Wastewater Resources. *Bioresour. Technol*, 102: 17-25.

Prassad, S., Singh, A., Joshi, H. C. 2007. Ethanol as an Alternative Fuel from Agricultural, Industrial and Urban Residues. *Resour. Conserv. Recycl*, 50: 1-39.

Quiroga, G., Castrillon, L., Fernandez-Nava, Y., Maranon, E., Negral, L., Rodríguez-Iglesias, J., and Ormaechea, P., 2014. Effect of Ultrasound Pre-Treatment in the Anaerobic Co-Digestion of Cattle Manure with Food Waste and Sludge. *Bioresource Technology,* 154: 74-79.

Rafique, R., Poulse, T. G., Nizami, A. S., Asam, Z. Z., Murphy, J. D., and Kiely, G., 2010. Effect of Thermal, Chemical and Thermo-Chemical Pretreatments to Enhance Methane Production. *Energy,* 35: 4556-4561.

Richards, B. K., Herndon, F. G., Jewell, W. J., Cummings, R. J., and White, T. E. 1994. In Situ Methane Enrichment in Methanogenic Energy Crop Digesters. *Biomass Bioenergy,* 6(4): 275-282.

Rittmann, BE. 2008. Opportunities for Renewable Bioenergy Using Microorganisms. *Biotechnol. Bioeng,* 100: 203-212.

Rodriguez, C., Alaswad, A., Benyounis, K. Y., and Olabi, A. G. 2017. Pretreatment Techniques Used in Biogas Production from Grass. Renew. Sustain. *Energy Rev*, 68: 1193-1204.

Romano, R. T., Zhang, R., Teter, S., and McGarry, J. A. 2009. The Effect of Enzyme Addition on Anaerobic Digestion of Jose Tall Wheat Grass. *Bioresour. Technol*, 100: 4564-4571.

Rozzi, A., and Remigi, E. 2004. Methods of Assessing Microbial Activity and Inhibition under Anaerobic Conditions: A Literature Review. *Reviews in Environmental Science and Biotechnology,* 3(2): 93-115.

Rusten, B., and Sahu, A. K. 2011. Microalgae Growth for Nutrient Recovery from Sludge Liquor and Production of Renewable Bioenergy. *Water Sci. Technol*, 64(6): 1195-1201.

Saharan, B.S., Sharma, D., Sahu, R., Sahin, O., and Warren, A. 2013. Towards algal biofuel production: a concept of green bioenergy development. *Innov. Rom. Food. Biotechnol,* 12: 1-21.

Sandino, J., Santha, H., Rogowski, S., Anderson, W., Sung, S., and Isik, F. 2005. Applicability of Ultrasound Pre-Conditioning of WAS to Reduce Foaming Potential in Mesophilic Digesters. *Residuals and Biosolids Management,* 17: 819-835.

Schink, B. 1997. Energetics of Syntrophic Cooperation in Methanogenic Degradation. *Microbiol. Mol. Biol. Rev,* 61: 262-280.

Shafiei, M., Kabir, M. M., Zilouei, H., Sarvari Horvath, I., and Karimi, K. 2013. Techno-economical Study of Biogas Production Improved by Steam Explosion Pretreatment. *Bioresource Technology*, 148: 53-60.

Shahriari, H., Warith, M., Hamoda, M., and Kennedy, K. J. 2012. Anaerobic Digestion of the Organic Fraction of Municipal Solid Waste Combining Two Pretreatment Modalities, High Temperature Microwave and Hydrogen Peroxide. *Waste Manag*, 32(1): 41-52.

Skiadas, I. V., Gavala, H. N., Lu, J., and Ahring, B. K. 2005. Thermal Pre-Treatment of Primary and Secondary Sludge at 70°C Prior to Anaerobic Digestion. *Water Science and Technology,* 52: 161-166.

Stephens, E., Ross, I. L., King, Z., Mussgnug, J. H., Kruse, O., and Posten. C. 2010. An Economic and Technical Evaluation of Microalgal Biofuels. *Nat. Biotechnol,* 28:126-128.

Subramani, T., and Ponkumar, S. 2012. Anaerobic Digestion of Aerobic Pretreated Organic Waste. *International Journal of Modern Engineering and Research Technology,* 2: 607-611.

Taherdanak, M., and Zilouei, H. 2014. Improving Biogas Production from Wheat Plant Using Alkaline Pretreatment. *Fuel*, 115: 714-719.

Tampio, E., Ervasti, S., Paavola, T., Heaven, S., Banks, C., and Rintala, J., 2014. Anaerobic Digestion of Autoclaved and Untreated Food Waste. *Waste Management,* 34: 370-377.

Tiehm, A., Nickel, K., Zellhorn, M., and Neis, U. 2001. Ultrasonic Waste Activated Sludge Disintegration for Improving Anaerobic Stabilization. *Water Res*, 35: 2003-2009.

Vavouraki, A. I., Angelis, E. M., and Kornaros, M. 2012. Optimization of Thermo-chemical Hydrolysis of Kitchen Wastes. *Waste Management*, 33(3): 740-745.

Wang, Q., Peng, L., and Su, H. 2013. The Effect of a Buffer Function on the Semi-Continuous Anaerobic Digestion. *Bioresource Technology*, 139: 43-49.

Weiland, P. 2009. Biogas Production: Current State and Perspectives. *Applied Microbiology and Biotechnology*, 85(4): 849-860.

Weiland, P. 2010. Biogas Production: Current State and Perspectives. *Appl. Microbiol. Biotechnol,* 85: 849-860.

Weinberg, Z. G., Muck, R. E., and Weimer, P. J. 2003. The Survival of Silage Inocculent Lactic Acid Bacteria in Rumen Fluid. *J. Appl. Biochem*, 93: 1066-1071.

Xiao, W., and Clarkson, W. W. 1997. Acid solubilization of lignin and bioconversion of treated newsprint to methane. *Biodegradation*, 1997, 8(1): 61-66.

Zhang, L., Lee, Y. W., and Jahng, D. 2011. Anaerobic Co-Digestion of Food Waste and Piggery Wastewater: Focusing on the Role of Trace Elements. *Bioresource Technology*, 102: 5048-5059.

Zhang, C., Su, H., Baeyens, J., and Tan, T., 2014. Reviewing the Anaerobic Digestion of Food Waste for Biogas Production. *Renewable & Sustainable Energy Reviews*, 38: 383-392.

Zheng, Y., Zhao, J., Xu, F., and Li, Y. 2014. Pretreatment of Lignocellulosic Biomass for Enhanced Biogas Production. *Progress in Energy and Combustion Science*, 42: 35-53.

Zhong, W., Zhang, Z., Qiao, W., Pengcheng, F., and Man, L. 2011. Comparison of Chemical and Biological Pretreatment of Corn Straw for Biogas Production by Anaerobic Digestion. *Renewable Energy*, 36(6): 1875-1879.

Zhu, B., Gikas, P., Zhang, R., Lord, J., Jenkins, B., and Li, X. 2009. Characteristics and Biogas Production Potential of Municipal Solid Wastes Pretreated with Rotary Drum Reactor. *Bioresource Technology*, 100: 1122-1129.

Zubr, J. 1986. Methanogenic Fermentation of Fresh and Ensiled Plant Materials. *Biomass*, 11: 159–71.

Chapter 4

BIOGAS FROM LANDFILLS

1. INTRODUCTION

A landfill site is an area of land that has been specifically engineered to allow for the deposition of waste on to and into it. Historically, a landfill has been called a *dump* or a *midden* and has been the most convenient method of waste disposal for millennia, or at least since the concept of communal living became a lifestyle. Landfills may include internal waste disposal sites (where a producer of waste carries out their own waste disposal at the place of production) as well as sites used by many producers. Many landfills are also used for other waste management purposes, such as the temporary storage, consolidation and transfer, or processing of waste material (sorting, treatment, or recycling). A landfill also may refer to ground that has been filled in with soil and rocks instead of waste materials, so that it can be used for a specific purpose, such as for building houses.

The current municipal solid waste (MSW) landfill sites are a large source of human-related methane emissions and, in some countries such as the United States, can account for up to a substantial portion of these emissions (Johnston et al., 2000; Pawlowska, 2014). At the same time, methane emissions from landfills represent a lost opportunity to capture and use a significant energy resource. Some landfills are also used for waste management purposes, such as the temporary storage, consolidation and transfer, or processing of waste material (sorting, treatment, or recycling). In addition, landfill gas can be upgraded to natural gas and another advantage manifests itself by having a specific location for disposal that can be monitored, where waste can be processed to remove all recyclable materials before tipping.

Generally, waste can be classified in several ways and typical examples of such classifications are: (i) biodegradable waste, (ii) recyclable materials such as paper, cardboard, glass, metals, tires, and batteries. (iii) electronic waste, and (iv) discarded clothing. Waste material such as toxic waste and biomedical waste are regulated to be sent for disposal in specified landfills and through high temperature combustion, such as in a cement kiln. The collection of such wastes not only includes the gathering of the waste and recyclable materials, but also the transport of these materials to the location where the collection vehicle is emptied and the waste discharged into the landfill. This location may be a materials processing facility, a transfer station or a landfill disposal site.

Table 4.1. General Composition of Landfill Gas

Component	Composition, %v/v
Methane (CH_4)	35-55
Carbon dioxide (CO_2)	30-44
Nitrogen (N_2)	5-25
Oxygen (O_2)	0-6
Hydrogen sulfide (H_2S)	0-5
Hydrogen	0-0.2
Carbon monoxide	0-0.2
Non-methane organic compounds	0.01-0.6
Siloxane derivatives	Trace-0.2
Heavy metals	Trace-0.2
Halogenated hydrocarbon compounds	Trace-0.2
Water vapor	Saturated

Table 4.2. Some Physical Characteristics of the Constituents of Landfill Biogas

Characteristics	CO_2	CH_4*	H_2S**
Molecular weight	44.1	16.04	34.08
Boiling point, °C (°F) (at 14.7 psi)	-164.0 (-263)	-161.61 (-258)	-59.6 (-75)
Freezing point, °C (°F) (at 14.7 psi)	-78.0 (-108)	-182.5 (-296)	-82.9 (-117)
Relative density (15°C, 59°F) (air = 1.0)	1.53	0.555	1.189
Density at 0°C, 32°F	1.85	0.719	1.539
Flammable limits in air	-	5.3-14	4.3-45
Solubility in water	4.0	24.0	3.4

*Combustion: $CH_4 + 2O_2 \rightarrow CO_2 + 2H_2O$. **Combustion: $H_2S + 2O_2 \rightarrow SO_3 + H_2O$.

Typically, in the landfill, the compacted waste is covered with soil or alternative materials daily. Alternative waste-cover materials include (i) chipped wood, (ii) any waste, (iii) sprayed-on foam products, (iv) chemically fixed bio-solids, and (v) temporary blankets. Blankets can be lifted into place at night and then removed the following day prior to waste placement. The space that is occupied daily by the compacted waste and the cover material is called a daily cell. Waste compaction is critical to extending the life of the landfill. Factors such as waste compressibility, waste-layer thickness and the number of passes of the compactor over the waste affect the waste densities. In spite of these precautions, there is always (through rainfall and snow melt) the potential for materials to be leached from the landfill. In some places, efforts are made to capture and treat leachate from landfills before it reaches groundwater aquifers. However, liners always have a lifespan, though it may be 100 years or more. Eventually, every landfill liner will leak, allowing pollutants to contaminate groundwater. In the current context, other disadvantages such as leachate production not being ignored, the real prize is the gas produced at the landfill.

Table 4.3. General Properties of Methane

Chemical formula	CH_4
Molar mass	16.04 $g \cdot mol^{-1}$
Appearance	Colorless gas
Odor	Odorless*
Density	0.656 g/L (gas, 25°C, 1 atm/14.7 psi) 0.716 g/L (gas, 0°C/32°F, 1 atm/14.7 psi)
Melting point	-182.5°C; -296.4°F
Boiling point	-162°C; -260°F
Solubility in water	22.7 mg/L
Solubility	Soluble in ethanol, diethyl ether, benzene, toluene, methanol, acetone
Molecular shape	Tetrahedron
Flash point	−188 °C (-306.4°F)
Autoignition temperature	537°C (999°F)
Explosive limits	4.4 to 17% v/v in air

*Natural gas which is predominantly methane has an odor because of the addition of odiferous compounds (such as t-butyl thiol, also known as t-butyl mercaptan that has a skunk-like odor) that is added by the seller so that inherently dangerous gas leaks can be detected immediately.

Table 4.4. General Properties of Carbon Dioxide

Chemical formula	CO_2
Molar mass	44.01 $g \cdot mol^{-1}$
Appearance	Colorless gas
Odor	Low concentrations: barely detectable High concentrations: sharp; acidic
Density	1.84 g/liter (NTP*); 1.96 g/liter (STP*)
Melting point	-56.6°C; -69.8°F
Solubility in water	1.45 g/liter at 25°C (77°F)

*NTP: 0°C (32°F) and 1 atm (14.696 psi); STP: 15.6°C (60°F) and 1 atm (14.696 psi).

In the landfill, the decaying-induced decomposition of organic materials such as food and other organic waste creates gases which include methane, carbon dioxide, and other constituents including hydrogen sulfide (Table 4.1, Table 4.2) (Farquhar and Rovers,

1973). The gases are produced by the anaerobic digestion process (Chapter 3) and, in fact, a landfill is considered by some observers to be a large anaerobic digester. The product gases can seep out of the landfill and into the surrounding air and soil. Methane (Table 4.3), a greenhouse gas, is flammable and potentially explosive at certain concentrations, which makes it perfect for burning to generate electricity cleanly. While carbon dioxide (Table 4.4) is also a greenhouse gas, it is produced not only not by the burning of fossil fuel by also by the decomposition of organic waste, such as food waste (Farquhar and Rovers, 1973). However, because of the dangers from such gases, there is the need for gas monitoring to detect the presence of a build-up of gases to a harmful level. In a managed landfill, the gas is collected, cleaned (Chapter 7), and put to a variety of uses (Chapter 8) while caution is taken not to release any greenhouse gases to the atmosphere (EPA, 2011).

Thus, a monitoring program must be an integral part of the operations; it should be site specific and provide the following data for efficient operations: (i) liquid injection volumes, (ii) temperature, (iii) moisture content, (iv) cellulose/lignin content, (v) leachate yield and quality, (vi) waste density, (vii) settlement, (viii) gas flow/quality, (ix) leachate levels, (x) volatile solids, and (xi) gas production. It is only through the installation of a monitoring procedure that the true condition of the landfill can be assessed.

2. Landfill Classification

Historically, a landfill has been called a *dump* or a *midden* and has been the most convenient method of waste disposal for millennia. In the modern sense, a landfill site is an area of land that has been specifically engineered to allow for the deposition of waste on to and into it. A well-developed approach to solid waste management usually incorporates landfilling as an essential element for the final disposal of waste. Municipal solid waste landfill sites are a large source of human-related methane emissions and, in some countries such as the United States, can account for up to 25% v/v of these emissions. At the same time, methane emissions from landfills represent a lost opportunity to capture and use a significant energy resource.

Landfills may include internal waste disposal sites (where a producer of waste carries out their own waste disposal at the place of production) as well as sites used by many producers. Many landfills are also used for other waste management purposes, such as the temporary storage, consolidation and transfer, or processing of waste material (sorting, treatment, or recycling). A landfill also may refer to ground that has been filled in with soil and rocks instead of waste materials, so that it can be used for a specific purpose, such as for building houses.

The need for new landfills cannot be overstated as waste quantities are increasing rapidly and large amounts of waste require disposal in landfills. In keeping with the expansion of landfill numbers and operations, a well-developed approach to solid waste management usually incorporates landfilling as an essential element for final disposal. In parallel with this increase, regulations by various levels of government have been adopted to improve landfill operations. Poor solid waste management practices in the past has made landfill permitting and siting a daunting task due to limited availability of suitable sites and challenges mounted from residents who have been (and remain) opposed to constructing a landfill in their neighborhood.

Landfilling represents disposal of all types of waste into engineered facilities that are specific to the type of disposed waste. Thus, not all landfills are suitable for the production of landfill gas and it is well worthwhile to understand the nature and constituents of the various landfills areas. Landfill gas is the result of the following processes: (i) evaporation of volatile organic compounds, such as solvents, (ii) chemical reactions between waste components, and (iii) microbial action. The first two categories depend strongly on the nature of the waste and the type of landfill. However, because gases produced by landfills are both valuable and sometimes hazardous, monitoring techniques have been developed to ensure the safety of the workers and inhabitants of nearby homes. Flame ionization detectors can be used to measure methane levels as well as total levels of volatile organic compounds (VOCs). Surface monitoring and sub-surface monitoring as well as monitoring of the ambient air is essential.

There are five types of waste disposal facilities: (i) bioreactor landfills, (ii) conventional municipal solid waste landfills – landfills built to accommodate municipal solid waste, (iii) construction and debris landfills – landfills built to accommodate construction waste and construction debris (iv) hazardous waste landfills, and (v) surface impoundments. Furthermore, as presented alphabetically below (and not in any order of preferences), it must be recognized that not all landfills are suitable for the production of landfill gas (collectable gas) and the components of each landfill must be understood given serious attention before embarking on a gas collection project.

Landfilling represents disposal of all types of waste in engineered facilities. There are 6 types of engineered waste disposal facilities: (i) conventional municipal solid waste landfills, MSW landfills (ii) hazardous waste landfills (iii) construction and debris, C&D landfills, (iv) surface impoundments (v) bioreactor landfills, and (vi) small landfills.

2.1. Bioreactor Landfills

One innovative approach to municipal solid waste disposal, which actually encourages rapid waste decomposition and speeds stabilization, is the bioreactor landfill. A bioreactor landfill is any permitted landfill where liquid or air, in addition to leachate and landfill gas (landfill gas) condensate, is injected in a controlled fashion into the waste mass in order to accelerate or enhance bio-stabilization of the waste (Mehta et al., 2002). In this type of landfill, liquid or air, in addition to leachate and landfill gas (landfill gas) condensate, is injected in a controlled fashion into the waste mass in order to accelerate or enhance bio-stabilization of the waste. The bioreactor landfill significantly increases the extent of waste decomposition, conversion rates and process effectiveness over what would otherwise occur within the landfill.

In contrast to conventional municipal solid waste landfills, bioreactors are designed and operated to enhance microbial processes significantly, to degrade and stabilize the organic constituents of the waste (Warith, 2002). In fact, the operation of a bioreactor landfill is treated as a large biological waste digester, and its operation should be closely monitored. The operation should be such that biological degradation of the biodegradable organic fraction of municipal solid waste is achieved in an efficient manner.

The enhanced waste degradation and stabilization, carried out by the indigenous microbial populations within the waste, is accomplished through control of moisture, temperature, pH, and nutrients. The enhanced microbiological processes within a bioreactor can transform and stabilize the decomposable organic waste within 5 to 10 years of implementation, compared with many decades for conventional landfills where wastes are essentially sealed off from air and moisture. All bioreactor landfills require specific management activities and operational modifications to enhance microbial decomposition of waste during their operational life and this results in higher initial capital costs. However the bioreactor-type landfill operations typically involve less monitoring over the duration of the post closure period than conventional dry tomb landfills.

In a bioreactor landfill, any leachate is returned to the landfill until the liquid holding capacity of the waste is achieved (Mehta et al., 2002). Leachate drained from the waste is recovered and recirculated to maintain the moisture content of the waste and distribute nutrients for microbial degradation. A distinction should be made between the bioreactor landfill operation and the conventional leachate recirculation performed at conventional municipal solid waste landfill sites. The bioreactor process requires significant liquid addition to maintain optimal conditions. However, leachate alone is usually insufficient to sustain the bioreactor needs. Water or non-toxic, non-hazardous liquids and semi-liquids are suitable amendments to supplement leachate. At the closure of a bioreactor landfill (i) leachate quantity will be a finite amount, amenable to on-site treatment with limited need for off-site transfer, treatment, and/or disposal, (ii) landfill gas generation will be in decline and (iii) long-term environmental risk will be minimized.

Moisture control, approximately 35 to 65% w/w of the field capacity, is critical for effective operation and it is the easiest to manipulate. It also influences the distribution of nutrients throughout the landfill. Other activities include shredding of waste, pH adjustment, nutrient addition and balance, waste predisposal and post-disposal conditioning, and temperature management; these all serve to optimize the bioreactor process. At a minimum, leachate is injected into the bioreactor to stimulate the natural biodegradation process.

Generally, a bioreactor landfill often needs other liquids such as storm water, wastewater, and wastewater treatment plant sludge to supplement leachate which enhance the microbiological process by purposeful control of the moisture content, and differs from a landfill that simply recirculates leachate for liquids management. A landfill that merely recirculates the leachate may not necessarily operate as an optimized bioreactor. In fact, the moisture content is the single most important factor that promotes the accelerated decomposition. The moisture content, combined with the biological action of naturally occurring microbes, decomposes the waste. The microbes can be either aerobic or anaerobic. A side effect of the bioreactor is that it produces landfill gas (LFG) like methane in an anaerobic unit at an earlier stage in the life of the landfill at an overall much higher rate of generation than traditional landfills.

Potential advantages of bioreactors include the following: (i) decomposition and biological stabilization in years versus decades in dry tombs, (ii) lower waste toxicity and mobility due to both aerobic and anaerobic conditions, (iii) reduced leachate disposal costs, (v) a 15 to 30 percent gain in landfill space due to an increase in density of waste mass, (vi) significant increased generation of the landfill gas that, when captured, can be used for energy use onsite or sold.

Due to degradation of the organic constituents of the landfill material and the sequestration of inorganic constituents, municipal solid waste can be rapidly degraded and made less hazardous by enhancing and controlling the moisture within the landfill under aerobic and/or anaerobic conditions. Leachate quality in a bioreactor rapidly improves, which leads to reduced leachate disposal costs. Landfill volume may also decrease with the recovered airspace offering landfill operators the full operating life of the landfill.

The landfill gas that is emitted by a bioreactor landfill consists primarily of methane and carbon dioxide, as well as lesser amounts of volatile organic chemicals and/or hazardous air pollutants, such as methane, carbon dioxide, and hydrogen sulfide (Table 4.1). Also, the operation of a bioreactor may generate landfill gas earlier in the process and at a higher rate than the traditional landfill. The bioreactor landfill gas is also generated over a shorter period of time, because the landfill gas emissions decline as the accelerated decomposition process depletes the source waste faster than in a traditional landfill. The net result appears to be that the bioreactor produces more landfill gas than the traditional landfill.

Thus, in contrast to conventional municipal solid waste landfills, bioreactor landfills are designed and operated to enhance microbial processes significantly, to degrade and stabilize the organic constituents of the waste. The enhanced waste degradation and stabilization, carried out by the indigenous microbial populations within the waste, is accomplished through control of moisture, temperature, pH, nutrients etc. The enhanced microbiological processes within a bioreactor can transform and stabilize the decomposable organic waste within 5 to 10 years of implementation, compared with many decades for conventional landfills where wastes are essentially sealed off from air and moisture.

Table 4.5. Comparison of Bioreactors

Bioreactor type:	Conventional	Anaerobic	Aerobic
Typical settlement after			
2 Years	2-5%	10-15%	20-25%
10 Years	15%	20-25%	20-25%
Anticipated waste stabilization time	30 to 100 years	10 to 15 years	2 to -4 years
Relative rate of methane generation	1.0	2.0	0.1 to 05
Liquid evaporation	Negligible	Negligible	50 to 80%

Table 4.6. Benefits of Bioreactors

1	Predictable long-term performance. Reduced risk of groundwater contamination. Leachate stored within the waste mass.
2	Leachate and landfill gas condensate re-injected. Leachate requires reduced treatment. Rapid decrease in leachate strength, measured by chemical oxygen demand.
3	Controlled landfill gas generation. Reduces the incidence of landfill gas emissions. Reduces the potential for the greenhouse effect. Gas generation rates higher than conventional municipal solid waste landfills.
4	Longer landfill lifespan through efficient use of air space. Increased rate of settlement (15 to 30% gain in landfill space). Reduces the need/urgency to find new sites.
5	Better waste stabilization. Improved value of the land after closure.
6	Potential for mining the waste to recover humic material. Allows inspection of the liner.

Depending on the microbial process that degrades and stabilizes waste, bioreactor landfills can be categorized into three types, (i) anaerobic bioreactor landfills (ii) aerobic bioreactor landfills, and (iii) hybrid (anaerobic-aerobic) bioreactor landfills. In each case, the bioreactor accelerates the decomposition and stabilization of waste (Table 4.5, Table 4.6).

Finally, the landfill gas emitted by a bioreactor landfill consists primarily of methane and carbon dioxide, as well as lesser amounts of volatile organic chemicals and/or hazardous air pollutants. There are also indications that the operation of a bioreactor may generate landfill gas earlier in the process and at a higher rate than the traditional landfill. The bioreactor landfill gas is also generated over a shorter period of time, because the landfill gas emissions decline as the accelerated decomposition process depletes the source waste faster than in a traditional landfill. As stated above this contributes to the result that the bioreactor produces more LFG overall than the traditional landfill.

2.1.1. Aerobic Bioreactor Landfills

The aerobic bioreactor landfill is analogous to a composting operation in which the organic fraction of municipal solid waste is rapidly biodegraded using air, moisture, and increased temperature (internal heat generated by the aerobic bacterial activity during waste decomposition). The aerobic bioreactor operates by the controlled injection of liquid and air into the waste, which is pumped into the waste until a sufficiently high moisture content (50 to 70% w/w) is achieved. When the optimal moisture of the waste is achieved, air is introduced through vertical or horizontal perforated wells using mechanical blowers. Any leachate produced is drawn off and stored for reinjection in a controlled manner into the landfill.

More specifically, in an aerobic bioreactor landfill, leachate is removed from the bottom layer, piped to liquids storage tanks, and re-circulated into the landfill in a controlled manner. Air is injected into the waste mass using vertical or horizontal wells to promote aerobic activity and accelerate waste stabilization. The leachate is formed when rain water filters through wastes placed in a landfill. When this liquid comes in contact with buried wastes, it leaches, or draws out, chemicals or constituents from those wastes. In the aerobic bioreactor landfill, moisture is also added to the waste mass in the form of re-circulated leachate and other sources to obtain optimal moisture levels. Biodegradation occurs in the absence of oxygen (anaerobically) and produces landfill gas.

Optimum temperatures for waste degradation are on the order of 60 to 72°C (140 to 162°F). The aerobic process continues until most of the readily degradable compounds have decomposed and the landfill temperature gradually decreases during the maturation phase. Also, the heat produced during decomposition causes large quantities of leachate to evaporate; in a study of two bioreactor landfills leachate volume was reduced by 86% and 50%, respectively. Changing the rate of air and liquid injection will alter the landfill temperature, and fires are prevented by wetting and uniform distributed air injection in the waste mass.

Because of the higher reaction rates, aerobic biodegradation is more rapid than anaerobic digestion; aerobic landfills have the potential to achieve waste stabilization in 2 to 4 years as opposed to decades or longer for conventional municipal solid waste landfill sites. The rapid rate of waste stabilization also offers the potential for 'mining.' Aerobic bioreactors do not produce significant quantities of methane and there is little potential for the sale of landfill gas for energy production. The benefits of aerobic bioreactor landfills are: (i) more rapid waste and leachate stabilization, (ii) the increased rate of landfill settlement, (iii) the reduction of methane generation, (iv) the capability of reducing leachate volumes by up to 100% due to evaporation, (v) the potential for landfill mining, and (vi) a reduction in the of environmental liability.

The municipal solid waste in a conventional landfills undergo four phases of decomposition, starting with a brief aerobic phase, through two additional anaerobic phases followed by a methane generation phase and finally maturation. In the aerobic bioreactor landfill, the Phase I activity (aerobic conditions) is sustained over a longer period of time than what occurs in a conventional municipal solid waste landfill site. In contrast, anaerobic bioreactor landfills reduce significantly the time involved for Phase IV (methane generation) to possibly 5-10 years (a 75% reduction), with 5-7 years for Phase IV considered optimum.

2.1.2. Anaerobic Bioreactor Landfills

An anaerobic bioreactor landfill seeks to accelerate waste degradation by anaerobic digestion of biodegradable organic fraction of municipal solid waste. The four processes: (i) hydrolysis, (ii) acidogenesis, (iii) acetogenesis, and (iv) methanogenesis are responsible for the conversion of organic wastes into organic acids and ultimately into methane and carbon dioxide. Since oxygen is naturally deficient in most landfills, the anaerobic conditions exist without intervention. However, the degradation of municipal solid waste in a landfill is frequently is rate-limited by the overall lack of sufficient moisture. The moisture content of solid waste ranges from 10 to 20% (wet weight) and to optimize the anaerobic digestion process, moisture at or near field capacity (45 to 65%, wet weight) is required.

Under optimal conditions, waste stabilization can be accomplished in a period on the order of 6 to 7 years. However, anaerobic bioreactor landfills require careful monitoring and start up and, if the addition of moisture takes place too rapidly, a buildup of volatile organic acids (volatile fatty acid derivatives, VFAs) typically the lower molecular weight acid derivatives with a chain length greater than that of acetic acid (i.e., up to an including butyric acid, Table 4.7) have the capability of lowering the pH Into the acid region (<pH 7.0) thereby inhibiting the action of the methanogenic microbes and reduce the rate of gas production. Optimal conditions for methanogenic bacteria are a neutral pH (pH = 7.0). Leachate parameters such as pH, volatile organic acids, alkalinity, and methane content are direct indicators to the health of the methanogenic bacterial population. An indicator, such as a high volatile organic acid to alkalinity ratio (>0.25), can show that the leachate might possess a low buffering capacity and such conditions could inhibit methane generation.

When the methane content of landfill gas >40% v/v, the methanogenic bacterial populations are considered established. On the other hand, when the methane contact of the landfill gas is <40% v/v, the indication is that the waste has too much moisture or insufficient moisture. Once the methanogenic bacteria have become established, the rate of leachate recirculation may be increased.

Table 4.7. Examples of Mono-carboxylic Acid Derivatives

Volatile fatty acids (VFAs)	Formic acid HCOOH
	Acetic acid CH_3COOH
	Propionic acid C_2H_5COOH
	Butyric acid C_3H_7COOH
Long chain fatty acids (LCFAs)	Lauric acid $C_{11}H_{23}COOH$
	Palmitic acid $C_{15}H_{31}COOH$
	Stearic acid $C_{17}H_{35}COOH$
	Oleic acid $C_{17}H_{33}COOH$ (one double bond)
	Linoleic acid $C_{17}H_{31}COOH$ (two double bonds)
	Linolenic acid $C_{17}H_{29}COOH$ (three double bonds)

2.1.3. Dry Tomb Landfill

The *dry tomb landfill* is, essentially, an open dump in which the waste from each day is covered by a several inches of soil (classical sanitary landfill) where compacted soil (clay) and plastic sheeting (flexible membrane liners, FMLs) are used to try to isolate the untreated municipal solid waste from moisture. This type of containment system is designed to collect and manage the leachate (often referred to as garbage juice) generated within the dry tomb that results from the entrance of moisture into the landfill site. Some governments have decided not to adopt the dry tomb method of municipal solid waste disposal because of the likelihood of the ultimate failure of the dry tomb containment system (the liner) to prevent moisture from entering the landfill and to collect all leachate generated in the landfill. Also, the decomposition and biological stabilization of the waste in a bioreactor landfill can occur in a much shorter time than in a traditional dry tomb landfill which can provide a potential decrease in long-term environmental risks and landfill operating and post-closure costs.

This containment system also is designed to collect and manage the leachate (also called garbage juice) generated within the dry tomb. Other countries and geographical areas in areas have chosen not to adopt the dry tomb method of landfilling, typically because of the likelihood of the ultimate failure of the dry tomb containment (liner) system to prevent moisture from entering the landfill and also the failure to collect all leachate generated in the landfill (Lee and Jones-Lee, 1993, 1996).

Since, with few exceptions, both of the types of landfills (classical and dry tomb sanitary landfills) will pollute groundwater and the aquifer system hydraulically connected to the landfill, the key to public health and environmental protection is the establishment of a leak-detectable cover that prevents moisture from entering the landfill after closure (Lee and Jones-Lee, 1994).

2.1.4. Hybrid Bioreactor Landfills

A hybrid bioreactor landfill (aerobic-anaerobic landfill) combines both aerobic microbes and anaerobic microbes in order to provide an optimum approach for effective biodegradation. This type of bioreactor landfill accelerates waste degradation by employing a sequential aerobic-anaerobic treatment to rapidly degrade organics in the upper sections of the landfill and collect gas from lower sections. Operation as a hybrid results in the earlier onset of methanogenesis compared to aerobic landfills.

One of the implementation schemes involves treating the uppermost layer of waste aerobically for a period on the order of 30 to 60 days before it is buried by another layer at which time it is treated anaerobically. This approach is operationally simple and can be effective, because during the initial phase of aerobic treatment, organic wastes are quickly converted into the acid phase and are then effectively treated by the anaerobic methanogens.

The hybrid bioreactor landfill accelerates waste degradation by employing a sequential aerobic-anaerobic treatment to rapidly degrade organics in the upper sections of the landfill and collect gas from lower sections. Operation as a hybrid results in the earlier onset of methanogenesis compared to aerobic landfills.

2.2. Landfill Design

In the past, many of the problems associated with landfills occurred as a result of non-engineered – in other words, the landfill was an unmanaged dumping area. If gas is to be the product assigned for collection, as well as to protect the environment, the problems (both short-term and long-term) must be carefully assessed so that issues related to (i) gas collection, (ii) the uncontrolled migration of landfill gas, (iii) possible contamination of the groundwater and surface water regimes, and (iv) the generation of odor, noise and visual nuisances are mitigated.

Good design of a landfill site will prevent, or reduce as far as possible, negative effects on the environment, as well as the risks to human health arising from the landfilling of waste. The design process should be consistent with the need to protect the environment and human health. Thus, in modern landfills, it is necessary to consider the design of landfills using the major elements of a landfill system include, but is not limited to: (i) cells, (ii) a liner and leachate collection system, (iii) a liquid injection system, (iv) a gas collection system, (v) a cover, and (vi) slope stability and settlement. Other aspects to be considered are the bottom liner and a storm water drainage system. Each of these parts is designed to address specific problems that are encountered in a landfill (Table 4.8).

The conventional landfills are essentially dry while the bioreactor landfills are wet. The performance standards for the bioreactor landfill types often conflict and pose challenges to designers. Design considerations to ensure satisfactory performance of bioreactor landfills are (i) the cell size, (ii) the liner and leachate collection system, (iii) the liquid injection system, (iv) the gas extraction system, (v) the final cover system, (vi) the stability of the slope, (vii) settlement, and (viii) the metals content of the waste.

Table 4.8. The Components of a Landfill and Their Function

Component	Purpose
Bottom liner system	Separates waste from the groundwater
Cells (old and new)	Separate leachate from the groundwater
	Where the waste is stored within the landfill
Storm water drainage system	Collects rain water that falls on the landfill
Leachate collection system	Collects water that percolating through the landfill
	Contains contaminating substances (leachate)
Methane collection system	Collects methane gas that is formed
Covering or cap	Seals off the top of the landfill

2.2.1. Cells

In a landfill, the waste is compacted into specific areas (cells) that contain waste from a specific time period (such as the waste from one day or from one week). Typically, a cell may be approximately 50 feet long by 50 feet wide by 14 feet high and the amount of waste within the cell is several tone to several thousand tons. Compression of the waste if achieved using heavy equipment (tractors, bulldozers, rollers, and graders) and the cells are arranged in rows and layers of adjoining cells (strata or lifts). Once the cell is constructed, it is covered with six inches of soil and compacted further. In addition to compressing the waste into cells, space is conserved by excluding bulky materials, such as carpets, mattresses, foam, and yard waste, from the landfill.

For economic and regulatory reasons, an emerging trend in traditional landfill design is to construct deep cells that are completed in 2 to 5 years. Once a landfill is closed, methanogenic conditions are optimized and gas generation and extraction facilitated. Extremely deep landfills might be so dense in lower strata that the permeability of the waste will inhibit the flow of leachate and gas. Therefore, it may be necessary to limit addition and recirculation to the upper strata or to develop the capability for internal drainage.

As part of the cell structure, bioreactor operation is most efficient when the waste has a high biodegradable organic fraction and large exposed specific surface areas. For this reason, bioreactor operations should be concentrated on waste segregated to maximize its organic content and shredded or flailed to maximize its exposed surface area. However, shredded waste may become exceedingly dense, limiting moisture penetration.

Waste segregation should include separation of non-biodegradables (glass, plastics, metals, tires, construction and demolition debris) from municipal solid waste before disposal in bioreactor landfills. There have been some concerns regarding the absence of decomposition in bioreactor landfills when municipal solid waste is placed in plastic bags, which may or may not be broken open during conventional compaction with heavy equipment. It may be feasible to open plastic bags to expose the contents or use degradable bags for optimum bioreactor performance.

In addition, there is also the requirement that the microbial population in a bioreactor receive suitable nutrients. Nutrients are normally supplied in the waste but other biological and chemical supplements may enhance biological activity. As with waste segregation or shredding the cost of nutrients need to be justified. Optimum pH for methanogens is in the range 6.8-7.4, buffering of leachate in this range improves gas production in laboratory tests. pH and buffering must be considered during early stages of leachate recirculation, careful operation through slow introduction of liquids should minimize the need for buffering.

Another aspect of cell structure relates to the heavy metals content of the waste. Heavy metals tend to concentrate during wastewater bio-solids treatment; similar effects could be anticipated in bioreactor landfills during decomposition and such changes in heavy metal concentrations have been detected in leachate. The reduced pH in the acidic stages of the waste decomposition process will mobilize metals that may leach out of the landfill or become toxic to the bacteria generating the gas. Other issues are (i) Microorganisms may concentrate heavy metals, (ii) the presence of sulfide derivatives may affect metal mobilization, (iii) there is the potential for remobilization of metals if landfill conditions allow.

2.2.2. Liner and Leachate Collection System

No system to exclude water from the landfill is perfect and water does get into the landfill. The water percolates through the cells and soil in the landfill much as water percolates through ground coffee in a drip coffee maker. As the water percolates through the waste, it picks up contaminants (organic and inorganic chemicals, metals, biological waste products of decomposition) just as water picks up coffee in the coffee maker. This water with the dissolved contaminants (the leachate) is typically acidic.

To collect leachate, perforated pipes run throughout the landfill which drain into a leachate pipe, which carries leachate to a leachate collection pond. The leachate can be pumped to the collection pond or flow to it by gravity. The leachate in the pond is tested for acceptable levels of various chemicals (biological and chemical oxygen demands, organic chemicals, pH, calcium, magnesium, iron, sulfate and chloride) and allowed to settle. After testing, the leachate must be treated like any other sewage/wastewater; the treatment may occur on-site or off-site. Some landfills recirculate the leachate and treat it later – this method reduces the volume of leachate from the landfill, but increases the concentration of the various contaminants in the leachate.

Since, with few exceptions, both of the types of non-bioreactor landfills (classical and dry tomb sanitary landfills) will pollute groundwater systems and the aquifer system hydraulically connected to the landfill, the key to public health and environmental protection is the establishment of a leak-detectable cover that prevents moisture from entering the landfill after closure of the landfill. US federal government regulations prescribe a 0.3 m maximum allowance leachate head on the bottom liner; the same criteria are used for conventional municipal solid waste landfill sites. This can be achieved through appropriate liner, leachate collection and removal system design. Adding liquid to the waste will increase its density by as much as 30% because of moisture uptake and settlement. Fouling of leachate collection and removal pipes must be considered during initial design, and provision for cleanouts included for periodic maintenance and inspection. Stability of waste mass and liner slopes are critical during and after landfill operations. Leachate-head predictions are validated using mathematical models, laboratory performance testing in accordance with regulatory design codes, standards and recommended practice (Pacey and Augenstein, 1990; Alexander et al., 2005; Kamalan et al., 2011).

On the other hand, there is also the need to ensure that any liquid does not escape from the landfill. In fact, one of the biggest changes is to contain the waste trash. The bottom liner prevents the trash from coming in contact with the outside soil, particularly the groundwater. In municipal solid waste landfills, the liner (usually 30-100 mils thick) is a type of durable, puncture-resistant synthetic plastic (polyethylene, high-density polyethylene, HDPE, polyvinylchloride, PVC). The plastic liner may be also be combined with compacted clay soils as an additional liner. The plastic liner may also be surrounded on either side by a fabric mat (geotextile mat) that will help to keep the plastic liner from tearing or puncturing from the nearby rock and gravel layers.

Finally, leachate seeps/breakouts may result from adding excess liquid to waste mass, leachate must be precluded from contaminating storm water runoff. Monitoring for leachate seeps is mandatory and the operation plan must consider a response. Measures such as the installation of slope and toe drains, surface grading, filling and sealing cracks to reduce surface water infiltration and reducing liquid addition are some of the standards measures used to prevent an occurrence.

2.2.3. Liquid Injection System

Moisture content (leachate and water) close to the field carrying capacity of the waste mass must be injected into the landfill to optimize decomposition. The volume of liquid needed for the waste to reach field capacity can be based on prior field studies, model predictions or landfill specific measurement. Reinhart and Townsend [27] and Phaneuf [21] note that moisture content > 25% (wet weight basis) is a minimum for bioreactor effectiveness; better results are obtained for moisture content 40 -70% as complete saturation is not conducive to methanogenesis. Achieving a moisture condition close to field capacity requires the addition of significant volume of liquid.

If enough leachate is not produced, clean water or contaminated sources (surface runoff from siltation ponds, contaminated groundwater, and certain municipal wastewater treatment plant effluents) could be utilized. Biosolids considered suitable are those in liquid form that typically undergo land application rather than dewatered sludge. [38] Use of the liquid form avoids dewatering costs. The contaminant loading of these sources should be evaluated to ensure that they will not increase the pollution potential of the bioreactor landfill, and they will be compatible with the bioreactor's microbiology. [21] There are concerns regarding operational health and safety issues about use of such liquids as it potentially exposes workers and nearby residential areas to pathogens and aerosols resulting from bio-solids application to the landfill surface. The addition of significant quantities of liquid is critical for bioreactor operation and landfills sited in arid regions may require significant amounts of liquid. In addition more liquid may be needed to sustain bioreactors after a low permeability cap is installed because the landfill moisture may be removed via the gas collection system. Thus, the leachate generated in the landfill should not be considered sufficient to support the bioreactor.

Methods for injecting liquids into landfills are: (i) the working face application, (ii) irrigation of the surface, (iii) infiltration ponds, (iv) vertical injection wells and manholes, and (v) horizontal trenches. In the *working face application,* liquid is applied directly to the waste to yield target moisture levels during active landfilling. The benefits are its simplicity and direct access to the waste. The drawbacks are one-time application and odor problems due to increased gas generation shortly after application. In the *surface irrigation method, t*anker trucks directly apply liquid to the surface of the waste, similar to *working face application. Infiltration ponds* are constructed using waste as berms to store leachate. Infiltration into the waste is by gravity. As the moisture in the waste attains its field capacity, infiltration continues and the liquid level in the ponds decreases. Both surface irrigation and infiltration ponds are simple and cost effective. Odors, vectors and litter make them unattractive. *Vertical wells and manholes* are easy to install during landfill operation or on a retrofit basis and have a relatively low cost. They are spaced 30-60 m apart, and injection rates are high if wells/manholes are not located close to the waste slopes. The major shortcoming is limited ability to distribute liquid laterally; liquid tends to accumulate around the well/manhole and result in dry zones.

Horizontal trenches are the most effective means to distribute liquid to the waste during landfilling operations. Trenches are constructed as the waste accumulates. Trenches comprise horizontal pipes embedded in an impervious media (gravel, cullet, tire chips). However, certain types of tire chips compressed under vertical pressure from thick waste layers (>50 feet) become less permeable and inhibit movement of liquid. Pervious cover layers can be designed into waste bodies and integrated with constructed injection trenches, vertical wells/manholes to form a distribution system.

Choice of a particular liquid injection system is site specific and guided by factors such as climate, mal-odors, worker exposure, environmental impacts, evaporative losses, reliability, uniformity and aesthetics. Buried trenches/vertical wells offer advantages of minimum exposure to trafficked areas, good all weather performance and favorable aesthetics; however they could be adversely affected by differential settlement. In cases of bioreactor landfills, the injection of liquid causes complex leachate flows, including heterogeneous moisture distribution. Free moisture may be generated by waste decomposition.

2.2.4. Gas Collection System

Landfill gas, which contains approximately 50% v/v methane and 50% v/v carbon dioxide with small amounts of nitrogen and oxygen, presents a hazard because the methane can explode and/or burn. So, the landfill gas must be removed and, to achieve collection of the gas, a series of pipes are embedded within the landfill. In some landfills, this gas is vented or burned but landfill gas represents a usable energy source.

The extraction system (often referred to as the extraction system) is a split system in which methane gas can go to the boilers and/or the methane flares that burn the gas. The reason for the split system is that the landfill will increase its gas production over time (from 300 cubic feet per minute to 1,250 cubic feet per minute) and exceed the capacity of the boilers at the chemical company.

A bioreactor landfill will generate more landfill gas in a much shorter time than a dry conventional municipal solid waste landfill site. To control the gas efficiently and avoid odor problems, the extraction system warrants installation of an active gas collection system (AGCS) (pipes, blowers and related equipment) early in its operational life. Horizontal trenches, vertical wells, near surface collectors and hybrid systems may be used for gas gathering and extraction. Greater gas flow is accommodated by increased pipe diameter because of the higher gas generation rates or closer spacing of collection systems may be required. Liquid addition systems have in the past been kept separate from gas gathering systems to avoid impedance; however dual function systems are becoming the preferred choice in modern landfill design. The porous leachate removal system underlying the waste should be considered for integration with the gas extraction system.

Enhanced gas production can affect side slope and cover negatively, if an efficient collection system is not installed during the active landfill phase. Uplift pressure on geomembrane cover can cause ballooning of the membrane and lead to local instability and soil loss. Temporary venting or aggressive extraction of gas during cover installation might facilitate cover placement. Once the final cover is in place, venting should be adequate to resist the uplift force created by landfill gas pressure buildup. The designer must consider the pressure buildup condition on slope stability when the collection system is shut down for any significant length of time.

2.2.5. Cover

Generally 150 mm of daily soil cover at conventional municipal solid waste landfill sites is necessary to control vectors, fire, blowing litter, odor and scavenging. Statutory regulations allow using alternative daily cover material such as relatively permeable waste materials (foundry sands, shredded tires, contaminated soils, incinerator ash residue, compost (green waste) and auto fluff). Many manufactured cover materials (sprayed on slurries, polymer foams and removable tarps) have also been developed and marketed on the basis that they help conserve air space.

A cover material more permeable than the waste can direct leachate to the sides, where the leachate must be collected and drained. A cover material less permeable than the waste can create barriers to the effective percolation of liquids and cause internal leachate mounding that promote surface seeps and contribute to waste mass instability. It can impede leachate distribution and landfill gas flow to collection and distribution systems. Its ability to serve as a barrier should be reduced through scarifying, or partial removal, prior to placing solid waste over it. When placed within 50 feet of slopes, it should be graded to drain back into the landfill to preclude leachate from reaching the slope and emerging as a seep. Use of alternative covers that do not create such barriers can mitigate these effects. Therefore the characteristics of the daily and intermediate cover materials used need to be factored into the evaluation of all bioreactor landfill operations.

When a section of the landfill is finished, it is covered permanently with a polyethylene cap (40 mil). The cap is then covered with a 2-foot layer of compacted soil. The soil is then planted with vegetation to prevent erosion of the soil by rainfall and wind. The vegetation consists predominantly of grass and kudzu and trees, shrubs, or plants with deep penetrating roots are not used so that the plant roots do not contact the underlying waste and allow leachate out of the landfill. Occasionally, leachate may seep through weak point in the covering and come out on to the surface. These seepages are promptly repaired by excavating the area around the seepage and filling it with well-compacted soil to divert the flow of leachate back into the landfill.

Conventional municipal solid waste and bioreactor landfills are designed to minimize infiltration and to help contain landfill gas. The final cover layers for long term closure consists of foundation, gas collection, barrier, surface water drainage, cover soil and topsoil with vegetative growth. For conventional landfills, the final cover system is constructed within one year upon the landfill attaining its final height. The final cover construction for bioreactor landfills should only commence when the majority of settlement has occurred (> 5 years); an interim closure measure is therefore required to optimize the disposal capacity of bioreactor landfills. To combat odors a temporary geotextile liner is needed, otherwise soil cover is sufficient once it does not impact the gas extraction system.

2.2.6. Slope Stability and Settlement

Injection of excess liquid into the bioreactor landfill will affect the geotechnical properties of the waste. The designer must consider the changes in properties of the waste and evaluate slope stability. The stability analysis procedures used in standard geotechnical engineering can also be used for the analysis of bioreactor landfill slopes. Seismic effects should be considered in the geotechnical analysis as required. The geotechnical properties considered for stability analyses include unit weight of municipal solid waste in conventional landfills which is dependent on (i) the composition of the waste, (ii) soil content, and (iii) waste compaction, as well as other environmental factors.

The other important geotechnical property needed for stability analysis is the shear strength of the waste, because of the low degree of saturation and relatively high permeability of municipal solid waste, drained shear strength of municipal solid waste is used in the analysis of waste mass stability in conventional landfills. In the case of bioreactor landfills, initially the waste conditions are similar to that of a conventional landfill; therefore, the drained shear strength of non-degraded waste in bioreactor landfills is the same as that in conventional landfills. However the liquid injected in bioreactor landfills, renders the waste fully saturated and degraded waste may have low permeability and under such conditions undrained shear strength of the degraded waste is of great engineering significance.

For stability analysis, both undrained and drained shear strengths of the degraded waste are required. Unfortunately, no data is available on undrained shear strength of degraded municipal solid waste and such data is crucial in evaluating stability of bioreactor landfills in seismic regions. Information on the drained shear strength of degraded waste is also very limited. For design purposes the drained shear strength of degraded waste in bioreactor landfills may be assumed equal to the drained shear strength of waste in conventional landfills.

A bioreactor landfill will experience more rapid and complete settlement than a drier conventional municipal solid waste landfill site. Accelerated settlement results from the increased rate of waste decomposition and compression of lower lifts through higher specific weights. Conventional landfill settlement is typically on the order of 10% of landfill height and generally occurs over a number of years as the waste decomposes. Aerobic bioreactors might achieve settlement within 2 to 4 years, while anaerobic bioreactors might require 5 to 10 years.

Settlement is dependent upon type of waste, amount of cover and compaction and does not occur uniformly across the landfill surface. This will affect the final grade, drainage, roads, gas collection, leachate collection and recirculation piping systems (shift to some extent with settlement); as such it is beneficial to overfill, above design grade before placement of final cover. A significant benefit will be if the final cover and final site improvement installations are postponed and the rapid settlement used to recapture airspace, reducing the need for new landfills by conserving volume. Settlement impacts can be accommodated in the project design and will be largely completed soon after landfill closure; long term maintenance costs and the potential for fugitive emissions will be avoided.

2.3. Construction and Debris Landfills

Construction and debris landfills are devoted exclusively to construction and debris materials, which is generated from construction, renovation, repair and demolition structures such as residential and commercial buildings, roads and bridges. The composition of Construction and debris waste varies for these different activities and structures.

Construction and debris (C&D) landfills are devoted exclusively to construction and debris waste which is generated from construction, renovation, repair and demolition structures such as residential and commercial buildings, roads and bridges. The composition of construction and debris waste varies for these different activities and structures. This type of landfill site typically receives construction and demolition debris, roadwork material, excavated material, demolition waste, construction/renovation waste, and site clearance waste. However, construction and debris landfills do not receive hazardous waste or industrial solid waste unless those landfills meet certain standards and are permitted to receive such wastes.

Overall, construction and debris waste is composed mainly of wood products, asphalt, drywall and masonry; other components often present in significant quantities are metals, plastics, earth, shingles, insulation, paper and cardboard. Construction and debris also contains wastes that may be hazardous, potentially toxic and problematic (Young and Parker, 1984), such as adhesives, leftover paint, excess roofing cement, waste oils, grease and fluids, batteries, fluorescent bulbs, appliances, carpet and treated or untreated wood. It is not prudent to dispose of such waste in municipal solid waste landfill sites. Construction and debris landfills are easier to site because demolition materials are for the most part inert unless contaminated from an external source introduced at the construction/ demolition site; therefore the environmental impact is considered minimal.

2.4. Conventional Municipal Solid Waste Landfills

A municipal solid waste landfill (MSWLF) is a discrete area of land or excavation that receives household waste. A municipal solid waste landfill may also receive other types of nonhazardous wastes, such as commercial solid waste, nonhazardous sludge, conditionally exempt small quantity generator waste, and industrial nonhazardous solid waste. Some materials may be banned from disposal in municipal solid waste landfills, including common household items like paints, cleaners/chemicals, motor oil, batteries and pesticides – leftover portions of these products are referred to as household hazardous waste. If these products, or mishandled there can be serious consequences for human health and the environment. Many municipal solid waste landfills have a household hazardous waste drop-off station for these materials.

More to the point of this section, municipal solid waste (MSW) is a type of waste consisting of everyday items that are discarded by the public. The term can also refer specifically to food waste. Within the waste, there are various types of organic materials that serve as sustenance for a variety of microbes. However, there are vast compositional differences between the types of solid waste that are sent to a landfill and the composition of the waste varies greatly from municipality to municipality and also, once discarded, the waste can change significantly with time. In municipalities which have a well-developed waste recycling system, the waste stream mainly consists of intractable wastes such as plastic film and non-recyclable packaging materials. At the start of the 20th Century, the majority of domestic waste in many countries consisted of coal ash from open fires. As the century progressed, in areas without significant recycling activity the waste also included food waste and product packaging materials (typically plastics), and other miscellaneous solid wastes from residential, commercial, institutional, and industrial sources.

Conventional municipal solid waste landfills (MSWLF) generally contain (less toxic) wastes from sources such as private homes, institutions, schools, and businesses without hazardous wastes. Modern sanitary landfills are engineered to protect public health and the environment and usually include detection and prevention of hazardous waste from entering the landfill, appropriate cover material for active cells and closed landfill, disease vector and explosive gas control, air monitoring, facility access, run-on control, run off control and surface water requirements, restrictions on liquids entering the cells and record keeping requirements. An engineered municipal solid waste landfill site comprise a liner, leachate collection and removal system, final cover, gas management and a groundwater monitoring system.

2.5. Hazardous Waste Landfills

Such facilities are disposal facilities for the more toxic chemicals and dangerous by-products. These landfills must be extremely well engineered to reduce any chance of the escape of hazardous compounds into the environment. These landfills are restricted, by permit or law, to the types of waste that they may handle (chemical vs. radioactive, liquid vs. dry). Double liner systems are the norm for hazardous waste landfills.

In terms of hazardous waste, a landfill is defined as a disposal facility or part of a facility where hazardous waste is placed in or on land and which is not a pile, a land treatment facility, a surface impoundment, an underground injection well, a salt dome formation, a salt bed formation, an underground mine, a cave, or a corrective action management unit. A hazardous waste landfill is a treatment, storage, and disposal facility (TSDF) and as such must be appropriately permitted and the permit will specify all design and operating practices necessary to ensure compliance.

All hazardous waste landfills are required to have a run-on control system, a runoff management system, and control the wind dispersal of particulate matter. The run-on control system must have the capacity to prevent flow onto the active portion of the landfill during peak discharge of a 25-year storm and the runoff management system must have an adequate capacity to collect and control water from a 24-hour, 25-year storm and the contents tested to determine correct disposal methodology. The collection and holding tanks or basins for run-on and runoff control systems must be emptied expeditiously after storms. For each hazardous waste landfill there must be a map with the exact location and dimensions, including depth of each cell with respect to permanently surveyed benchmarks. The contents of each cell and the approximate location of each hazardous waste type within the cell must be recorded.

In the landfill, bulk or non-containerized liquid waste or waste containing free liquids must not be placed in the landfill. The exemption to this rule are that containers holding free liquids may only be placed in a landfill if they meet one of the following standards: (i) all free-standing liquid has been removed by decanting or mixed with sorbent or solidified so that free-standing liquid is no longer observed and has been otherwise eliminated, (ii) the container is very small, such as an ampule, the container is designed to hold free liquids for use other than storage such as, for example, a battery or capacitor, and (iii) the container is a lab pack. In order to dispose of the sorbents used to treat free liquids in a hazardous waste landfill, the sorbent must be non-biodegradable.

Approved sorbents include: (i) inorganic minerals, other inorganic materials, and elemental carbon, (ii) high molecular weight synthetic polymers, except for polymers derived from biological material or polymers specifically designed to be degradable, (iii) mixtures of non-biodegradable materials. Unless they are very small (such as ampules), containers must be at least 90 percent full and crushed, shredded, or similarly reduced in volume to the maximum practical extent.

The final cover of a hazardous waste landfill must be designed and constructed for the long term to: (i) minimize migration of liquids through the closed landfill, (ii) function with minimum maintenance, (iii) promote drainage and minimize erosion or abrasion of the cover, (iv) tolerate settling and subsidence so that the cover's integrity is maintained, (and (v) have a permeability less than or equal to the permeability of the liner at the base of the system or any natural subsoil that is present.

During post-closure the integrity and effectiveness of the final cover must be maintained, including making repairs to the cap (cover) as necessary to correct the effects of settling, subsidence, erosion, or other disruptive events. Also, during post-closure all leak detection systems and groundwater monitoring system are required to be maintained. Post-closure care usually lasts 30 year after closure. The removal of any landfill gas taken from such a landfill site must be properly analyzed by standard (acceptable) analytical procedures, permitted, and authorized.

2.6. Manual Landfills

The manual sanitary landfill is a technically and economically feasible alternative, benefiting urban and rural populations of less than 30,000 inhabitants who have no way of acquiring the heavy equipment they would need for constructing and operating a conventional sanitary landfill. This type of landfill is also a good alternative for the marginal areas of some cities

The manual operation technique requires heavy equipment only to prepare the site, that is, for the construction of the internal road, the preparation of the supporting base or the digging of trenches and the extraction of cover material in accordance with the progress made and the fill method. The rest of the work can be carried out without heavy machinery, which means that small communities with scanty resources, which are unable to acquire and maintain the necessary heavy equipment, are able to dispose hygienically of the small amount of waste they produce. All of the key steps for landfill preparation are carried out manually including cell preparation, compaction and daily cover.

In terms of gas collection and/or gas venting, a gas venting system consists of stone-filled ditches or perforated concrete piping lined with stone that function as chimneys or vents, penetrating the whole landfill vertically. These vents are built to link with the leachate drainage at the bottom of the landfill, and they are projected to the surface, for more efficient drainage of both leachate and gases.

2.7. Surface Impoundments

Surface impoundments (lagoons or ponds) deal with liquid waste disposal. These impoundments must be properly lined to prevent infiltration of chemical constituents into the soil and protect groundwater quality. The liner is considered as part of the impoundment and waste migration into the liner is allowed. The liner must therefore have sufficient strength and thickness to prevent failure due to pressure gradients, be chemically compatible with the migrating waste constituents, be protected from climatic conditions, installation and operation stresses, and prevents gradients caused due to settlements, compression or uplift.

Surface impoundments are a lot like landfill cells insofar as both units are either a natural topographic depression, manmade excavation, or diked area formed primarily of earthen materials, such as soil. In addition, both types of site may be lined with manmade materials but it is the use of each site that makes one different from the other. Surface impoundments are generally used for temporary storage or treatment, whereas a landfill is an area designated for final waste disposal. Because of this, the closure and post-closure standards are very different.

In addition to the liner system, the surface impoundment unit has to have a leachate collection and removal system (LCRS) which also serves as a leak detection system. The leachate collection and removal system, along with the leak detection system drainage layers, must be designed with a bottom slope of at least one percent, be made of materials chemically resistant to the wastes placed in the unit, and be able to remove the liquids at a specified minimum rate. The leachate collection and removal system itself must be designed to collect liquids in a sump area and subsequently pump out those liquids. In addition to the performance and design requirements, the leachate collection and removal system must be located between the liners immediately above the bottom composite liner, enabling the leachate collection and removal system to collect the largest amount of leachate, while also representing the most efficient place to identify leaks. Ensuring that the material cannot leak back into the earth is not the only consideration that must be made. The surface impoundment must also be designed to prevent liquids from flowing over the top (called overtopping) and to ensure the structural integrity of any dikes.

Because of the importance of these design regulations (and since none of the aforementioned technologies will work if the impoundment is not installed correctly or made of quality materials) the should be a construction quality assurance (CQA) program to make sure that an impoundment meets all technical criteria. The construction quality assurance program requires a plan that identifies how construction materials and their installation will be monitored and tested and how the results will be documented. The construction quality assurance program is developed and implemented under the direction of a registered professional engineer, who must also certify that the construction quality assurance plan has been successfully carried out and that the unit meets all specifications before any waste is received.

3. Landfill Gas Generation

Paper waste, yard waste, kitchen and food waste are constituents of the principal biodegradable organic fraction in municipal solid waste (Mehta et al., 2002). While such organic waste may contain a constant ultimate biodegradable organic fraction; practical biodegradability varies substantially. The organic component of municipal solid waste can be divided into two categories: (i) waste that will decompose rapidly – 3 months to 5 years, (ii) waste that will decompose slowly – up to or greater than 50 years (Table 4.9) (Speight, 2008).

Landfill gas composition and production rates are primarily affected by the waste that has been deposited in the landfill site. Municipal solid waste contains organic carbon which microorganisms convert to gas by way of anaerobic processes. The gas formation is influenced by a number of factors such as waste composition, landfill storage height and density, air temperature, atmospheric pressure and precipitation levels. Landfill gas production starts one to two years after the waste is deposited in the landfill and can lasts 15 to 25 years. The continuously decreasing landfill gas volume can be compensated by the disposal of additional waste during this period.

Table 4.9. Rapidly and Slowly Biodegradable Organic Constituents in Municipal Solid Waste

Organic waste component	Rapidly biodegradable	Slowly biodegradable
Food wastes	x	
Newspaper	x	
Office paper	x	
Cardboard	x	
Yard wastes	x	x
Textiles		x
Rubber		x
Leather		x
Wood		x

Factors such as size, time, environmental conditions (moisture, temperature, nutrient requirements, pH, atmospheric conditions) will influence the final outcome of biodegradation. For example, because favorable conditions do not exist in most landfills, the biodegradability estimated using modeling software and analytical testing will usually be greater than the biodegradation that actually occur. Landfill gas is generated by one or more of three mechanisms: (i) evaporation/volatility, (ii) biological decomposition, and (iii) chemical reactions.

3.1. Evaporation/Volatilization

Evaporation is a type of vaporization that occurs on the surface of a liquid as it changes into the gas phase. The surrounding gas must not be saturated with the evaporating substance otherwise the evaporation; process will be kinetically inhibited. Typically, the evaporation process continues until an equilibrium is reached when the evaporation of the liquid is the equal to its condensation. In an enclosed environment, a liquid will evaporate until the surrounding air is saturated. On the other hand, volatilization is the process whereby a dissolved sample is vaporized. Volatilization is the change from a liquid to a vapor. Sublimation is the change from a solid to a vapor when the intermediate change from solid to liquid does not occur.

Evaporation and volatilization occur as a result of the chemical phase equilibrium that exists within a landfill. Volatile organic compounds (VOCs) and non-methane organic compounds (NMOCs) in landfill gas may be the result of volatilization of certain chemicals disposed of in the landfill and vaporization until the equilibrium vapor concentration is reached. This process is accelerated when the waste becomes biologically active as a result of heat which can be generated by the activity of the landfill microbes. The rate at which compounds are evolved depends on its physical and chemical properties. Henry's law is used to determine the extent of volatilization of a contaminant dissolved in water, and its constant describes the equilibrium partitioning between the vapor and the aqueous phases at given temperature and pressure.

3.2. Biodegradation

Biodegradation is the dominant mechanism in the landfill and generates methane and carbon dioxide along with traces of carbon monoxide, hydrogen, hydrogen sulfide, ammonia and other miscellaneous compounds (depending on the composition of the waste), typically in trace amounts (<1% v/v) (Allen et al., 1997).

Municipal solid waste comprises biodegradable organic fractions that rapidly or slowly decompose over time and contribute to landfill gas production. The bacteria involved in biological decomposition exist in the municipal solid waste, soil used in landfill operations, wastewater treatment plant sludge (WWTPS) addition and leachate. A generalized scheme for the biological decomposition of solid waste and occurs in 4 distinctive phases, appropriately named Phases I, II, III, and IV:

Phase 1: aerobic digestion begins shortly after waste is placed in the landfill and continues until all of the entrained oxygen is depleted (voids and within the organic waste). Heterotrophic aerobic bacteria produce a gas of high carbon dioxide content (approximately 30% v/v), water, NO_3, other oxygenated compounds and low methane content (approximately 2 to 5% v/v). Once the oxygen level drops below 10 to 15% v/v, the anaerobic microorganisms are activated and the aerobic decomposition may last for 6 to 18 months for waste in the bottom lifts of a landfill.

Phase II: this is the first anaerobic phase in which the waste decomposition changes from aerobic to anaerobic. The extent of anaerobic conditions can be monitored and measured by the redox potential of the waste. The pH of leachate decreases due to the formation of organic acids and the elevated concentrations of carbon dioxide which may partly dissolve and form carbonic acid (Tchobanoglous et al., 1993).

Phase III: this is the 2nd anaerobic phase (*acidogenesis*) in which the anaerobic microbial activity is accelerated with the concomitant production of copious amounts of organic acids and modest amounts of hydrogen. Anaerobes hydrolyze complex organic compounds such as cellulose, fats and proteins into simpler compounds and subsequently organic acids as acetic acid (CH_3COOH), butyric acid ($CH_3CH_2CH_2COOH$), lactic acid ($CH_3CH (OH) COOH$) and small concentrations of fulvic acid and other complex organic acids.

The pH of landfill liquids drops to approximately 5 due to the presence of the organic acids and the relatively high concentrations of carbon dioxide (principal gas generated in this phase). There is no methane production as methanogenic bacteria cannot tolerate acid conditions. The five-day biological oxygen demand (BOD), chemical oxygen demand (COD) and conductivity of leachate increase significantly due to the dissolution of the organic acids in the leachate. Because of the low pH, metals and other inorganic constituents are solubilized during this phase.

As a recap (Chapter 3), Phase 1 of the process is the hydrolysis reaction in which high-molecular weight naturally- occurring compounds (cellulose, hemicellulose and lignin) are converted to lower molecular weight products. Typically sugars and oxygenated aromatic derives. During acidogenesis, soluble monomers are converted into small organic compounds, such as short chain (volatile) acids (propionic, formic, lactic, butyric, succinic acids), ketones (glycerol, acetone), and alcohols (ethanol, methanol). Generally, the chemistry of Phase 2 (acidogenesis) and Phase 3 (acetogenesis) can be represented by a series of simple equations:

Acidogenesis:

$$C_6H_{12}O_6 + 2H_2 \rightarrow 2CH_3CH_2COOH + 2H_2O$$
$$C_6H_{12}O_6 \rightarrow 2CH_3CH_2OH + 2CO_2$$

Acetogenesis:

$$CH_3CH_2COO^- + 3H_2O \rightarrow CH_3COO^- + H^+ + HCO_3^- + 3H_2$$
$$C_6H_{12}O_6 + 2H_2O \rightarrow 2CH_3COOH + 2CO_2 + 4H_2$$
$$CH_3CH_2OH + 2H_2O \rightarrow CH_3COO^- + 2H_2 + H^+$$
$$2HCO_3^- + 4H_2 + H^+ \rightarrow CH_3COO^- + 4H_2O$$

Phase IV: is the phase of methane formation (methanogenesis) in which the concentrations of the intermediate acids are usually small in proportion to their production and degradation rates, and they are quickly transformed to methanogenic substrates, including acetate, methanol, and formate. Thus, the last phase of anaerobic digestion is the methanogenesis phase. Several reactions take place using the intermediate products from the other phases, with the main product being methane. The reactions are commonly represented as:

$$2CH_3CH_2OH + CO_2 \rightarrow 2CH_3COOH + CH_4$$
$$CH_3COOH \rightarrow CH_4 + CO_2$$
$$CH_3OH \rightarrow CH_4 + H_2O$$
$$CO_2 + 4H_2 \rightarrow CH_4 + 2H_2O$$

$CH_3COO^- + SO_4^{2-} + H^+ \rightarrow 2HCO_3 + H_2S$

$CH_3COO^- + NO^- + H_2O + H^+ \rightarrow 2HCO_3 + NH_4^+$

Several bacterial contribute to methanogenesis, including: *Methanobacterium, methanobacillus, methanococcus,* and *methanosarcina.*

Any kind of organic matter can be fed to an anaerobic digester, including manure and litter, food wastes, green wastes, plant biomass, and wastewater sludge. The materials that compose these feedstocks include polysaccharides, proteins, and fats/oils. Some of the organic materials degrade at a slow rate; hydrolysis of cellulose and hemicellulose is rate limiting. There are some organic materials that do not biodegrade: lignin, peptidoglycan (a polymer consisting of sugar derivatives and amino acid derivatives that forms a mesh-like layer outside the plasma membrane of most bacteria, forming the cell wall), and membrane-associated proteins (Amani et al., 2010). The organic residues contain water and biomass composed of volatile solids and fixed solids (minerals or ash after combustion) and it is the volatile solids that can be non-biodegradable and biodegradable. However, unlike the chemistry of the anaerobic digester, the organic waste in the landfill is not available for separate pretreatment unless there is a form of pretreatment before he material is discharged into the landfill.

Methanogenic bacteria degrade the volatile acids, primarily acetic acid and use the hydrogen to generate methane (45 to 57% v/v) and carbon dioxide (40 to 48% v/v). Both methane and some acid formation proceed simultaneously and the methane bacteria can use only limited ranges of substrates for growth and energy production. As the acids decompose the pH rises and stabilizes at approximately 6.8 to 8 (Tchobanoglous et al., 1993). Consequently metals which were previously soluble now precipitate and the concentration of biochemical oxygen demand, the chemical oxygen demand, and the conductivity also decline.

During anaerobic digestion of municipal solid waste, a varied mixture of complex compounds is converted to a very narrow range of simple compounds, mainly methane and carbon dioxide. The anaerobic bacteria are responsible for the biochemical transformation of the biodegradable organic fraction (BOF); these transformations are involved in the breakdown of complex polymers, such as cellulose, fats and proteins to long and short chain fatty acids and finally methane, carbon dioxide and water (Table 3.9) (Kreith and Goswami, 2007; Valdez-Vazquez and Poggi-Varaldo, 2009).

To some observers, maturation is considered to be a 5th phase and occurs after the readily available biodegradable organic fraction has been converted to methane and carbon dioxide. The rate of gas generation declines significantly because most of the nutrients have been removed with the leachate during the previous phases and the substrates that remain are only slowly biodegradable. The principal landfill gas constituents are now methane and carbon dioxide. During the maturation phase, the leachate contains humic acids and fulvic acids, which are complex and highly stable compounds.

Table 4.10. Summary of the Common Biochemical Reactions in Anaerobic Digestion Process

Substrate	Biochemical Reactions
H_2 and CO_2	$4H_2 + CO_2 \rightarrow CH_4 + 2H_2O$
CO	$4CO + 2H_2O \rightarrow CH_4 + 3CO_2$
Alcohols	$4CH_3OH \rightarrow 3CH_4 + CO_2 + 2H_2O$ $CH_3OH + H_2 \rightarrow CH_4 {}^+ H_2O$ $4HCOO^- + 2H^+ \rightarrow CH_4 + 2H_2O$
Monosaccharides	$C_6H_{12}O_6 + 2H_2O \rightarrow 2H_2 + CH_3CH_2CH_2COO^- + 2HCO_3^- + 3H^+$ $C_6H_{12}O_6 + 4H_2O \rightarrow 4H_2 + 2CH_3COO^- + 2HCO_3^- + 4H^+$
Organic Acids	Butyrate $+ 2H_2O \rightarrow 2H_2 + 2CH_3COO^- + H^+$ Propionate $+ 3H_2O \rightarrow 3H_2 +$ acetate $+ HCO_3^- + H^+$ $HCOO^- + 3H_2 + H^+ \rightarrow CH_4 + 2H_2O$ $CH_3COO^- + H_2O \rightarrow CH_4 + HCO_3^-$ $4CH_2NH_2 + 2H_2O + 4H^+ \rightarrow 3CH_4 + CO_2 + 4NH_4^+$
Methanogenic substances	$2(CH_3)_2NH + 2H_2O + 2H^+ \rightarrow 3CH_4 + CO_2 + 2NH_4^+$ $4H_2 + HCO_3^- + H^+ \rightarrow CH_4 + 3H_2O$ $4H_2 + 2HCO_3^- + H^+ \rightarrow CH_3COO^- + 4H_2O$
Sulfate	$4H_2 + SO_4^{2-} \rightarrow HS^- + 3H_2O + OH^-$

Table 4.11. Examples of the Variability of Landfill Gas Composition over Time

Time after closure (months)	N_2, % v/v	CO_2, % v/v	CH_4, % v/v
0-3	5.2	88	5
3-6	3.8	76	21
6-12	0.4	65	29
12-18	1.1	52	40
18-24	0.4	53	47
24-30	0.2	52	48
30-36	1.3	46	51
36-42	0.9	50	47
42-48	0.4	51	48

The duration for each phase is a function of the distribution of the organic components in the waste, the availability of nutrients, moisture content and compaction of the waste mass, increasing bulk density may prevent movement of moisture to all parts of the waste thereby reducing the rate of biological reactions and subsequent gas production (Table 4.10, Table 4.11) (Pichtel, 2007). The chemical reactions within the waste are likely to occur when the waste materials that are prone to reaction are co-disposed. Heat generated by biological decomposition may accelerate chemical reaction rates.

3.3. Factors Impacting Landfill Gas Generation

With time, the ability of a landfill to generate gas decreases, thus landfills closed for less than 5 years are typically ideal for landfill gas recovery because of relatively fresh and moist municipal solid waste. Under optimum conditions, a landfill might produce gas for 15 years or more depending on the rate of gas generation, the water content of the waste and the manner in which the landfill was closed.

Current closure requirements for landfills are intended to restrict the entry of moisture; this will lead to greatly reduced gas generation after closure. Landfill gas generation is impacted by the following factors: (i) availability of nutrients in the landfill are and in the waste, (ii) temperature, (iii) moisture, and (iv) acidity or alkalinity, i.e., pH.

3.3.1. Nutrients

Bacteria in a landfill require various nutrients for growth, primarily carbon, hydrogen, oxygen, nitrogen and phosphorous (macro-nutrients), but also trace elements, such as sodium, potassium, sulfur, calcium and magnesium (micronutrients). For stable anaerobic decomposition, these nutrients must be present in the waste in the correct ratios and concentrations. The availability of macronutrients in the waste has an effect on both the volume of water generated from microbial processes and gas composition. Landfills that accept municipal solid waste and use daily soil cover have an adequate nutrient supply for most microbiological processes.

If the occurrence of nutrients is lacking, nutrient supplementation can be improved by the addition of wastewater treatment plant sludge (WWTPS), farm and agricultural waste.

3.3.2. Temperature

Landfill temperatures can influence the types of bacteria that are predominant and the level of gas production. The optimum temperature for aerobic decomposition is 54 to 71°C (129 to 160°F), for anaerobic bacteria 30 to 41°C (86 to 106°F. Landfill temperatures are usually in the range 29 to 60°C (84 to 140°F) for anaerobic decomposition, a temperature rise from 35 to 65°C (95 to 149°F) leads to an increase in population of thermophilic bacteria and an increase in methanogenic activity.

3.3.3. Moisture Content

Moisture is the most important parameter affecting waste decomposition and gas production. A moisture content of between 50 to 60% w/w favors maximum methane production. In general, the moisture content of municipal solid waste varies from 10 to 20% to a high of 30 to 40% with an average of 25% w/w.

Low moisture content, as is the case in arid regions, may prevent decomposition and limit gas production. For example, in a landfill holding waste having 15% w/w moisture, the waste will be *fossilized*, i.e., it will not decay and therefore produce very little methane.

Leachate circulation in bioreactor landfills allows control of moisture within the landfill. Typically when waste achieves 50% w/w moisture, it has reached the field capacity and will leach continuously downward thereafter with additional moisture added to lower layers. In-situ moisture content as high as 70% w/w is possible. At this level, a decrease in the efficiency of gas collection can be expected.

3.3.4. Acidity and Alkalinity

Acidity and alkalinity in a landfill is variable, usually falling with the ranges from 5 to 9 and may change depending on the biological processes. For example, a pH change from 5 to 7 leads to an enrichment of acidophiles and a reduction in methanogenesis. Most landfills have an acidic environment initially, but when the aerobic and acidic anaerobic stages have been completed, the methanogenic processes returns the pH to neutral (7-8) due to the buffering capacity of the systems pH and alkalinity. Important hydrogen-producing (e.g., *C thermocellum*) and hydrogen consuming (e.g., *M. thermoautitrophicum*) aerobes do not grow at pH values below 6. One concern during the acidic stages of the biological process is that the reduced pH will mobilize metals that may leach out of the landfill or become toxic to the bacteria generating the gas. In some cases, the addition of sewage sludge, manure or agricultural waste during waste placement will improve methane gas generation.

3.3.5. Atmospheric Conditions

Temperature, barometric pressure and precipitation, affect landfill conditions. Atmospheric temperature and pressure affect the surface layer of waste by reducing the surface concentrations of gas components and creating advection near the surface. Precipitation affects gas generation significantly by supplying water to the process and carrying dissolved oxygen into the waste. For example, the methane emission rate is strongly dependent on changes in barometric pressure, i.e., rising barometric pressure suppressed the emission, while falling barometric pressure enhanced the emission (Xu et al., 2014).

Temperatures as well as methane and carbon dioxide concentrations vary with seasonal fluctuations. At shallow depths, decomposition of wastes is inhibited during cold months of the year and is manifested by low methane and carbon dioxide concentrations occurring at low temperatures. Waste temperatures increase during the aerobic phase of the digestion process and continue to increase during all of the decomposition stages (Yeşiller et al., 2005; Hanson et al., 2010).

3.4. Gas Migration

Landfill gas migration is a complex process of the gas moving from the site of original deposition to other places by means of diffusion as well as other variables (Table 4.12).

Typically, the gas moves from areas of high concentration to areas of low gas concentration in, or in the area around a landfill. The process is also affected by the permeability of the ground and factors such as pressure differences in the soil, cavities, pipes, and tunnels. Furthermore, changes in the atmospheric pressure and the water table can encourage this migration. Containing the gas can be achieved by means of combinations of geomembranes and clay-based products.

Table 4.12. Summary of the Conditions that Can Affect Landfill Gas Migration

Conditions	Result
Cover type	If the landfill cover consists of relatively permeable material, such as gravel or sand, the gas will likely migrate up through the landfill cover. If the landfill cover consists of silt minerals and clay minerals, it is not very permeable and the gas will tend to migrate horizontally underground. If one area of the landfill is more permeable than the rest, gas will migrate through that area.
Pathways	Drains, trenches, and buried utility corridors (such as tunnels and pipelines) can act as conduits for gas movement. The natural geology often provides underground pathways, such as fractured rock, porous soil, and buried stream channels, where the gas can migrate.
Wind speed/ direction	Landfill gas naturally vented into the air at the landfill surface is carried by the wind. The wind dilutes the gas with fresh air as it moves it to areas beyond the landfill. Wind speed and direction determine the concentration of the gas in the air, which can vary considerably in one day. Gentle winds may provide the least dilution and dispersion of the gas to other areas.
Moisture	Wet surface soil conditions may prevent landfill gas from migrating through the top of the landfill into the air above. Rain and moisture may also seep into the pore spaces in the landfill and expel the gas from these spaces.
Groundwater levels	Gas movement is influenced by variations in the groundwater table. If the water table is rising into an area, it will force the landfill gas upward.
Temperature	Increases in temperature stimulate gas movement with an increase gas diffusion, so that landfill gas might spread more quickly in warmer conditions. Although the landfill itself generally maintains a stable temperature, freezing and thawing cycles can cause the soil's surface to crack, causing landfill gas to migrate upward or horizontally. Frozen soil over the landfill may provide a physical barrier to upward landfill gas migration, causing the gas to migrate further from the landfill horizontally through soil.
Barometric pressure	The difference between the soil gas pressure and barometric pressure allows gas to move either vertically or laterally, depending on whether the barometric pressure is higher or lower than the soil gas pressure. When barometric pressure is falling, landfill gas will tend to migrate out of the landfill into surrounding areas. As barometric pressure rises, gas may be retained in the landfill temporarily as new pressure balances are established.

In an ideal situation landfill gas would simply diffuse to the surface of the landfill and disperse into the atmosphere (O'Leary and Tansel, 1986). However there are a number of circumstances that will force landfill gas to migrate or be transported laterally. The directions are controlled in part by the permeability of the soil and fill material; this is especially relevant in landfills that lack a complete liner.

Coarse, porous soils will allow greater lateral movement of landfill gas than fine grained soils. If a cell is closed and the final cover dense and impermeable and if the side slopes do not contain a gas barrier the gases will migrate laterally. If the soil near the surface has high moisture content, upward movement is inhibited; similarly a cell with a frozen surface will promote lateral migration. Lateral migration is more common in older facilities that lack both liners and gas control systems. The mechanisms by which landfill gas migrate are: (i) molecular effusion (ii) molecular diffusion (iii) convection.

3.4.1. Molecular Effusion and Diffusion

Effusion is the process in which a gas escapes from a container through a hole of diameter considerably smaller than the mean free path (the average distance travelled by a moving particle such as an atom or a molecule between successive impacts or collisions), of the molecules whereas diffusion refers to the process of particles moving from an area of high concentration to one of low concentration.

Molecular effusion occurs at the air-landfill interface, where the material has been compacted but not yet covered; effusion. For dry solids, the principal release mechanism is direct exposure of the waste vapor phase to the ambient atmosphere. Raoult's law predicts the release rate based on the vapor pressure of the compounds present.

The diffusive flow is in the direction of decreasing concentration and occurs where a concentration difference in landfill gas exists. The concentration of landfill gas will be higher in the landfill than that of the surrounding atmosphere, so its constituents will tend to migrate out of the landfill. Constituents of landfill gas exhibit various diffusion coefficients.

3.4.2. Convection

Convection is the heat transfer due to the bulk movement of molecules within fluids, such as gases and liquids.

In the landfill, the gas will move from higher to lower pressure and out of the landfill into the atmosphere in response to pressure gradients and the rate of gas movement is generally sundtantially higher for convection-type movement than for diffusion-type movement. The source of the pressure may be the production of landfill gas from (i) the biodegradation process, (ii) chemical reactions within the landfill, (iii) compaction, or (iv) methane generation at the lower regions of the landfill which drive vapors toward the surface. For most cases of landfill gas recovery, diffusive and convective flows occur in the same direction.

Also, the permeability of the system (or the cell) is an indication of the ease with which the landfill gas, water or other fluid travels through a porous media. The permeability distribution to landfill gas has a profound influence on gas flow and recovery rates. Coarse waste typically exhibits higher permeability than fine waste. The water table acts as a no flow boundary for gas flow. The presence of the water table beneath the landfill inhibits the depth of gas migration and increases lateral gas movement.

Condition of waste, spatial variations, layering, compaction, porosity, and moisture affect landfill gas migration, flow patterns and recovery rates. The subsurface soils, daily cover soil, and the final cover system can influence vertical and lateral migration of landfill gas. Manmade features as sewers, drainage culverts, and buried utility right of ways adjacent to landfills can provide corridors for gas migration and potential for pockets of gas to accumulate. Additionally natural features as gravel and sand lenses, void spaces, cracks and fissures resulting from landfill differential settlement can lead to landfill gas migration.

4. Gas Collection Systems

In order to use the landfill gas, it must first be collected from the source using a functional collection system. A collection system usually consisting of vertical extraction wells gas pipeline, blower units and possibly a flare to burn off gases safely is used. These collection wells are placed strategically throughout the landfill site. If the gas production is significant, each extraction well is connected to the main gas line, which carries the gas to a station; landfill operators can either sell the gas or use it for energy.

The gas is recovered from the landfill by means of a gas pump or a compressor leading the gas to a utilization plant by means of pressure in the transmission pipe. The connection of the single wells to the pump and utilization system can be done in different ways. The oldest and maybe most common way, is to connect the wells to a main collection pipe which goes around on the landfill. The main problem with this system is the difficulty involved with the regulation of both the quality and quantity of the gas. Another issue related to finding the location of the leak(s) when all the wells are connected in an integrated system. To prevent explosions, reduce landfill gas odor and migration, and adverse effects on human health and environment, control systems are installed to collect and remove landfill gas. From a regulatory perspective landfill gas should be controlled so that methane concentration at the landfill property line is <5% v/v (LEL).

Landfill gas control systems can be grouped into two types: (i) the passive system or (ii) the active system. Passive systems function on convection (pressure gradient), diffusion (concentration) and density gradients to move the landfill gas to extraction wells. Passive systems provide corridors to intercept lateral gas migration and direct to a collection point. These systems employ barriers to prevent migration outside the perimeter of the landfill.

On the other hand, active systems move landfill gas using mechanical equipment such as vacuum pumps, blowers, compressors. Factors considered in the choice of passive or active systems are (i) design, (ii) age of the landfill, (iii) soil conditions of the facility, (iv) hydrogeologic conditions, (v) the surrounding environment, (vi) potential for gas hazards, and (vii) parameters such as temperature and the corrosion resistance of materials used in the construction of the system.

4.1. Passive Gas Collection Systems

Passive gas collection systems use existing variations in landfill pressure and gas concentrations to vent landfill gas into the atmosphere or a control system. Passive collection systems can be installed during active operation of a landfill or after closure. Passive systems use collection wells, also referred to as extraction wells, to collect landfill gas. These systems control convective gas flow; however they are less effective in controlling diffusive gas flow. Passive systems involve ‘high permeability’ or ‘low permeability’ techniques. High permeability systems incorporate well vents, trench vents, and perforated vent pipes surrounded by coarse material to vent landfill gas to the surface.

To control lateral migration, a low permeability barrier (synthetic membranes and clayey soils) is incorporated in the landfill design; this needs to be deeper than the landfill to intercept all lateral gas movement. Passive systems may be incorporated at the design stage, later for corrective purposes or at the time of landfill closure (incorporated into the final cover system and may consist of perforated collection pipes and high permeability soils located directly beneath the impermeable layer). They may also be installed along the perimeter or between the landfill and the facility property boundary. The passive gas collection system may be connected to header pipes along the perimeter of the landfill unit.

4.1.1. Well Vents

Well vents are used to control lateral and vertical migration of landfill gas and are located at points where gas accumulates. Lateral gas migration can be controlled only if the vents are located in a tightly bunched cluster. Preliminary sampling is an essential requirement to determine gas collection points for proper vent placement. For optimum effectiveness, well vents are placed at maximum concentration and/or pressure contours and vent depth should exceed to the bottom of the waste. Atmospheric well vents are generally not recommended for the control of lateral gas migration, if used however they must be placed very close together (<50 feet apart).

The vents are constructed from perforated polyvinyl chloride pipes and a gravel packed layer is placed around the vent to prevent clogging. The pipe vent should be sealed off from the atmosphere and the passive wells should generally be located approximately 30 to 50 feet from the edges of the waste and typically not more than one well per acre). Additional wells may be needed further within the body of the waste to intercept their full depth if the site is benched. The well vents can be incorporated into the final cover systems and the geo-membrane in the landfill cover should be properly sealed at the well vent locations.

4.1.2. Trench Vents

Trench vents are used primarily to control lateral gas migration. Such systems are suitable where the depth of gas migration is limited (impervious formation or presence of groundwater). If trenches can be excavated to such depths, trench vents can offer full containment and control of lateral landfill gas migration. Passive open trenches can be used to control gas migration; however, their effectiveness is low. An impervious liner can be added to the outside of the trench to increase control efficiency.

Open trenches are suited for sparsely populated areas, not likely to be covered or planted over. Passive trench vents may be covered by clay or any impervious material and vented to the atmosphere. Such a system ensures adequate ventilation and prevents infiltration of rainfall into the vent. Also an impermeable clay layer can be used as an effective seal against the escape of landfill gas. Open trenches are prone to runoff and clogging, and the surrounding ground should be sloped to direct runoff away. The gravel pack in the trench should be sufficiently permeable to transport the gas adequately. A barrier may be installed on the outside of the trench to prevent bypass of landfill gas. Three types of barriers are commonly used: geosynthetic liners, natural clays or admixed materials, these are similar to the liner materials used in the base and side slope liner systems.

4.1.3. Closed Trenches

Closed trenches comprise laterals and risers that lay in gravel packed trenches. An impermeable clay layer is used for sealing to protect against escape of landfill gas. Gas collection trenches can be used where vertical extraction wells are not practical, such as in areas where the waste depth is shallow or where the leachate levels are high. A drawback of trenches is their tendency to draw air if the seal over each trench is inadequate. Care must be taken in the design of vent systems to prevent them from being a source of infiltration through the cover. Advantages for trench systems include ease of construction and relatively uniform withdrawal influence areas. Trenches are susceptible to damage (crushing, severing) as a result of differential settlement due to placement of subsequent lifts of waste. When placed below groundwater levels, the trenches are susceptible to flooding, measures should be taken to avoid drawing water or leachate into the gas collection system.

The trenches may be either vertical or horizontal at or near the base of the landfill. For new sites, horizontal trenches are installed within a landfill cell as each layer of waste is applied. The distance between layers should be no greater than 15 feet which allows for gas collection as soon as possible after gas generation begins and avoids the need for aboveground piping, which can interfere with landfill maintenance equipment. Additional legs of the system are connected to the manifold as the landfill grows in size and height.

The horizontal trench pipes are constructed of perforated polyvinyl chloride (PVC), high density polyethylene (HDPE) or other nonporous material of suitable strength and corrosion resistance. The trench should be about 1 m wide filled with gravel of uniform size and extend into the waste approximately 5 feet below the landfill cap layer. Trenches should be located between the waste fill and the gas barrier or side of the site. The side of the trench nearest the property boundary should be sealed with a low-permeability barrier material, such as a geomembrane, to prevent gas migration. The remainder of the trench should be lined with a fabric filter to prevent clogging of the permeable medium.

The gas collection piping is connected to surface vent pipes. Spacing is determined from monitoring and site investigation data (generally 50 m apart). Passive vents can be used in combination with horizontal trenches by connecting vents to the pipes with flexible hosing. The flexible hose between the extraction well or trench and the collection header system allows differential movement. Because of its horizontal layout, the collection header system would be expected to settle more than a vertical extraction well. This flexible connection allows more movement than would be possible if the two pipes were rigidly connected. Sampling ports can be installed, allowing monitoring of pressure, gas temperature, concentration and liquid level.

4.2. Active Gas Collection Systems

Active gas collection systems include vacuums or pumps to move gas out of the landfill and piping that connects the collection wells to the vacuum. Vacuums or pumps pull gas from the landfill by creating low pressure within the gas collection wells. The low pressure in the wells creates a preferred migration pathway for the landfill gas. The size, type, and number of vacuums required in an active system to pull the gas from the landfill depend on the amount of gas being produced.

Thus active gas collection systems (AGCS) use mechanical means to remove landfill gas by air injection (positive pressure) or vacuum extraction (negative pressure). Vacuum extraction systems are more commonly used. Gas extraction wells may be installed within the landfill cells or beyond the landfill, in nearby extraction trenches. Active systems are not as sensitive to freezing or saturation of cover soils as passive systems. The capital, operating and maintenance costs of an active gas collection system are higher than for passive systems. These costs continue throughout the post-closure period. It is possible to convert an active gas collection system to a passive gas collection system when gas production decreases.

The effectiveness of an active gas collection system depends on its ability to collect, handle and monitor max landfill gas flows and adjust operation of individual extraction wells and trenches. Often, for gas extraction, several wells are connected on a pilot basis to determine the radius of influence; they are fitted with valves, condensate traps and connected to the suction manifold of the blower/compressor/vacuum pump. The gas is either vented to the atmosphere, flared or processed. The system includes a series of gas well vents, manifold to a blower and subsequent processing plant and distribution to customer, this arrangement can be used to prevent atmospheric emissions and monetize landfill gas. Active gas collection systems haves four major components (Figure 2.6) and these are: (i) gas extraction wells or horizontal trenches, (ii) a gathering, collecting and gas moving system, and (iii) a gas treatment and processing facility.

Gas extraction wells or horizontal trenches (the number of wells is determined by the radius of influence, spacing, landfill geometry). Some overlap of influence zone is desirable for the perimeter wells, designed for the control of gas migration between wells along the landfill boundary. The gas extraction rate and radius of influence are interrelated and individual well flow rates can be adjusted to provide effective migration control and/or efficient methane recovery (Young, 1990; Young et al., 1993). The *gathering, collecting and gas moving system* such as pipeline, headers, mechanical blower a typical header pipe is made of polyvinyl chloride (high density polyethylene with a diameter that is determined by the anticipated gas flow. The size and type of blower is a function of flow rate, system pressure drop and vacuum to induce pressure gradient at the landfill.

The *treatment and processing facility* should be equipped for carbon dioxide removal (Chapter 7). Processing increases the calorific value (CV, also called heat content) value of the gas. In large municipal solid waste landfill sites, landfill gas is treated by water and carbon dioxide removal through scrubbing and gas polishing with carbon or polymer adsorption and is used as a low calorific value fuel source for boilers, gas and steam turbines and internal combustion engines for electricity generation, process heating or it is upgraded to pipeline quality gas (water, carbon dioxide, hydrogen sulfide, oxygen, nitrogen removal) and distributed to local utilities.

5. Upgrading

After collecting the gas from the wells, the methane rich gas is flared or pre-treated before being used in an energy recovery system – the treatment requirements depend on the end use application. Landfill gas can be upgraded to fuels or chemicals by a variety of thermal and catalytic processes (Chapter 7). In many cases most of the carbon dioxide must be removed from the methane before upgrading. Steam or oxidative reforming can be used to convert the biogas to syngas, from which hydrogen may be separated, if desired.

Flaring is used when the gas cannot be put to useful means and has to be released into the environment in a safe manner. Flaring is the process of burning the LFG in a controlled environment so as to destroy harmful constituents and dispose the products of combustion safely to the atmosphere. The operating temperature is a function of gas composition and flow rate (Young, 1990). Some of the factors that need to be considered in the design of a landfill gas flare are: residence time, operating temperature, turbulence and delivery of combustion air. Adequate time must be available for complete combustion, the temperature must be high enough to ignite the gas and allow combustion of the mixture (air/fuel).

The operating temperature of the combustor is approximately 760ºC (140ºF; the temperature at which methane ignites). Flares are also used as a backup to an energy recovery system. The gas on its way to the flare passes a water seal; this prevents flashback and the possibility of explosions and fires. Landfill gas flows through the flame arrestor and burner tips, the flare stack drafts air through dampers and around the burner tips. The stack acts as a chimney in developing sufficient draft and residence time for efficient operation.

When the landfill gas is required for energy or any other use a rigorous pre-treatment of the gas is required. Treatment systems can be divided into primary treatment processing and secondary treatment processing. Most primary processing systems include de-watering and filtration to remove moisture and particulates. Dewatering can be as simple as physical removal of free water or condensate in the landfill gas. However it is common to remove water vapor using gas cooling and compression.

Secondary treatment systems are designed to provide much greater gas cleaning than primary gas cleaning systems. These systems may employ multiple clean up processes depending on the gas specifications of the end use. Such processes include both physical and chemical treatments. The most common technologies used for secondary treatment are absorption and adsorption. A large portion of the landfill gas consists of carbon dioxide which is non-combustible and therefore reduces the calorific value of the gas. Furthermore, in the presence of ambient moisture, acid gases produce acid rain, which is a well-known global environmental menace. The removal of these toxic compounds can be done through dry or wet scrubbing processes. The scrubbing process is an operation in which one or more components of a gas stream are selectively absorbed into an absorbent.

The term scrubbing is often used (but often not quite correctly) interchangeably with absorption when describing this process. In wet scrubbing a relatively non-volatile liquid is used as the absorbent. In dry scrubbing a dry powder or semidry slurry are possible absorbents. Adsorption involves the binding of particle to the surface. In this case the surface is usually carbon or silica gel. There also exist advanced treatment technologies that remove carbon dioxide, non-methane organic compounds (NMOCs) and a variety of other contaminants in landfill gas to produce a high-Btu gas.

As part of the cleaning process, landfill gas can be upgraded to natural gas quality so that it can be used in the existing natural gas distribution network. One of the indices used to evaluate the quality of gas for comparative use the Wobbe Index. The Wobbe Index allows for the comparison of the volumetric energy content of different gas fuels at different temperatures. Gases can possess varying heating values and relative densities but have the same Wobbe Index. i.e., they will deliver the same heat and yield the same performance for a given process. Gas containing methane with a Wobbe Index of 13-15 kWh/m^3 can be considered comparable to natural gas and can be used in any standard equipment designed for natural gas without any additional treatment.

The upgrading process involves a series of complex processes which will result in the removal of elements such as, hydrogen sulfide, carbon dioxide, heavy hydrocarbon, siloxanes, nitrogen and oxygen. Though the process used to remove such gasses is not new when applied to landfill gas, it may not be a cost effective option. Upgraded Landfill gas can also be used as chemical feedstock in the process industry such as in the production of diesel fuel and methanol. The process involes the upgrading of the landfill gas to chemical feedstock quality hence rending this option economically unattractive.

Low grade heat recovered from flaring of landfills can also be used in organic Rankine cycle and Stirling cycle for the generation of electrical power (Chapter 8). It can also be used directly to evaporate leachate in landfill, powering and heating greenhouses and heating water for aquaculture (fish farming) operation. Harnessing the power of landfill gas energy provides environmental benefits to landfills, energy providers and end-users through emerging technologies by: (i) offsetting the use of non-renewable resources; coal, oil and natural gas and (ii) reducing greenhouse gas emissions due to the escape of the methane rich landfill gas into the atmosphere (EPA, 2011).

However a process utilizing waste produces substances of concern from two major products of the process: (i) solids, such as ash or slag, heavy metals and dust, and (ii) gases, such as flue gas and other gaseous chemical constituents. Ash or slag originates from the incineration or gasification of the waste and consists of solid incombustible material combined with traces of metals which can either be toxic or benign in nature. Apart from the traces of metal ash also contains toxic organic compounds (TOCs), such as dioxin derivatives. These metals can be made harmless through the sintering of the ash although extremely toxic ash remains buried in many landfills. Dioxins are formed during waste combustion and the formation of these chemicals must be controlled as a lack of control can lead to serious environmental issues.

Heavy metals such as mercury released from coal-fired power plants, cadmium and lead must be monitored. However, new technologies exist today to mitigate the negative effects of the metals. Another solid element of concerned is fly ash. Fly ash consists of fine solid particles which often contain high levels of toxic metals. This ash escapes with the flue gas or exhaust gas from the waste-to-energy plants. Fabric or bag filters and electrostatic precipitators can be used to capture the ash which can then be buried in the landfill.

REFERENCES

Alexander, A., Burklin, C., and Singleton, A. 2005. *Landfill Gas Emissions Model (LandGEM)* Version 3.02 User's Guide. EPA-600/R-05/047. United States Environmental Protection Agency, Eastern Research Group Morrisville, North Carolina. May; or United States Environmental Protection Agency, Washington, DC.

Allen, M. R., Braithwaite, A. and Hills, C. C. 1997. Trace Organic Compounds in Landfill Gas at Seven UK Waste Disposal Sites. *Environmental Science and Technology,* 31(4): 1054-1061.

Amani, T., Nosrati, M., and Sreekrishnan, T. R. 2010. Anaerobic Digestion from the Viewpoint of Microbiological, Chemical, and Operational Aspects – A Review. *Environ. Rev.,* 18: 255-278.

EPA. 2011. *Available and Emerging Technologies for Reducing Greenhouse Gas Emissions from Municipal Solid Waste Landfills.* Office of Air and Radiation, Sector Policies and Programs Division, Office of Air Quality Planning and Standards, United States Environmental Protection Agency, Research Triangle Park, North Carolina. June.

Farquhar, G. J. and Rovers, F. A. 1973. Gas Production during Refuse Decomposition. *Water, Air and Soil Pollution*, 2: 483-493.

Hanson, J. L., Yeşiller, N., and Oettle, N. K. 2010. Spatial and Temporal Temperature Distributions in Municipal Solid Waste Landfills. *J. Environmental Engineering,* 136(8): 1095-1102.

Johnston, A. G., Edwards, J. S., Owen, J., Marshall, S. J. and Young, S. D. 2000. Research into Methane Emissions from Landfills. *Proceedings. Waste 2000: Waste Management at the Dawn of the Third Millennium.* Stratford-upon-Avon, United Kingdom. Page 13-20.

Kamalan, H., Sabour, M., and Shariatmadari, 2011. A Review on Available Landfill Models. *Journal of Environmental Science and Technology*, 4(2): 79-92.

Kreith, F., and Goswami, Y. D. 2007. *Handbook of Energy Efficiency and Renewable Energy,* CRC Press, Boca Raton, Florida.

Lee, G. F. and Jones-Lee, A. 1993. Landfills and Groundwater Pollution Issues: Dry Tomb vs F/L Wet-Cell Landfills. *Proceedings. Sardinia '93 IV International Landfill Symposium*, Sardinia, Italy. Page 1787-1796.

Lee, G. F., and Jones-Lee, A. 1996. Dry Tomb Landfills: Nothing Stays Buried Forever. *MSW Management,* 6(1): 82-89.

Lee, G. F., and Jones-Lee, A. 1994. A Groundwater Protection Strategy for Lined Landfills. *Environmental Science Technology*, 28(13): 584-585.

Lee, U., Han, J., and Wang, M. 2017. Evaluation of Landfill Gas Emissions from Municipal Solid Waste Landfills for the Life-Cycle Analysis of Waste-To-Energy Pathways. *Journal of Cleaner Production*. 166(10): 335-342.

Mehta, R., Barlaz, M. A., Yazdani, R., Augenstein, D., Bryars, M., and Sinderson, L. 2002. Refuse Decomposition in the Presence and Absence of Leachate Recirculation, *Journal of Environmental Engineering*, 128(3): 228-236.

O'Leary, P., and Tansel, B. 1986. Landfill gas Movement, Control and Uses. *Waste Age,* page 104-115.

Pacey, J. and Augenstein, D. 1990. Modelling landfill Methane Generation. *Proceedings. International Landfill Gas Conference, Energy and Environment '90.* Energy Technology Support Unit. Department of the Environment. London, UK. Page 223-267.

Pawlowska, M. 2014. *Mitigation of Landfill Gas Emissions.* CRC Press, Taylor & Francis Group, Boca Raton, Florida.

Pichtel, J. 2007. *Waste Management Practices*, CRC Press, Taylor & Francis Group, Boca Raton, FL, 2007.

Speight, J. G. 2008. *Synthetic Fuels Handbook, Properties, Process, and Performance.* McGraw Hill Inc, New York.

Tchobanoglous, G., Theisen, H., and Vigil, S. 1993. *Integrated Solid Waste Management: Engineering Principles and Management Issues,* McGraw Hill, Inc., New York.

US EPA. 2005. *Landfill gas Emissions Model*, Version 3.02, User's Guide, EPA-600/R-05/047. United States Environmental Protection Agency, Office of Research and Development Washington, DC. http://www.epa.gov/ttncatc1/dir1/landgem-v302-guide.pdf) and http://www.epa.gov/landfill/res/handbook.htm.

Valdez-Vazquez, I., and Poggi-Varaldo, H. M. 2009. Hydrogen Production by Fermentative Consortia. *Renewable and Sustainable Energy Reviews*, 13: 1000-1013.

Warith, M. 2002. Bioreactor Landfills: Experimental and Field Results. *Waste Management,* 22: 7-17.

Xu, L., Lin, X., Amen, J., Welding, K., and McDermitt, D. 2014. Impact of Changes in Barometric Pressure on Landfill Methane Emission. *Global Biogeochemical Cycles,* 28(7): 679-695.

Yeşiller, N., Hanson, J. L., and Liu, W. L. 2005. Heat generation in municipal solid waste landfills. *J. Geotech. Geoenviron. Eng.*, 131(11): 1330-1344.

Young, A. 1990. Volumetric Changes in Landfill Gas flux in Response to Variations in Atmospheric Pressure. *Waste Management and Research,* 8: 379-385.

Young, A., Latham, B. and Graham, G. 1993. Atmospheric Pressure Effects on Gas Migration. *Wastes Management*. Page 44-46.

Young, P G. and Parker, A. 1984. *Vapors, Odors and Toxic Gases from Landfills. Proceedings. Hazardous and Industrial Waste Management and Testing: Third Symposium.* American Society for Testing and Materials (ASTM) Special Technical Publication No. 851. American Society for Testing and Materials, Philadelphia, Pennsylvania. Now: ASTM International, West Conshohocken, Pennsylvania.

Chapter 5

BIOGAS BY PYROLYSIS

1. INTRODUCTION

Biomass is one of the first sources of energy used by mankind and is still the major source of energy in developing countries. Charcoal has been the major commercial product of biomass pyrolysis for a long time and is the largest single biofuel produced today. Both in historical times, and in developing countries today, relatively simple charcoal kilns have provided charcoal for use as a high quality fuel and as a reducing agent in the winning of ores. The Iron Age was characterized by the use of charcoal, and in medieval times the increasing industrialization of society resulted first, in the over consumption of wood for charcoal production for use in iron production, and then charcoal was replaced with coke produced from coal.

The actual extent of the Iron Age is difficult to define with any degree of accuracy – by convention, the Iron Age in the Ancient Near East is taken to last from c. 1200 BC (the end of the Bronze Age) to approximately 550 BC (or the end of the proto-historical period). In Central and Western Europe, the Iron Age is taken to last from approximately 800 BC to approximately 1 BC, in Northern Europe from approximately 500 BC to approximately 800 AD, in China from approximately 500 BC to approximately 100 BC during which time ferrous metallurgy was present even if not dominant.

Whatever the approximately date of the extent of the Iron Age, in the western world, a renewed interest in biomass started in the nineteen-seventies. Charcoal, which is a smokeless fuel used for heating purpose, has been produced from wood biomass for thousands of years. An early use of biomass as a source of energy can be dated to the Iron Age when charcoal was used in ore melting to produce iron.

More than 5500 years ago in Southern Europe and the Middle East, pyrolysis technology was used for charcoal production (Antal and Grönli, 2003). Pyrolysis has also been used to produce tar for caulking boats and certain embalming agents in ancient Egyptian (Mohan et al., 2006). Since then, use of pyrolysis processes has been increasing and is widely used for charcoal and coke production. This is because only the burning of charcoal allowed the necessary temperatures to be reached to melt tin with copper to produce bronze. Pyrolysis technology has the capability to produce bio-fuel with high fuel-to-feed ratios. Therefore, pyrolysis has been receiving more attention as an efficient method in converting biomass into bio-fuel during recent decades (Demirbaş, 2002). The ultimate goal of this technology is to produce high-value bio-oil for competing with and eventually replacing non-renewable fossil fuels.

The disadvantages of early pyrolysis technology include slow production, low energy yield and excessive air pollution. However, of late, pyrolysis has become of major interest due to the flexibility in operation, versatility of the technology and adaptability to a wide variety of feedstocks and products. Pyrolysis operates in anaerobic conditions where the constituents of biomass are thermally cracked to gases and vapours which usually undergo secondary reactions thereby giving a broad spectrum of products. There are a number of conditions and circumstances that have a major impact on the products and the process performance. These include feedstock, technology, reaction temperature, additives, catalysts, hot vapor residence time, solids residence time, and pressure.

Biomass is now recognized as a renewable resource for energy production and is abundantly available around the world and is receiving increased attention due to the real potential for the depletion of conventional (fossil fuel) energy sources as well as stricter environmental regulations Chapter 1) (Zhang et al., 2007). Although complex in nature, biomass is not a stable energy source in terms of composition and properties and contains varying amounts small amount of sulfur, nitrogen, and mineral matter (which is manifested as ash after thermal processing (Speight, 2008, 2011).

As a result of the variability in composition of biomass, combustion of biomass as fuel (like fossil fuel products) produces gas emissions such as nitrogen oxides (NOx), sulfur dioxide (SO_2), and particulate matter (soot). However, one of the advantages of using biomass as a fuel is the so-called zero or negative carbon dioxide (CO_2) emissions that is possible (from theoretical calculations) from biomass combustion because released carbon dioxide from the combustion process can be recycled into the plant by photosynthesis (Tsai et al., 2007).

Biomass can also be converted into a variety of bio-fuels by way of different thermal, physical, and biological processes. Among the biomass-to-energy (BTE) conversion processes, pyrolysis of the bio-oil (the liquid product from) thermal conversion processes and gasification of the biochar (the solid products from thermal conversion processes) has attracted more interest in producing fuel product because of the advantages in storing the oil and char, transport, and versatility of the bio-oil in applications such as combustion engines, boilers, and turbines. Not surprisingly, the bio-liquids and other products (char and gas) by pyrolysis of different types of biomass – such as beech wood, woody biomass, straws, seedcakes, and municipal solid waste (MSW) – has been extensively investigated

(Karaosmanoglu and Tetik, 1999; Jensen et al., 2001; Pütün et al., 2006; Asadullah et al, 2007; Aho et al., 2008).

The pyrolysis process is not new to industrial operations and is commonly used to the treatment of organic feedstocks. In general, the pyrolysis of organic feedstocks produces volatile products and leaves a solid residue enriched in carbon (the char). If the pyrolysis temperature parameters involve high temperature (>500°F, >930°F), the predominant product (to the exclusion of liquid products) is the char. In the coal industry, this process is frequently referred to as carbonization (Speight, 2013). The process is also used during crude oil refining to produce ethylene, various forms of carbon, and other chemicals or chemical intermediates which serves as feedstock to the petrochemical industry (Speight, 2019).

In a modern setting, there are mainly three ways frequently used to extract energy from biomass and these are (i) combustion, which is an exothermic process (ii) gasification, which is an exothermic process, and (iii) pyrolysis, which is an endothermic process. Combustion is the oxidation of fuel in which biomass can be completely oxidized and transferred into heat. Gasification is a partly oxidizing process that converts a variety of solid fuels (such as coal, petroleum residua, and biomass) into a gaseous fuel, while pyrolysis is the first stage of both combustion and gasification processes (Speight, 2013, 2014a, 2014b). Therefore pyrolysis is not only an independent conversion technology, but also a part of gasification and combustion, which consists of a thermal degradation of the initial solid fuel into gases and liquids without an oxidizing agent.

However, in the current context, application of the pyrolysis process would convert biomass to bio-oil and then to biogas and in the context of a later chapter (Chapter 6), into synthesis gas (often called bio-synthesis gas or bio-syngas). Thus, the focus of this chapter is to present an examination of the production of bio-oil and biochar and the ways by which bio-oil can be converted to biogas by the use of pyrolytic methods, especially when the product balance involved the production of high amounts of biogas from biomass feedstocks.

2. Pyrolysis Chemistry

Pyrolysis is a chemical reaction process that is represented by the chemical changes that occur when heat is applied to an organic feedstock in the absence of oxygen. In the current context, pyrolysis is the thermal decomposition of biomass occurring in the absence of oxygen. It is the fundamental chemical reaction that is the precursor of both the combustion and gasification processes and occurs naturally in the first two seconds. The products of biomass pyrolysis include (i) gas, often referred to as biogas because of the portion of the gas from a biomass sources, which includes methane, hydrogen, carbon monoxide, and carbon dioxide, (ii) bio-oil, and (iii) biochar. Depending on the thermal environment and the final temperature, pyrolysis will yield mainly biochar at low temperatures, less than 450°C (840°F), when the heating rate is quite slow, and mainly gases at high temperatures (>800°C, >1470°F), with rapid heating rates. At an intermediate temperature and under relatively high heating rates, the main product is bio-oil – the yield and properties of the bio-oil are dependent upon the feedstock (Table 5.1).

Biomass pyrolysis provides product options such as fuels or green chemicals. In the process, lignocellulosic materials undergo thermal degradation in the absence of oxidative environment at atmospheric pressure during a definite residence time, which produces biogas, bio-oil, and biochar. If the bio-oil is going to be used as a fuel source or to be processed for producing chemicals, it requires an upgrading of which the catalytic pyrolysis is the most promising method.

Generally, biomass is usually made up of three main components – cellulose, hemicellulose and lignin with water and ash. Cellulose (Figure 5.1) is a polymer of glucose, a six-carbon molecule, which can be thermally and catalytically cracked to monomers and decomposition products. Hemicellulose (Figure 5.2) is a polymer of five-carbon rings that can also be cracked to smaller organic molecules. Lignin (Figure 5.3) is a complex polymer built up of phenolic units that can be cracked to a wide range of phenolic products – the actual structure is not known with any degree of certainty. Other components of biomass include water (up to 60% w/w in freshly grown biomass) and mineral matter that occurs as mineral ash during and after thermal processing. The mineral ash is predominantly alkali metals from nutrients, which is catalytically active and causes cracking of the organic constituents. This is beneficial in gasification (Chapter 6) where they help to crack tars, but not beneficial in pyrolysis where they crack the organics in the vapor resulting in lower liquid yields with an adverse effect on liquid properties.

The alkali metals that form ash are significant in fast pyrolysis and the most active alkali metal potassium followed by sodium and calcium. The alkali metals act by causing secondary cracking of low-boiling products thereby reducing yield and quality of the bio-oil. When gas is the main objective of the process, the presence of these metals can be a benefit. The vast majority of these alkali metals end up in the char which results in the char byproduct acting as a cracking catalyst thus requiring rapid and effective removal of char within the fast pyrolysis process. The mineral matter in the feedstock can be reduced by water-washing or dilute acid-washing but using these methods to the extreme can cause loss of hemicellulose and cellulose through hydrolysis (Banks and Bridgwater, 2016).

The products of biomass pyrolysis include gas (including methane, hydrogen, carbon monoxide, and carbon dioxide), water, oil, tar products, and a carbonaceous solid). While the bio-oil can be analyzed and many of the constituents identified by the application of a suite of analytical test methods (ASTM, 2019). The chemical composition of the tar product is unknown and difficult to define. Physically, tar is a dark brown or black viscous liquid containing various aromatic constituents (especially polynuclear aromatic constituents) that is obtained from a wide variety of organic feedstocks through destructive distillation. By way further clarification, tar may also be composed to two fractions, often referred to as tar liquid and pitch. Treatment of the distillate (boiling up to 250°C, 480°F) of the tar with caustic soda causes separation of a fraction known as *tar acids*; acid treatment of the distillate produces a variety of organic nitrogen compounds known as *tar bases*. The residue left following removal of the distillate, is *pitch*, a black, hard, and highly ductile material. In the chemical-process industries, pitch is the black or dark brown residue obtained by distilling coal tar and biomass tar.

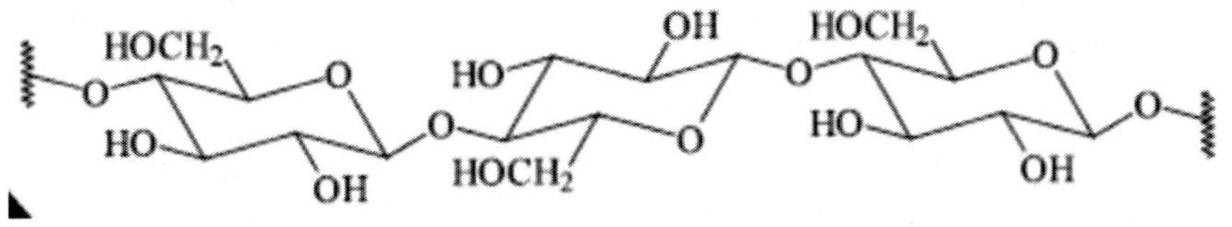

Figure 5.1. Generalized Structure of Cellulose.

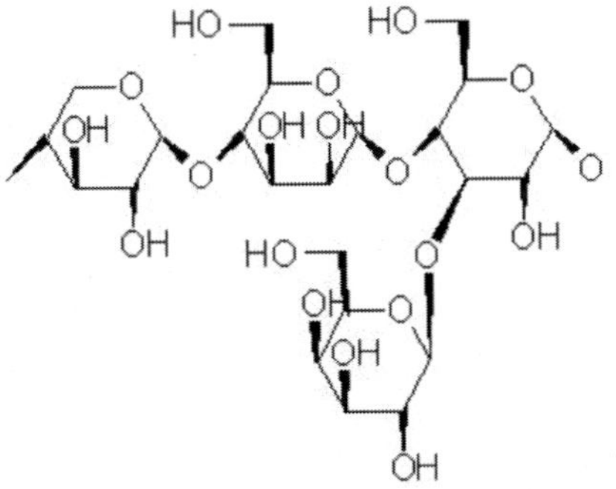

Figure 5.2. Representation of a Portion of the Hemicellulose Macromolecule.

The nature of the changes in pyrolysis depend on the character of the biomass feedstock being pyrolyzed, the final temperature of the pyrolysis process and the rate of heating to the maximum temperature. As typical lignocellulosic biomass feedstocks such as wood, straws, and stalks are poor heat conductors, management of the rate of heating requires that the size of the particles being heated be quite small. Otherwise, in massive feedstocks such as wood logs, the heating rate is very slow, and this determines the yield of pyrolysis products.

Pyrolysis can be performed at relatively small scale and at remote locations which enhance energy density of the biomass resource and reduce transport and handling costs. Heat transfer is a critical area in pyrolysis as the pyrolysis process is endothermic and sufficient heat transfer surface has to be provided to meet process heat needs. Pyrolysis offers a flexible and attractive way of converting solid biomass into an easily stored and transported liquid, which can be successfully used for the production of heat, power and chemicals.

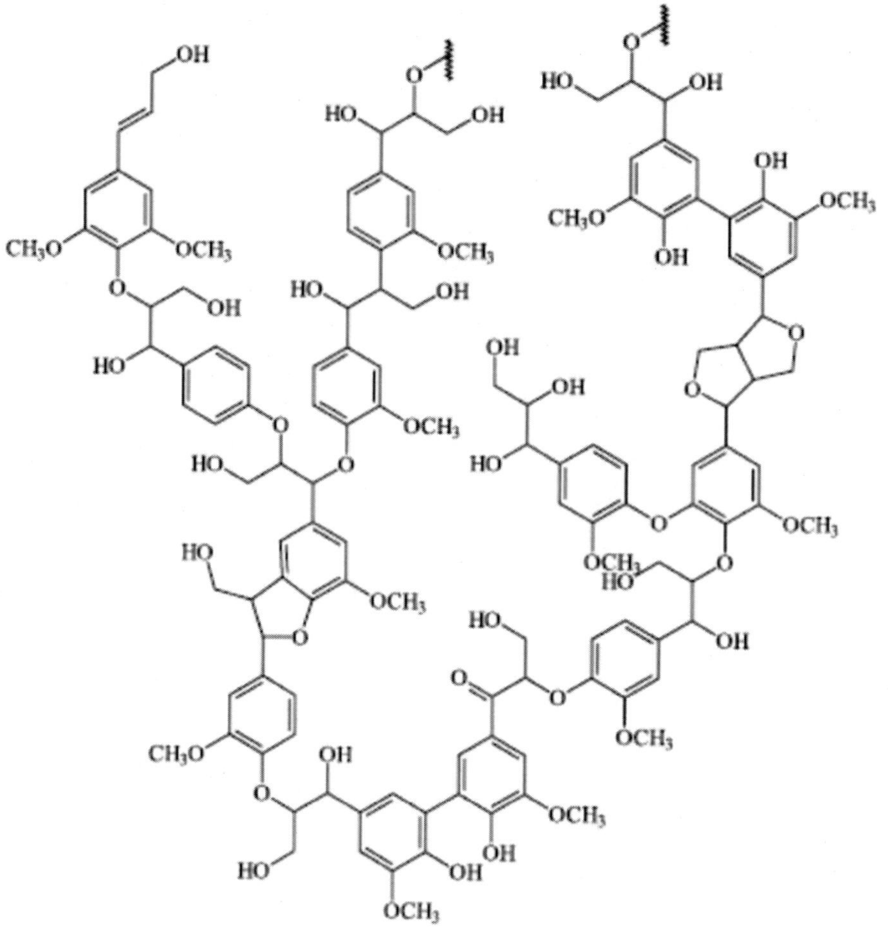

Figure 5.3. Hypothetical Structure of Lignin to Illustrate the Complexity of the Molecule.

The process of pyrolysis of organic matter is a complex process that consists of both simultaneous and successive reactions when organic feedstock is heated in a non-reactive atmosphere. In this process; thermal decomposition of organic components in biomass starts at 350 to 550°C (650 to 1000°F) and may require temperatures as high as 700 to 800°C (1290 to 1470°F) in the absence of air/oxygen (Fisher et al., 2002). The long chains of carbon, hydrogen and oxygen compounds in biomass break down into lower molecular weight products in the form of gases, condensable vapors (oil and tar) and solid charcoal under pyrolysis conditions (Table 5.2). The reaction rate and the extent of the decomposition of the feedstock components depends on the process parameters such as: (i) reactor temperature, (ii) heating rate of the feedstock, (iii) pressure, (iv) reactor configuration, and (v) the composition of the feedstock (Table 5.3) which is in keeping with observations made on a variety of processes and feedstocks (Fegbemi et al., 2001; Demirbaş, 2007). In terms of the mechanism of the pyrolysis, the pyrolysis process cannot be limited to a single reaction path because of widely varying structure and compositional properties of biomass (Demirbaş, 2000b; Speight, 2008, 2011).

Table 5.1. Properties of Bio-oil from Various Wood Feedstocks

Property	Birch	Pine	Poplar	Various
Carbon, % w/w	44.0	45.7	48.1	3249
Hydrogen, % w/w	6.9	7.0	5.3	6.9-8.6
Nitrogen % w/w	<0.1	<0.1	0.14	0.0-0.2
Sulfur, % w/w	0.00	0.02	0.04	0.0-0.05
Oxygen, % w/w	49.0	47.0	46.1	44-60
Sodium plus potassium, ppm, w/w	29	22	2	5-500
Calcium, ppm w/w	50	23	1	4-600
Magnesium, ppm w/w	12	5	0.7	2-11
Ash, % w/w	0.004	0.03	0.007	0.004-0.3
Conradson carbon residue, % w/w	20	16	18	14-23
Density	1.25	1.24	1.20	1.2-1.3
Viscosity, cSt @ 50°C (122°F)	28	28	13.5	13-80
Flash Point, °C (°F)	62 (144)	95 (203)	64 (147)	50-100 (122-212)

Table 5.2. Representation of the Reaction Paths Occurring During Biomass Pyrolysis

Phase:	Primary phase	Secondary phase	Final phase
Temperature, °C	450-550	400-500	variable
Temperature, °F	840-1020	750-930	variable
Reactions	Decomposition	Cracking and condensation	Condensation
Gaseous products	Carbon monoxide	Carbon monoxide	Carbon monoxide
	Carbon dioxide	Carbon dioxide	Carbon dioxide
	Methane	Methane	Methane
Liquid products		Bio-oil	Bio-oil
Solid products		Char	Char

Table 5.3. Range of Operating Parameters and Product Yields for Pyrolysis Processes

Pyrolysis process	Residence time (seconds)	Heating rate (degrees/second)	Particle size (mm)	Temp (°C)	Product	Yield % w/w
Slow	450-550	0.1–1	5-50	675	30	35
Fast	0.5-10	10–200	<1	575-975	50	20
Flash	<0.5	>1000	<0.2	775-1025	75	12

Most of the constituents of biomass have a cellular structure with extensive voids, such that while the density of the lignocellulosic cell wall feedstock typically is between 1.5 to 2 g cm^{-3} while more typical softwoods, such as pine, have values on the order of 0.4 g cm^{-3} and typical hardwoods, such as oak, the density is on the order of 0.6 g cm^{-3}. As a result, the heat transfer characteristics are very poor and, consequently, there are large temperature gradients within the solid wood feedstock. Thus, at high external heat fluxes with large particles of >2 cm thick, the surface rapidly reaches the external temperature, while the center of the particle is still cold. However, the net chemical result is that the primary products of pyrolysis have close contact with char and can react with the solid char matrix, modifying both it and the composition of the tars and gases. For thermally thick samples in high external heat flux regimes, the processes of drying and pyrolysis travel together as a wave through the feedstock. This pyrolysis wave is an exotherm and augments the rate of heat transfer through massive feedstocks.

3. Pyrolysis Classification

Pyrolysis is a thermochemical conversion process in the absence of an oxidizing agent and can be regarded as the initial stage of gasification and combustion. Solid char, liquid pyrolysis oil, and gas are the main products of biomass pyrolysis. The amount of products and their fractions are influenced by many factors such as heating rate, pyrolysis temperature, biomass composition, and catalyst effect. Since biomass pyrolysis is feedstock composition-dependent process, finding suitable catalysts to regulate pyrolysis processes is an alternative way to reduce overall energy consumption [15]. Also, catalytic pyrolysis of biomass provides, with optimum catalyst/biomass ratio, chemically more homogenous fractions of pyrolysis products.

Depending on the operating condition, pyrolysis can be classified into three main categories: (i) slow pyrolysis, (ii) fast pyrolysis, and (iii) flash pyrolysis which can be classified on the basis of temperature, residence time, and heating rate (Table 5.4). These processes differ in parameters such as (i) temperature, (ii) heating rate, (iii) feedstock residence time, and (iv) feedstock particle size. However, the product type and the relative distribution of the products (the product mix) is dependent on pyrolysis type and pyrolysis operating parameters.

In the process, the thermal decomposition of the biomass leading to fragmentation of the biomass constituents generates lower molecular weight products that, upon cooling, result in (i) gaseous products, (ii) condensable liquids such has bio-oil and some products classed as tars, and (iii) residual solids (bio-carbon or bio-char). The gas consists mainly of carbon monoxide, hydrogen and methane, with lower amounts of carbon dioxide and ethane. The bio-oil is a mixture of many chemicals, including acid derivatives, aldehyde derivatives, ketone derivatives, furfural derivatives, anhydrosugar derivatives, phenol derivatives, and water. The biochar consists of the mineral matter of the original biomass entrapped into a porous carbon structure.

Table 5.4. Classification of Pyrolysis Methods on the Basis of Temperature, Residence Time, and Heating Rate

Method	Temperature (°C)	Residence Time	Heating rate (°C/s)
Conventional/slow pyrolysis	Med-high 400-500	Long 5-30 min	Low 10
Fast pyrolysis	Med-high 400-650	Short 0.5-2 s	High 100
Ultra-fast/flash pyrolysis	High 700-1000	Very short < 0.5 s	Very high >500

However, the slow and fast pyrolysis generate solid bio-carbon products with different characteristics, even when produced from the same raw biomass material. The most significant differences include (i) the evolution of the specific surface area resulting from the development of a porous structure during the pyrolysis process, and (ii) the average pore size and pore size distribution, which is the fraction of micro-pores and meso/macro-pores).

3.1. Slow Pyrolysis

The slow pyrolysis process is an example of the pyrolysis wave phenomena in massive biomass, such that the only requirement is for the initiation of the process in which a small proportion of the wood charge is burnt to provide the initial heat in a well-insulated reactor.

In this process, the vapor residence time is high (5 min to 30 min) and components in the vapor phase continue to react with each other which results in the formation of solid char and other liquids (Bridgwater et al., 2001; Bridgwater, 2004). However, slow pyrolysis has some technological limitations which made it unlikely to be suitable for good quality bio-oil production. Cracking of the primary product in the slow pyrolysis process occurs due to high residence time and could adversely affect bio-oil yield and quality. For example, slow pyrolysis at low to moderate temperatures (approximately 300°C, 570°F) and long reaction times (up to days) has been used for thousands of years for the conversion of wood into high yields of charcoal (bio-carbon). The slow pyrolysis process generates also lower yields of bio-oil and gaseous products.

3.2. Fast Pyrolysis

In the past several decades, fast pyrolysis (not to be confused with flash pyrolysis, see Section 3.3, below) is carried out at intermediate temperatures (around 500°C, 930°F) and very short reaction times (1 to 5 seconds) has become of considerable as a method for producing higher yields of bio-oil (normally around 65% w/w) with significantly higher energy density than the original biomass, in addition to bio-carbon (20%) and gas (15%). Depending on the pyrolysis process and on the biomass material utilized, both the yields as well as the physical and chemical characteristics of the products, and consequently, the performance of these reactors varies considerably (Bridgwater and Peacock, 1999).

The essential features of a fast pyrolysis process are: (i) very high heating and heat transfer rates, which often require a finely ground biomass feed, (ii) a carefully controlled reaction temperature of around 500°C (930°F) in the vapor phase and residence time of pyrolysis vapors in the reactor less than 1 second, and (iii) quenching – i.e., rapid cooling – of the pyrolysis vapor to yield the bio-oil product. Typically the majority of the feedstock is converted to a liquid that contains about 10% w/w water. At 455°C (850°F) the yield of liquid can be as high as 90% w/w, and as the temperature is increased to 900°C (1650°F) , there is hardly any solid char and the liquid is converted into gas.

Thus, in the fast pyrolysis process, biomass is rapidly heated to a high temperature in the absence of oxygen. Typically on a weight basis, fast pyrolysis produces 10 to 20% w/w of gases phase products, 60 to 75% w/w of oily products (which may include some tar) with 15 to 25% w/w of solids (mainly biochar) and which, like many processes and the variety of feedstock employed, is dependent on the composition and properties of the feedstock (Fegbemi et al., 2001). The basic characteristics of the fast pyrolysis process are high heat transfer and heating rate, very short vapor residence time, rapid cooling of vapors and aerosol for high bio-oil yield and precision control of reaction temperature (Demirbas and Arin, 2002; Czernik and Bridgwater, 2004; Dobele et al., 2007).

Fast-pyrolysis technology is can be used to produce liquid fuels and a range of specialty and commodity chemicals and the liquid product can be transported and stored relatively easily , thereby de-coupling the handling of solid biomass from utilization (Brammer et al., 2006). The process also has potential to supply a number of valuable chemicals that offer the attraction of much higher added value than fuels. In fact, the production of bio-oil through fast pyrolysis has received more attention in recent years due to the following potential advantages: (i) renewable fuel for boiler, engine, turbine, power generation and industrial processes, (ii) utilization of second generation bio-oil feedstocks and waste feedstocks, such as forest residue, municipal solid waste, and industrial waste, (iii) secondary conversion to motor-fuels, additives or special chemicals, and (primary separation of the sugar and lignin fractions in biomass with subsequent further upgrading (Czernik and Bridgwater, 2004; Chiaramonti et al., 2007; Speight, 2008, 2011; Lee et al., 2014).

Though often included in discussions of fast pyrolysis by virtue of its high liquid yields, vacuum pyrolysis is not a rapid heating technique and is in the same thermal regime of time and temperature as slow pyrolysis charcoal production. However, yields of liquid that are over 50% w/w of the original biomass are achieved by removing the vapors as soon as they are formed by operating under a partial vacuum – essentially the converse of work at high pressures, in which the liquids are held in the charring mass to increase the yield of char.

Using the fast pyrolysis technique, biomass decomposes very quickly to generate gas, liquid vapor and chars. After cooling and condensation, a dark brown homogenous mobile liquid is formed if wood or a low ash feed is used. The liquid, referred to as bio-oil (ASTM D7544), can be obtained in high yield of liquid is obtained from low mineral matter biomass. The essential features of a fast pyrolysis process for producing liquids are: (i) a low-moisture feedstock (<10% w/w) since all the feed water reports to the liquid phase along with water from the pyrolysis reactions, (ii) a carefully controlled fast pyrolysis reaction temperature of on the order of 500°C (930°F) for most biomass gives the high liquid yield.

3.3. Flash Pyrolysis

Flash pyrolysis (sometime referred to as ultra-fast pyrolysis) is an extremely rapid thermal decomposition pyrolysis, with a high heating rate that can vary from 100 to 10,000°C per second (180 to 18,000°F per second) with a residence time that is short in duration (Scott and Piskorz, 1984). Flash pyrolysis is an irreversible thermo-chemical process in which organic material is rapidly heated in the absence of oxygen, whereby the material is decomposed and can be separated into distinct fractions of tar, char, and gas. Ash is largely retained in the char, whereas the tar is a homogeneous mixture of organics/water commonly referred to as bio-oil. The main products from biomass feedstocks are (i) gases: 60 to 80% w/w, (ii) bio-oil: 10 to 20% w/w, and (iii) char: 10 to 15% w/w.

The flash pyrolysis of biomass is a promising process for the production of solid, liquid and gaseous fuel from biomass which can achieve up to 75% of bio-oil yield (Demirbaş, 2000a). This process can be characterized by rapid devolatilization in an inert atmosphere, high heating rate of the particles, high reaction temperatures between 450 °C and 1000 °C and very short gas residence time (less than 1 second). However this process has some technological limitations, for instance: poor thermal stability and corrosiveness of the oil, solids in the oil, Increase of the viscosity over time by catalytic action of char, alkali concentrated in the char dissolves in the oil and production of pyrolytic water (Cornelissen et al., 2008.).

The main issues for flash pyrolysis are the quality and stability of the produced oil, both of which strongly influenced by the char/ash content of the bio-oil. Char fines in the oil can catalyze the repolymerization reactions inside the oil resulting in a higher viscosity. The char can be removed by post-treatment of the condensed products, such as by filtration of the oil. The disadvantage of this is that the alkali, concentrated in the char, will dissolve in the bio-oil because of the high acidity of the oil (pH = 2 to 3). Another option to remove the char is hot gas cleaning of the oil vapor, but this will result in lower oil yield if the oil undergoes further cracking on the char.

4. Process Technology

The pyrolysis of biomass to produce charcoal from biomass (wood is the best example of the biomass feedstock) that can be used to produce pig iron and in metallurgical applications is still used in many parts of the world. Additional charcoal is derived from wood that is not from forests and from non-woody biomass. Non-woody biomass is biomass that has a lower lignin content than woody materials, is a common waste material found in agricultural processing plants and fields. Non-woody biomass is often bulky and has a comparatively low energy content and due to the heterogeneous nature of non-woody biomass, it is often critical to apply suitable pretreatment prior to use as a feedstock for thermal processes (pyrolysis and gasification).

As heat is applied to the biomass, the chemical bonds become thermally activated and eventually some bonds break. In cellulose the bonds are broken at random locations along the chain composed of thousands of glucan moieties. Thus, as heat continues to be applied, the macromolecule decomposes to lower molecular weight products which can (at typical pyrolysis temperatures between 400 and 600°C, 750 and 1110°F) evaporate from the solid mass. These small fragments (in the case of cellulose, they are called *anhydro sugars*) and if these volatile anhydro sugars are not quickly removed from the high temperatures of the pyrolysis, they will also undergo thermal fragmentation, producing highly reactive small intermediates. These intermediates will, in turn, if not removed from the original solid feedstock (as in vacuum pyrolysis), undergo chemical reactions with the remaining solid feedstocks and may result in the creation of new higher molecular weight species or accelerate the breakdown of the original chains. These exothermic reactions, unlike the endothermic chain-breaking reactions, can result in the propagation of a thermal wave that accelerates the overall pyrolysis reaction. The removal of the products of pyrolysis and quenching them by sweeping them from the pyrolysis reactor into a cold zone (a condensation zone), results in their capture for use as chemicals or fuels.

Hemicellulose rapidly loses the side chains, which are often acetyl groups that are condensed as acetic acid. Similarly lignin, which has methoxy substituents on the majority of the phenyl-propane monomer units, leads to the production of methanol. Prior to the advent of petrochemical synthesis, wood pyrolysis was a major source of acetic acid, acetone, and methanol; the latter, as a result, is still known as wood alcohol. The volatile feedstocks also undergo thermal rearrangements according to the temperature and the duration of exposure to that temperature.

Charcoal has value not only as fuel and metallurgical reducing agent, but is also the base for the production of special carbons in gas and liquid absorption applications. The product known as activated charcoal is charcoal that has been chemically treated at high temperature to produce an almost pure carbon product with a very high surface area. The absorptive capacity as a result of the large surface areas (e.g., $10^3 m^2 g^{-1}$) is extremely large and can trap a wide range of toxic substances and vapors.

Thus, depending on the thermal environment and the final temperature, pyrolysis will yield mainly biochar at low temperatures, less than 450°C (840°F), when the heating rate is quite slow, and mainly gases at high temperatures, greater than 800°C (1470°F), with rapid heating rates. At an intermediate temperature and under relatively high heating rates, the main product is bio-oil. The process can be performed at relatively small scale and at remote locations which enhance energy density of the biomass resource and reduce transport and handling costs. Heat transfer is a critical area in pyrolysis as the pyrolysis process is endothermic and sufficient heat transfer surface has to be provided to meet process heat needs. In fact, pyrolysis offers a flexible and attractive way of converting solid biomass into an easily stored and transported liquid, which can be successfully used for the production of heat, power and chemicals.

A wide range of biomass feedstocks can be used in pyrolysis processes and the process is very dependent on the moisture content of the feedstock, which is typically held at approximately 10% w/w. At a higher moisture content, a high level of water is produced and at a lower moisture content there is a risk that the process only produces dust instead of oil. Thus, a feedstock with a high moisture content, such as sludge and some food processing wastes, require drying before being used as a feedstock for the process. Also, the efficiency and nature of the pyrolysis process is dependent on the particle size of the feedstock. Most of the pyrolysis technologies can only process small particles (on the order of 2 mm or less) because of the need for rapid heat transfer through the particle. This criterion (i.e., the need for small particle size) is an indicator that the feedstock has to be pretreated in size-reduction equipment before being used in the process.

Ash management is another aspect of the process that needs attention (Bridgwater, 2018a, 2018b). Ash is the product of the thermal processing of mineral matter originally in the feedstock – for convenience here it will be referred to as mineral matter – which can be managed to some extent by selection of crops and harvesting time especially with rhizome crops such as Miscanthus (a genus of African, Eurasian, and Pacific Island plants in the grass family) which ages over winter with alkali metals returning to the rhizome, however it cannot be eliminated from growing biomass. Mineral matter in the feedstock can be reduced by washing in water or dilute acid or treatment with surfactants and the more extreme the conditions in temperature or concentration respectively, the more complete the removal of the mineral mater. However as washing conditions become more extreme, hemicellulose and then cellulose can be lost through hydrolysis which serves to reduce the yield of the bio-oil and the quality of the bio-oil. In addition, it is essential that the washed biomass needs to have any acid removed as completely as possible and recovered or disposed of and the wet biomass has to be dried. In some cases, washing may not be considered to be a viable pretreatment method unless the washing process is necessary to remove contaminants. Another consequence of the high mineral matter removal is the increased production of levoglucosan and levoglucosenone which can reach levels in bio-oil where recovery becomes an interesting proposition.

Levoglucosan ($C_6H_{10}O_5$) is an organic compound with a six-carbon ring structure formed from the pyrolysis of carbohydrates, such as starch and cellulose. The hydrolysis of levoglucosan generates the fermentable sugar glucose.

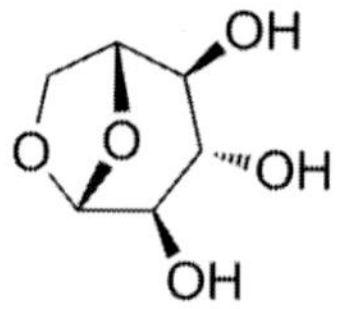

Levoglucosan

Levoglucosenone is a bridged, unsaturated heterocyclic ketone formed from levoglucosan by loss of two molecules of water – it is an anhydrosugar (Lakshmanan and Hoelscher, 1970). It is the major component produced during the acid-catalysed pyrolysis of cellulose glucose and levoglucosan.

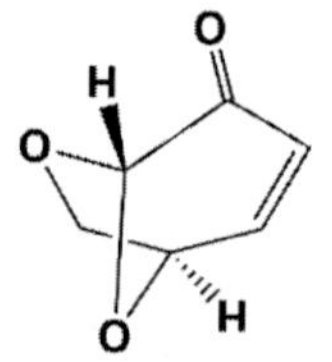

Levoglucosenone

The primary method for obtaining levoglucosenone is by the pyrolysis of carbohydrates, particularly cellulose (Halpern et el., 1973). Levoglucosenone can be derived from biomass or from other cellulosic materials including domestic/commercial waste paper. The availability of multiple sources for this compound is a key advantage when compared to other platform chemicals which are solely derived from biomass.

5. REACTORS

A chemical reactor is an enclosed volume in which a chemical reaction takes place and it is generally understood to be a process vessel used to carry out a chemical reaction. The design of a chemical reactor is to maximize the efficiency of a reaction. The reactor design must ensure that the reaction proceeds with the highest efficiency towards the desired output product, producing the highest yield of product while requiring the least amount of money to purchase and operate. Energy changes can come in the form of heating or cooling, pumping to increase pressure, frictional pressure loss or agitation.

The reactor is the heart of any pyrolysis process. Reactors have been the subject of considerable research, innovation and development to improve the essential characteristics of high heating rates, moderate temperatures and short vapor product residence times for liquids.

The reactors, in which chemicals are made in industry, vary in size from a few cubic centimeters to vast structures. For example, some reactors may be over 80 feet high and capable of holding, at any one time, several hundred tons of materials. The design of the reactor is determined by many factors but of particular importance are the thermodynamics and kinetics of the chemical reactions being carried out.

Initially, pyrolysis reactor developers had assumed that small biomass particles size (less than 1 mm) and very short residence time would achieve high bio-oil yield, however later research has found different results. With the continuation of pyrolysis technology development, a number of reactor designs have been explored to optimize the pyrolysis performance and to produce high quality bio-oil. However, each reactor type has specific characteristics, bio-oil yielding capacity, advantages and limitations (Bridgwater and Peacock, 1999). Of the various reactor designs, the most popular types are described in the following sub-sections. However, even though descriptions of the various reactors are presented below, not all of these reactors are in use on a commercial basis. The inclusion and descriptions of these various units is to give the reader a better presentation of the types of reactors that are available for use in the biomass industry.

5.1. Fixed Bed Reactor

The fixed bed pyrolysis system consists of a reactor with a gas cooling and cleaning system. The technology of the fixed bed reactor is simple, reliable and proven for fuels that are relatively uniform in size and have a low content of fines (Yang et al., 2006). In this type of reactor, the solids move down a vertical shaft and contact a counter-current upward moving product gas stream. Typically, a fixed bed reactor is made up of firebricks, steel or concrete with a fuel feeding unit, an ash removal unit and a gas exit.

The fixed bed reactors generally operate with high carbon conservation, long solid residence time, low gas velocity and low ash carry over. These types of reactor are being considered for small scale heat and power applications. The cooling system and gas cleaning consists of filtration through cyclone, wet scrubbers and dry filters. The major problem of fixed bed reactors is tar removal; however recent progress in thermal and catalytic conversion of tar has given viable options for removing tar.

5.2. Fluidized-Bed Reactor

A fluidized bed is a physical phenomenon occurring when a quantity of the solid particulate matter that forms the bed in the reactor vessel is placed under appropriate conditions to cause a solid/fluid mixture to behave as a fluid. This is usually achieved by the introduction of pressurized fluid through the particulate the solid particulate matter and, as a consequence, the bed has many of the properties that are characteristic of a fluid, such as the ability to free-flow under gravity, or to be pumped using fluid type technologies. The resulting phenomenon (fluidization) has found use in several industries, most notably the petroleum refining industry where fluidized bed reactors are common-place (Parkash, 2003; Gary et al., 2007; Speight 2014a; Hsu and Robinson, 2017; Speight, 2017).

In the fluidized-bed reactor, the solid substrate (the catalytic material upon which chemical species react) is typically supported by a porous plate (the distributor). The fluid is then forced through the distributor up through the solid material. At lower fluid velocities, the solids remain in place as the fluid passes through the voids in the material (the packed-bed reactor). As the fluid velocity is increased, the force of the fluid on the solids is enough to balance the weight of the solid material. This stage (the incipient fluidization) occurs at this minimum fluidization velocity and once this minimum velocity is exceeded, the contents of the reactor bed begin to expand and move and the reactor is, at this point, a fluidized-bed reactor. Depending on the operating conditions and properties of solid phase various flow regimes can be observed in this reactor.

Fluidized-bed reactors appear to be popular for fast pyrolysis as they provide rapid heat transfer, good control for pyrolysis reaction and vapor residence time, extensive high surface area contact between fluid and solid per unit bed volume, good thermal transport inside the system and high relative velocity between the fluid and solid phase (Wang et al., 2005). Different types of fluidized-bed reactors are described below.

5.2.1. Bubbling Fluidized-Bed Reactor

Bubbling fluidized-beds are simple to construct and operate insofar as they provide better temperature control, solids-to-gas contact, heat transfer, and storage capacity because of the high solids density in the bed. Heated sand is used as the solid phase of the bed which rapidly heats the biomass in a non-oxygen environment, where it is decomposed into char, vapor, gas and aerosols. After the pyrolytic reaction, the charcoal is removed by a cyclone separator and stored.

The remaining vapor is then rapidly cooled with a quenching system, condensed into bio-oil and stored. Bubbling fluidized-bed pyrolysis is very popular because it produces high quality bio-oil and liquid yield is approximately 70 to 75% w/w of the biomass on a dry basis. Char does not accumulate in the fluidized bed, but it is rapidly separated. The residence time of solids and vapor is controlled by the fluidizing flow rate. One important feature of bubbling fluidizing bed reactors is that they need small biomass particle sizes (less than 2–3 mm) to achieve high biomass heating rates.

The bubbling fluidized bed reactor has been used in the petroleum industry and in the chemical processing for more than six decades and, therefore, has a long operating history. As reactor designs, they are characterized as providing high heat transfer rates in conjunction with uniform bed temperatures, both being necessary attributes for fast pyrolysis. By selecting the appropriate size for the bed fluidizing media, the gas flow rate can be established such that gas/vapor residence time in the freeboard section above the bed can be set to a desired value, generally between 0.5 to 2.0 seconds. An operating temperature of on the order of 500 to 550°C (930 to 1000°F) in the bed is often sufficient to produce the optimal yield of liquid at (approximately) 0.5 sec residence time. The temperatures may also vary depending on the type of biomass being processed.

In principle, the bubbling bed is self-cleaning insofar as the byproduct char is carried out of the reactor with the product gases and liquid vapors. However, in practice this requires using carefully sized feedstock with a relatively narrow particle size distribution. If the biomass particles are too large, the remaining char particles (after pyrolysis) may have too much mass to be effectively entrained out of the reactor with the carrier gas and product vapors. Also, the density of this char will typically be less than that of the fluidizing media and, consequently, this char will be on top of the bed – the char is sometime said to float on top of the bed. In this location, the char will not experience enough turbulence with the bed media to undergo attrition into smaller particles that will eventually leave the reactor. Another issue with having the char on top of the bed is that there will be a catalytic effect on the on the vapors as they pass through the char on the way out of the bed. This can affect the yields and the chemical nature of the resulting liquid product. On the other hand, if fines are present in the feed, then the feed must be introduced lower in the bed otherwise the fines will be quickly entrained out of the bed before complete pyrolysis can occur.

Thus, some design considerations in bubbling fluidized bed systems include (i) heat can be applied to the fluid bed in a number of different ways that offer flexibility for a given process, (ii) the residence time of the vapor is controlled by the carrier gas flow rate, (iii) the particle size of the biomass feedstock should be on the order of 2 to 3 mm or less, (iv) the char can catalyze cracking reactions so it needs to be removed from the bed quickly, and (v) the char can accumulate on top of the bed if the biomass feed is not sized properly, provisions for removing this char may be necessary.

In summary, bubbling fluid beds, which have found wide use in the crude oil refining industry (Parkash, 2003; Gary et al., 2007; Speight 2014a; Hsu and Robinson, 2017; Speight, 2017) offer good temperature control and very efficient heat transfer to biomass particles arising from the high solids density. In terms of biomass operations, the usual fluidizing medium is sand, but catalysts are also being considered as the fluidizing medium. The residence tome of the vapor and solid product is controlled by the fluidizing gas flow rate. Liquid collection is either by indirect heat exchange or quenching in recycled bio-oil or an immiscible hydrocarbon.

5.2.2. Circulating Fluidized-Bed Reactors

Circulating fluidized-beds have similar features to bubbling fluidized-bed reactors except shorter residence times for chars and vapours. This results in higher gas velocity and char content in bio-oil than in bubbling fluidized bed reactors. One advantage is that this type of reactor is suitable for very large throughputs, even though the hydrodynamics are more complex. There are generally two types of circulating fluidized bed reactors: (i) the single circulating fluidized-bed reactor and (ii) the double circulating fluidized-bed reactor.

This reactor design also is characterized as having high heat transfer rates and short vapor residence times which makes it another good candidate for pyrolysis of biomass. It is somewhat more complicated by virtue of having to move large quantities of sand (or other fluidizing media) around and into different vessels. This type of solids transport has also been practiced for many years in refinery catalytic cracking units (Parkash, 2003; Gary et al., 2007; Speight 2014a; Hsu and Robinson, 2017; Speight, 2017), so it has been demonstrated in commercial applications. Various system designs have been developed with the most important difference being in the method of supplying heat. Earlier units were based on a single indirectly heated reactor, cyclone, and standpipe configuration, where char was collected as a byproduct. Later designs incorporated a dual reactor system in which the first reactor operates in pyrolysis mode while the second one is used to burn char in the presence of the sand and then transfer the hot sand to the pyrolysis vessel.

The size of the feedstock particles sized for a circulating bed system must be even smaller than those used in bubbling beds because, in this type reactor, the particle will only have 0.5 to 1.0 second residence time in the high heat transfer pyrolysis zone before it is entrained over to the char combustion section in contrast to the bubbling bed where the average particle residence time is 2 to 3 seconds. For relatively large particles this would not be sufficient time to allow heat to penetrate into the interior of the particle which is especially relevant when an insulating char layer develops on the outside surface thereby preventing further penetration of heat. As a result, the incompletely pyrolyzed larger particles will end up in the char combustor where they will simply be burned. Particles in the 1-2 mm are the desired size range.

5.3. Ablative Reactor

Ablative pyrolysis is fundamentally different from fluid bed processes with the mode of heat transfer being through a molten layer at the hot reactor surface and the absence of a fluidizing gas. Thus, the ablative pyrolysis is substantially different in concept compared with other methods of fast pyrolysis. In all the other methods, the rate of reaction is limited by the rate of heat transfer through the biomass particles, which is why small particles are required.

Mechanical pressure is used to press biomass against a heated reactor wall. Feedstock in contact with the wall essentially "melts" and, as it is moved away, the residual oil evaporates as pyrolysis vapors.

Advantages of ablative reactors are that feed feedstock does not require excessive grinding, and the process allows much larger biomass particle size than other types of pyrolysis reactors. These types of reactor can use particle sizes up to 20 mm in contrast to the 2 mm particle size required for fluidized bed designs. On the other hand, this configuration is slightly more complex due to mechanical nature of the process. Scaling is a linear function of the heat transfer as this system is surface area controlled. Therefore, the ablative reactor does not benefit from the same economies of scale as other reactor types. The commonly used ablative reactor types are ablative vortex and ablative rotating disk, which are described in the following paragraphs.

The rotating blade type of ablative reactor along with the rotating cone and vacuum pyrolysis reactors do not require an inert carrier gas for operation. When issues of product vapor collection and quality are considered, the lack of a carrier gas can be a real advantage because the carrier gas tends to dilute the concentration of bio-oil fragments and enhances the formation of aerosols as the process stream is thermally quenched. This, in turn, makes recovery of the liquid oil more difficult. Another disadvantage is that high velocities from the carrier gas entrain fine char particles from the reactor, which then are collected with the oil as it condenses which has a deleterious effect on the quality of the bio-oil.

5.3.1. Vortex Reactor

The vortex reactor was developed to exploit the phenomena of ablation. In this approach the biomass particle is melted/vaporized from one plane or side of its aspect ratio. This design approach had the potential to use particle sizes up to 20 mm in contrast to the 2 mm particle size required for fluidized bed designs.

In the process, biomass particles are accelerated to high velocity by an inert carrier gas (steam or nitrogen) and then introduced tangentially to the vortex (tubular) reactor and, under these conditions the particle is forced to slide across the inside surface of the reactor at high velocities. Centrifugal force at the high velocities applied a normal force to the particle against the reactor wall, which is temporarily maintained at 625°C (115°F), which effectively melts the particle. The vapor generated at the surface are immediately swept out of the reactor by the carrier gases to result in vapor residence times of 50 to 100 milliseconds.

In practice, a solids recycle loop close to the exit of the reactor has been incorporated into the unit in order to re-direct larger incompletely pyrolyzed particles back to the entrance to insure complete pyrolysis of the biomass. Particles could escape the reactor only when they were small enough to become re-entrained with the vapor and gases leaving the reactor. While the solids recycle loop was able to effectively address the issue of insuring all particles would be completely pyrolyzed it also resulted in a small portion of the product vapors being recycled into the high temperature zone of the reactor. This portion of the product vapor effectively had a longer residence time at the pyrolysis reactor temperature and most likely resulted in cracking of the product to gases thus resulting in slightly lower yields compared to other fluidized bed designs.

Other design issues with the vortex reactor are: (i) high entering velocities of particles into the reactor caused erosion at the transition from linear to angular momentum, (ii) the potential for excessive wear in the recycle loop which is increased when inert feedstock (such as stone particles) are introduced with the feed, and (iii) uncertainties about the scalability of the design related to maintaining high particle velocities throughout the length of the reactor since the high velocity is necessary for centrifugal force to maintain particle pressure against the reactor wall. The high sliding velocity and constant pressure of the particle against the 600°C (1110°F) reactor wall are necessary to achieve the high heat transfer requirements for fast pyrolysis.

In a vortex ablative pyrolysis reactor, biomass particles are entrained in a hot inert gas (steam or nitrogen) flow and then enter into the reactor tube tangentially. Biomass particles are then forced to slide on the reactor wall at high velocity by means of high centrifugal forces. The particles are melted on the hot reactor wall which is maintained at a temperature of about 625°C and leave a liquid film of bio-oil. Unconverted particles are recycled with a special solids recycle loop. Vapours generated on the reactor wall are quickly swept out by carrier gases in 50 to 100 milliseconds. This design is able to meet the requirements of fast pyrolysis and has demonstrated a bio-oil yield of 65% w/w.

5.3.2. Rotating Disk Reactor

The rotating disc reaction (sometimes referred to as the spinning disc reactor) is are capable of operating horizontally or vertically and is mounted on a rotating axle (Visscher et al., 2013). The feedstock is fed near the center and moves across the surface of a spinning disc under the influence of centrifugal force. This force stretches and spreads the feedstock which allows for high rates of mass transfer so that it favors unit operations such as absorption, stripping, mixing, and reactions. Residence time on the disc is in the range of 0.1 to 3 seconds. Both the thickness of the spread material and residence time are dependent on the physical properties, rotational speed, and radial location of the feedstock. On exiting the periphery (edge) of the disc, the product is thrown onto the enclosing wall which is heated or cooled depending upon the process requirements. Typically a 100-mm-diameter disc rotating at 600 to 1200 rpm and heated to 150°C (300°F) using a heat transfer fluid circulated through a chamber below the rotating disc. This leads to very high heat and mass transfer coefficients and there is no back mixing.

Thus, in the current context of a biomass feedstock, in the reactor the feedstock is forced to slide on a hot rotating disk. While under pressure and heat transfer from the hot surface, biomasses soften and vaporize in contact with the rotating disk which causes the pyrolysis reaction. The most important feature of this reactor is that there is no inert gas medium required, thus resulting in smaller processing equipment. But this process is dependent on the surface area, and hence scaling can be an issue for larger facilities.

The rotating disc reactor is an attractive and viable technology for the conversion of biomass to products operating in a continuous operation mode because of its unique characteristics and its capability of being tailored to a wide range of applications. The rotational speed confers an extra variable for reaction optimization, with higher speeds giving better mixing, heat/mass transfer, and shorter residence times.

The technology of the rotating cone reactor is analogous to the transported bed design (circulated fluidized bed) in that it co-mingles hot sand with the biomass feed to affect the thermal pyrolysis reactions. The primary distinction between the two is that centrifugal force resulting from a rotary cone is used for this transport instead of a carrier gas (Wagenaar et al., 1994). In the process, the biomass feedstock and sand are introduced at the base of the cone while spinning causes centrifugal force to move the solids upward to the lip of the cone. As the solids spill over the lip of the cone, pyrolysis vapors are directed to a condenser. The char and sand are sent to a combustor where the sand gets re-heated before introducing at the base of the cone with the fresh biomass feed. The advantage of this design are that the process does not require a carrier gas for pyrolysis which makes the bio-oil product recovery easier.

5.4. Vacuum Pyrolysis Reactor

Vacuum reactors perform a slow pyrolysis process with lower heat transfer rates which results in lower bio-oil yields of 35 to 50% w/w compared to the 75% w/w reported with the fluidized bed technologies. The pyrolysis process in vacuum reactor is very complicated mechanically and requires high investment and maintenance costs. A moving metal belt conveys biomass into the high temperature vacuum chamber. On the belt, biomasses are periodically stirred by a mechanical agitator. A burner and an induction heater are used with molten salts as a heat carrier to heat the biomass. Because of operating in a vacuum, these types of pyrolysis reactors require special solids feeding and discharging devices to maintain a good seal at all times. The main benefit of a vacuum reactor is that larger sized biomass particles (up to 5 cm) can be processed than can be accommodated in a fluidized bed reactors.

The vacuum pyrolysis process for converting biomass to liquids is a slow pyrolysis process (lower heat transfer rate) it generates a chemically similar liquid product because the shorter vapor residence time reduces secondary reactions. However, the slow heating rates also result in lower bio-oil yields of 30 to 45% w/w compared to the 70% w/w that has been reported using a fluidized bed technology. The process involves a moving metal belt that carries the biomass into the high temperature vacuum chamber. There are also mechanical agitators that periodically stir the biomass on the belt; all of this mechanical transport is being done at 500°C (930°F). The process does have several advantages: (i) a clean oil having little or no char without using hot vapor filtration, (ii) condensation of the liquid product is easier than for fluidized bed or entrained flow technologies, (iii) the process can accept larger feed particles than fluidized bed processes, (iv) the lignin-derived fraction of the oil can be of a lower molecular weight than that from fast pyrolysis processes, which may have advantages if extracting this component for phenol derivatives, (v) the need to sweep vapors out of the reaction vessel by a carrier gas is eliminated and aerosol formation is minimized.

On the other hand, vacuum pyrolysis technology is a slow pyrolysis process that will not be able to provide oil yields as high as fast heating rate processes. The process also generates more water than other fast pyrolysis processes and also generates liquid effluents as volatile feedstock that is not collected in the scrubbers but absorbed in the liquid ring compressor pump and need to be recycled back to the scrubbers.

5.5. Cyclone Reactor

In this reactor, pyrolysis is implemented in a cyclonic reactor with an integrated hot gas filter (the rotational particle separator) in one unit to produce particle free bio-oil. The biomass and the inert heat carrier are introduced as particles into the cyclone and the solids are transported by recycled vapours from the process. By centrifugal force the particles are moved downwards to the periphery of the cyclone.

During the transport downwards in the reactor, the biomass particles are dried, heated up and devolatilized. The average process temperature is 450 °C–550 °C. The typical gas residence time in the reactor is 0.5 to 1 s, so secondary cracking reactions of tars in the reactor can be reduced. Evolved vapours are transported rapidly to the center of the cyclone and leave the cyclone via the rotating filter. The remaining gases and char can be used to heat up the heat carrier and transportation gas. This reactor is comparatively compact and low cost, with 70 to 75% w/w bio-oil yield capability.

The flexibility of the cyclone reactor is such that, according to the operating conditions, it can be used either for the gasification or for the liquefaction of biomass. A fraction of the gaseous products can be used as the carrier gas (recycling process) without noticeable changes of the gas composition and with fast gasification yields close to 100%. The vapor-phase cracking reactions mainly occur inside a very thin and hot boundary layer close to the heated surface of the cyclone. Thus, the reactor is an efficient multifunctional reactor making it possible to perform in less than a second heating and pyrolysis of the reactants as well as the quenching and separation of the products.

5.6. Auger Reactor

In this type of reactor (sometimes referred to as a screw reactor), the auger is used to move biomass feedstock through an oxygen free cylindrical heated tube. A passage through the tube raises the feedstock to the desired pyrolysis temperature ranging from 400°C to 800°C (750 to 1470°F)) which causes it to devolatilize and gasify. Char is produced and gases are condensed as bio-oil, with non-condensable vapor collected as bio-gas. In this design the vapor residence time can be modified by changing the heated zone through which vapor passes prior to entering the condenser train.

In the process, the dried biomass is first added to the hopper from where it is fed with a volumetric feeder into the auger reactor at a feeding rate between 1 to 40 g/min. The auger is inside a stainless-steel tube with 58.5 cm length and 10 cm diameter heated by a furnace. The biomass is pushed through the hot zone of the reactor with a variable-speed auger screw driven by a 1 hp variable speed motor and nitrogen is typically used as the carrier gas. The temperature on the external wall is recorded as well as the temperature of the biomass bed at the exit of the heating zone and at the entrance into the collection pot. The pyrolysis vapors resulting are condensed in three condensation units. The pressure inside the reactor is typically kept close to atmospheric pressure (under a very slight vacuum) by drawing the pyrolytic vapors and the carrier gas through the condensers with a vacuum pump. The yield of liquid is typically determined by weighing the traps, the vacuum pump and the liquid collected in condensers. The non-condensable gases can also be trapped by use of low-temperature traps but, if bio-oil is the desired product the gases are calculated by difference.

In summary, this type of reactor is particularly suitable for feed materials that are difficult to handle or feed, or are heterogeneous. The liquid product yield may be lower than the yield of liquid product from a fluidized bed reactor and, also, the yield of the biochar yields are higher.

5.7. Plasma Reactor

The plasma reactor is based on the principle that once an atom is heated above the ionization energy, the electrons are stripped away (the atom is ionized), leaving just the bare nucleus (the ion). The result is a hot cloud of ions and the electrons formerly attached to them – this cloud is known as plasma. Because the charges are separated, plasmas are electrically conductive and magnetically controllable. Many fusion devices take advantage of this to control the particles as they are heated. Thus, plasma is an ionized gas that conducts electricity. Fusion exploits several plasma properties, including: (i) self-organizing plasma conducts electric and magnetic fields. Its motions can generate fields that can in turn contain it, (ii) diamagnetic plasma which can generate its own internal magnetic field – this can reject an externally applied magnetic field, making it diamagnetic, and magnetic mirrors which can reflect plasma when it moves from a low to high density field.

The plasma pyrolysis reactors are usually made with a cylindrical quartz tube surrounded by two copper electrodes. Biomass particles are fed at the middle of the tube using a variable-speed screw feeder located on the top of the tube. Electrodes are coupled with electrical power sources to produce thermal energy to gas flows through the tube. Oxygen is removed by an inert gas incorporated in the reactor. This inert gas also serves as working gas to produce plasma. The pyrolysis product vapours are evacuated from the reactor by means of a variable speed vacuum pump (Tang and Huang, 2005).

Although consuming high electrical power and having high operating costs, plasma reactors offer some unique advantages in biomass pyrolysis compared with conventional reactors. The high energy density and temperature produced in plasma pyrolysis corresponds with a fast reaction which provides a potential solution for the problems that occur in slow pyrolysis such as the generation of heavy tarry compounds and low productivity of syngas (Tang and Huang, 2005). In this type of reactor, tar formation is eliminated due to the cracking effects of the highly active plasma environment with a variety electron, ion, atom and activated molecule species. However, a significant proportion of heat from the thermal plasma is released to the surrounding environment by means of radiation and conduction.

5.8. Microwave Reactor

The microwave reactor is one of the recent research focuses in pyrolysis application in which energy transfer occurs through the interaction of molecules or atoms using a microwave-heated bed. The drying and pyrolysis processes of biomass are carried out in a microwave cavity oven powered by electricity. Inert gas is flowing continuously through the reactor to create an oxygen free atmosphere and to serve as the carrier gas as well. Microwave reactors offer several advantages over slow pyrolysis systems which make them an effective method of recovering useful chemicals from biomass.

These advantages include efficient heat transfer, exponential control of the heating process and an enhanced chemical reactivity that reduces the formation of undesirable species (Fernández and Menéndez, 2011). Additionally, unexpected physical behavior such as hot spots are appearing in microwave reactors which increases syngas yield (Tang and Huang, 2005). Therefore a wide range of biomasses and industrial wastes are possible to process in microwave reactors with high yields of desirable products such as syngas and bio-oil (Tang and Huang, 2005). The biomass types that are subjected to pyrolysis using microwave reactors include sewage sludge, coffee hulls, glycerol, rice straw, corn stalk, automotive waste oil, wood blocks, and wood sawdust (Domínguez et al., 2007; Lam et al., 2010.; Zhao et al., 2010, 2012).

In summary, microwave pyrolysis offers the advantage of avoiding or reducing the low thermal conductivity of biomass encountered in conventional thermal pyrolysis. The uniform heating resulting from use of microwaves is likely to reduce secondary reactions as reaction products are less likely to interact with pyrolyzed biomass (Bridgwater, 2018a).

5.9. Solar Reactor

The use of solar reactors in pyrolysis provides a suitable means of storing solar energy in the form of chemical energy. This type of reactor is usually made with a quartz tube which has opaque external walls exposed to concentrated solar radiation. A parabolic solar concentrator is attached with the reactor to concentrate the solar radiation. The concentrated solar radiation is capable of generating high temperatures (>700°C, >1290°F) in the reactor for pyrolysis processes (Antal et al., 1983; Rony et al., 2018). However, solar reactors have some advantages over slow reactors. In slow pyrolysis a part of the feedstock is used to generate the process heat. Therefore it reduces the amount of feedstock available and, at the same time, causes pollution. Hence utilization of solar energy in the pyrolysis process maximizes the amount of feedstock available and overcomes the prolusion problem. Moreover, solar reactors are capable of faster start up and shut down periods compared to slow reactors.

6. Pyrolysis Products

The three primary products obtained from pyrolysis of biomass are char, permanent gases, and vapours that at ambient temperature condense to a dark brown viscous liquid and the amounts is dependent upon the time of the biomass and the primary products in the hot zone (Table 5.5). Maximum liquid production occurs at temperatures between 350 and 500°C (660 and 930°F) because different reactions occur at different temperatures in pyrolysis processes. Consequently, at higher temperatures, the molecular constituents present in the liquid and residual solid are broken down to produce smaller molecules which enrich the gaseous fraction. The yields of products resulting from biomass pyrolysis can be maximized as follows: (i) charcoal, which is a low temperature, low heating rate process, (ii) liquid products which are due to a low temperature, high heating rate, short gas residence time process, and (iii) fuel gas which are due to a high temperature, low heating rate, long gas residence time process.

The products from pyrolysis processes also strongly depend on the water content in the biomass which produces large quantities of condensate water in the liquid phase. This contributes to the extraction of water-soluble compounds from the gaseous and tar phases, and thus a greater decrease in gaseous and solid products.

Table 5.5. Pyrolysis Products

Feedstock	Primary products	Secondary products
Biomass	Gas	Gas
	Light oil	Gas
		Light oil
		Heavy tar
	Heavy tar	Gas
		Light Oil
		Heavy tar
		Char

6.1. Biogas

The yield of biogas is similar to the yield of char – approximately 10 to 35% of biogas is produced in slow pyrolysis processes. However, a higher gas yield is possible in flash pyrolysis with high temperatures. For example, an investigation of syngas production from pyrolysis of municipal solid waste in a fixed-bed reactor over a temperature range of 750 to 900°C (1380 to 1650°F) using calcined dolomite as a catalyst achieved a high yield of gas (on the order of 79% gas yield at 900°C (1650°F). However, the yield of biogas is strongly influenced by the pyrolysis temperature.

The reactor temperature has a significant influence on pyrolysis processes and resulting product distribution. As the temperature is increased, the following events occur: first, the moisture inside the biomass evaporates and second, the thermal degradation and devolatilization of the dried portion of the particles take place (Demirbaş, 2004). At the same time, tar is produced and volatile species are gradually released from the surface of the particles. The volatile species and the tar then undergo a series of secondary reactions such as decarboxylation, decarbonylation, dehydrogenation, deoxygenation, and cracking to form the components of the biogas. Therefore higher temperatures favor tar decomposition and the thermal cracking of tar to increase the proportion of the biogas, resulting in decreased yields of the bio-oil and decreased yield of the biochar.

The moisture content of the feedstock influences the heat transfer process in pyrolysis, having an unfavorable effect on the production of the biogas. A high moisture content contributes to the extraction of water-soluble components from the gaseous phase, hence causing a significant decrease in gaseous products. For a given temperature, dry biomass produces the greatest quantity of gas at the early stage of pyrolysis, whereas with wet biomass the production of the greatest quantity occurs later in the process – an increase in the humidity leads to an increase in drying time.

The biogas mainly consists of hydrocarbons (of which methane, ethylene, and ethane are examples) as well as hydrogen, carbon monoxide, and small amounts of carbon dioxide, water, nitrogen, with some tar and mineral ash, depending on biomass feedstock and pyrolysis conditions. These components are obtained during several endothermic reactions at high pyrolysis temperatures: hydrogen is produced from the cracking of hydrocarbons at higher temperatures while carbon monoxide and carbon dioxide are the indicators of the presence of oxygen in the biomass. These latter constituents (carbon monoxide and carbon dioxide) mainly originate from the cracking of partially oxygenated organic compounds. Therefore, as a highly oxygenated polymer, the amount of cellulose present in the biomass is an important factor determining the amount of carbon oxides produced (Yang et al., 2007).

The presence low-boiling hydrocarbon derivatives such as methane, ethane, and ethylene may be due to the reforming and cracking of higher molecular weight (higher-boiling) hydrocarbon derivatives as well as the cracking of tar in the vapor phase. However, the composition of the biogas is clearly affected by the reactor temperature: the yield of hydrogen increases sharply while the yield of carbon monoxide increases slowly with the increase of the temperature – other constituents of the biogas show an opposite tendency. The molar ratio of hydrogen and carbon monoxide in the biogas is an important factor that determines its possible applications, as for example, in the Fischer-Tropsch process. For example, a higher hydrogen-carbon monoxide molar ratio is desirable to produce Fisher-Tropsch synthesis for the production of transportation fuel and to produce hydrogen for ammonia synthesis, so optimization of reaction temperature in pyrolysis is a critical issue for the production of bio-synthesis gas.

Synthesis gas from biomass pyrolysis could be a renewable source for an alternative fuel (bio-gasoline) for internal combustion (IC) engines and industrial combustion processes. Commercial gasoline engines and diesel engines can be converted to use gaseous fuel for the use of power generation, transportation and other applications.

6.2. Bio-Oil

Bio-oil (also often usually referred as pyrolysis-oil or bio-crude) is the liquid produced from the condensation of vapor of a pyrolysis reaction and has the potential to be used as a fuel oil substitute. Initially the interest of using bio-oil (often referred to as pyrolysis oil) was driven by concerns for potential shortages of crude oil, but in recent years the ecological advantages of biomass fuels have become an even more important factor.

The bio-oil from a pyrolysis process presents a much better opportunity for high efficiency energy production compared to traditional biomass fuel such as black liquor or hog oil. Therefore a significant effort has been made to upgrade bio-oil so that it is usable for the generation of heat and power and for use as a transport fuel. Unfortunately bio-oil has not reached commercial standards yet due to significant problems during its use as fuel in standard equipment such as boilers, engines and gas turbines constructed for operation with petroleum-derived fuels. The main reasons for this are poor volatility, high viscosity, coking, and corrosiveness of bio-oil. In this section, application of bio-oil in boilers, turbines and diesel engines is summarized. However, pyrolysis bio-oils have a good potential to replace conventional diesel fuel.

Furthermore, bio-oil can be converted to biogas and/or to biochar by the selection of the appropriate pyrolysis technology. It is a liquid product that can be converted with high efficiency into biogas using pyrolysis. The advantage of the using bio-oil is that it is well suited for integration to existing operations and it also makes possible the distributed production of high value biogas. The pyrolysis of bio-oil is the fundamental chemical reaction that is the precursor of the gasification process and occurs naturally in the first two seconds. The products of bio-oil pyrolysis include predominantly gases and some biochar. Hence, its inclusion here.

Bio-oil can be used as a substitute for fossil fuels to generate heat, power and chemicals. Short-term applications are boilers and furnaces (including power stations), whereas turbines and diesel engines may become users in the somewhat longer term. Upgrading of bio-oil to a transportation fuel quality is technically feasible, but needs further development. Transportation fuels such as methanol and Fischer-Tropsch fuels can be derived from bio-oil through synthesis gas processes. Furthermore, there is a wide range of chemicals that can be extracted or derived from bio-oil.

Typically, a bio-oil has a heating value on the order 40 to 50% of the heating value of a hydrocarbon fuel. The main advantages of pyrolysis liquid fuels are: (i) the carbon dioxide balance is clearly positive in biomass fuel, (ii) there is the possibility of utilization in small-scale power generation systems as well as use in large power stations, (iii) the storage stability of the liquid fuel, (iv) the fuel has a high-energy density compared to biomass gasification fuel, and (v) there is the potential of using pyrolysis liquid in existing power plants.

Bio-oil is composed of a complex mixture of oxygenated compounds and, thus, contains various chemical functional groups such as carbonyl derivatives, carboxyl derivatives, and phenol derivatives that provide both potentials and challenges for utilization. The total number of compounds in the bio-oil can be on the order of 300 to 400, even higher. Even if all of these constituents could be identified with a high degree of accuracy, numerous unknown factors are affecting the thermo-physical properties of pyrolysis bio-oil such as (i) limitations in fuel quality, (ii) the tendency for phase separation, (iii) stability/instability, and (iv) fouling issues during thermal processing of the bio-oil (Demirbaş, 2004).

The chemical constituents of bio-oil can be classified into five broad categories: (i) hydroxy-aldehyde derivatives, (ii) hydroxyketone derivatives, (iii) sugar derivatives, (iv) carboxylic acid derivatives, and (v) phenol derivatives. The phenol derivatives compounds are present as oligomers having molecular weights ranging from 900 to 2500 and which are primarily derived from the lignin component of biomass. Another classification can be employed in which the constituents of bio-oil are classed as specific organic compound types, thus: (i) acid derivatives, (ii) alcohol derivatives, (iii) aldehyde derivatives, (iv) ester derivatives, (v) ketone derivatives, (vi) phenol derivatives, (vii) furan derivatives, (viii) alkene derivatives, (ix) aromatic derivatives, (x) nitrogen compounds, and (xi) miscellaneous oxygenate derivatives. Thus, while there is a rich mixture of identifiable compound types in bio-oil, the vast majority of the compound types are in low concentrations. The highest concentration of any single chemical compound (after water) is hydroxy-acetaldehyde ($HOCH_2CHO$, also called glycolaldehyde) at levels up to 10% w/w. This is followed by acetic acid (CH_3CO_2H) and formic acid (HCO_2H), at approximately 5% w/w and 3% w/w, respectively and is a primary reason why bio-oils exhibit a pH in the range of 2.0 to 3.0.

Furthermore, the wide range of some of these properties is tied to certain processing methods employed by the particular organization producing the bio-oil. For example some producers may not have used bone dry feed as a starting feedstock and the additional moisture ends up in the oil which is clearly due to the ability of the sample to produce moisture as a result of chemical reactions during the process. A similar issue applies for the mineral matter (manifested as mineral ash or process ash), which is a function of the amount of char permitted to carry over to the condensation system where the bio-oil is recovered. Also, increasing the cracking severity (time/temperature relationship) is known to alter the chemical profile of the resulting oils and there is a relationship between compound classes in the bio-oil and the temperature to which the vapors were exposed to before quenching.

Bio-oil has a much higher density than woody feedstocks which reduces storage and transport costs. However, bio-oil is not suitable for direct use in standard internal combustion engines but, alternatively, the bio-oil can be upgraded to either a special engine fuel or through gasification processes to a synthesis gas and then bio-diesel. On the other hand, bio-oil is particularly attractive for co-firing because it can be more readily handled and burned than solid fuel and is cheaper to transport and store. It is in such applications that bio-oil can offer major advantages over solid biomass and gasification due to the ease of handling, storage and combustion in an existing power station when special start-up procedures are not necessary. In addition, bio-oil is also a vital source for a wide range of organic compounds and specialty chemicals.

During storage, the pyrolysis oil becomes more viscous due to chemical and physical changes as many reactions continue and volatiles are lost due to aging. These types of reaction than can render a reacted constituent incompatible with the unchanged boy of the oil occur faster at higher temperatures but the effects can be reduced if the pyrolysis oil is stored in a cool place. A more stable pyrolysis oil can be produced if an energy crop is used but this can affect the process optimization by lowering the organic yield due to the high level of mineral matter (manifested as process ash) in the feedstock and producing a high level of reaction water, resulting in a reduction of the heating value of the oil as well as risking phase separation. The feedstock content limits for metals, ash and lignin need to be identified and addressed in order to produce bio-oil which can be used for commercial applications, which does not change considerably over time, but is still produced at acceptable yield levels and with good heating values (Chen et al., 2007).

In terms of fuel use, the thermal efficiency of pyrolysis oils is identical to that of diesel fuel in combustion engine operations, but may exhibit excessive ignition delay. Therefore pyrolysis oil requires a moderate degree of preheated combustion air for reliable ignition. However the yield of the pyrolysis oil as well as the quality and stability of the oil can also be modified by process variables such as heating rate, pyrolysis temperature and residence times. Other factors such as different reactors (such as the ablative reactor and the fixed-bed reactor), particle size, and char accumulation can affect the yield and quality of the pyrolysis oil by varying its ash content and composition which affects the thermal degradation of these biomass (Ringer et al., 2006).

In fact, the properties of bio-oil can encompass a broad range of parameters because of the complex nature of the feedstock. Even if it is possible to perfectly reproduce all of the processing conditions necessary to produce bio-oil, the biomass feedstock, irrespective of an optimization of the process parameters, can influence the nature of the final product. Not only are there differences between types of biomass species but also where a particular species is grown can affect things such as the composition of mineral matter present. As a result of this non-uniformity in the starting feedstock and the high temperature reactive environment to which the prompt biomass vapor fragments are exposed during pyrolysis, it is not unusual for there to be considerable variations in many of the physio-chemical properties of bio-oil. For some applications, the small variations will be of little consequence, but in situations where it is desirable to use bio-oil in devices that have been designed to operate on hydrocarbon fuels, some of these properties will make operation difficult or simply not feasible.

Commercial pyrolysis bio-oil should maintain its chemical and physical properties such as stability and viscosity which can be achieved if the oil exhibits high inhomogeneity, processing lower molecular weight compounds. High molecular weight derived compounds present in pyrolysis oil typically arise as a result of lignin present in the biomass. Therefore, biomass with less lignin content is desirable to reduce the heavier molecular weight compounds present in pyrolysis oil and produce a more homogenous liquid. However, some properties such as moisture content and viscosity create issues when using the bio-oil as a fuel. Although the water content (hydrophilic) can exert a beneficial effect by lowering the nitrogen oxide (NOx) emissions and improving the flow characteristics of the bio-oil, the present of water in the bio-oil also means that it (the bio-oil) is not completely miscible with in petroleum fuels (which are hydrophobic) and the presence of water also lowers the heating value of the fuel. The solids entrained in the bio-oil principally contain fine char particles that are not removed by the cyclones. In addition, the viscosity of bio-oil may become problematic when the bio-oil is stored over time because unfavorable reactions (such as, for example, oxidation by acrial oxygen) take place that make the liquid too viscous to be a viable fuel. To address some of the obstacles of using bio-oil as a fuel, actions must be taken to modify and improve the quality of bio-oil and, thus, prevent these unwanted chemical changes to the oil.

It is also suggested that bio-oil or a slurry of bio-oil with char can be used as a feedstock for a gasification and synthesis of hydrocarbon transport fuels, for example by Fischer Tropsch synthesis or alcohols Bridgwater, 2018b).

6.3. Biochar

Biochar is the charcoal-like (carbonaceous) residue that is produced das a result of the pyrolysis of biomass. It is created using a pyrolysis process in which biomass is heated in a low oxygen environment. Once the pyrolysis reaction has begun, it is self-sustaining, requiring no outside energy input. Byproducts of the process include gas (variable yield and composition, depending on the feedstock and the process parameters) and bio-oil (variable yield and composition, depending on the constituents of the feedstock and the process parameters.

Thus, biochar is a high-carbon, fine-grained residue that is produced through a pyrolysis process by the direct thermal decomposition of biomass in the absence of oxygen (preventing combustion), which produces a mixture of gas (biogas), liquid (bio-oil), solid (biochar). The specific yield from the pyrolysis is dependent on process condition, such as temperature, and can be optimized to produce either energy or biochar. Temperatures of 400 to 500°C (750 to 930 F) produce more char, while temperatures above 700°C (1290 F) favor the yield of liquid and gas fuel components. Pyrolysis occurs more quickly at the higher temperatures, typically requiring seconds instead of hours – typical yields are biogas: 20% w/w, bio-oil: 60% w/w, biochar: 20% w/w. By comparison, slow pyrolysis can produce substantially more char (approximately 50% w/w). The gas and excess heat from the process can be used directly or employed to produce a variety of biofuels (Goyal et al., 2008).

Char acts as a vapor cracking catalyst – thus effective separation from the pyrolysis product vapor product is essential. Cyclones are the usual method of char removal and some success has been achieved with hot vapor filtration which is analogous to hot gas cleaning in gasification systems (Hoekstra et al., 2009). Finally, while biochar is often presented as a beneficial product to enhance the anaerobic digestion process (Chapter 3) and for use as a fertilizer (Chapter 8), it is also used as a feedstock for gasification to produce bio-synthesis gas (Chapter 6). Hence it inclusion here.

7. CATALYSTS

A catalyst increases the rate of a chemical reaction without consuming or changing itself during the reaction. Its plays an important role and is widely applied in biomass pyrolysis processes. In general; catalysts are used to enhance pyrolysis reaction kinetics by cracking higher molecular weight compounds into lighter hydrocarbon products. However; different catalysts have different product distributions in different operating conditions.

Depending on the application, the catalysts for pyrolysis process can be classified into three different groups: (i) the catalyst is added to the biomass before being fed into the reactor, (ii) the catalyst is added into the reactor; therefore permitting immediate contact with vapours; solid and tar, and (iii) the catalyst is placed in a secondary reactor located downstream from the pyrolysis reactor. Also, the catalysts can be divided into four groups depending on composition: (i) dolomite-type catalysts, (ii) nickel-based catalysts, (iii) alkali metal catalysts, and (iv) novel metal catalysts (French and Czernik, 2010).

Dolomite is an anhydrous carbonate mineral composed of calcium carbonate and magnesium carbonate [represented as $CaMg(CO_3)_2$ or $CaCO_3.MgCO_3$]. The term is also used for a sedimentary carbonate rock composed mostly of the mineral dolomite. An alternative name sometimes used for the dolomitic rock type is dolostone. Calcined dolomite as a catalyst is inexpensive, abundant and significantly reduces tar formation in the product gas. Therefore, dolomite was extensively investigated in different reactors such as fixed bed reactors and fluidized bed reactors. However calcined dolomite catalyst has some limitations including low melting point which makes it unstable at high temperatures, not effective in heavy tar cracking and difficult to achieve or exceed 90 to 95% w/w tar conversion.

Other types of catalysts have been used in biomass pyrolysis to enhance productivity include nickel (Ni), cerium oxide (CeO_2), alumina (Al_2O_3,)alumina, sodium feldspar ($NaAlSi_3O_8$), rhodium (Rh), silica (SiO_2), sodium carbonate (Na_2CO_3), potassium carbonate (K_2CO_3) lithium carbonate (Li_2CO_3,), zinc chloride ($ZnCl_2$), and zirconia (zirconium oxide, ZrO_2). The nickel-based catalyst ad the alkali metal-based catalysts have been proven to be effective for heavy tar elimination (by cracking to lower molecular weight products) but can be rendered inactive by carbon deposition.

REFERENCES

Aho, A., Kumar, N., Eranen, K., Salmi, T., Hupa, M., Murzin, D. Y. 2008. Catalytic Pyrolysis of Woody Biomass in a Fluidized Bed Reactor: Influence of the Zeolite Structure. *Fuel,* 87: 2493-2501.

Antal, M. J., 1983. Hofmann, L., Moreira, J. R., Brown, C. T., and Steenblik, R. 1983. Design and Operation of a Solar Fired Biomass Flash Pyrolysis Reactor. *Solar Energy,* 30(4): 299-312.

Antal, M. J., Jr., and Grönli, M. 2003. The Art, Science, and Technology of Charcoal Production. *Ind. Eng. Chem. Res.*, 42: 1619–1640.

Asadullah, M., Rahman, M. A., Ali, M. M., Motin, M. A., Sultan, M. B., and Alam, M. R. 2007. Production of Bio-Oil from the Fixed Bed Pyrolysis of Bagasse. *Fuel,* 86: 2514-2520.

ASTM. 2019. *Annual Book of Standards,* ASTM International, West Conshohocken, Pennsylvania.

ASTM D7544. 2018. Standard Specification for Pyrolysis Liquid Biofuel. *Annual Book of Standards.* ASTM International, West Conshohocken, Pennsylvania, USA.

Banks, S. W., Bridgwater, A. V. 2016. Catalytic Fast Pyrolysis for Improved Liquid Quality. In: *Handbook of Biofuels Production: Processes and Technologies.* R. Luque, C. S. K. Lin, K. Wilson and J. Clark (Editors). Elsevier BV, Amsterdam, Netherlands. Page 391-429.

Brammer, J. G., Lauer, M., and Bridgwater, A. V. 2006. Opportunities for Biomass-Derived Bio-Oil in European Heat and Power Markets. *Energy Policy*, 34: 2871-2880.

Bridgwater, A. V., and Peacock, G. V. C. 1999. Fast Pyrolysis Process for Biomass. *Sustainable and Renewable Energy Reviews*, 4: 1-73.

Bridgwater, A.V.; Czernik, S.; Piskorz, J. 2001. An Overview of Fast Pyrolysis. Prog. Thermochem. *Biomass Convers.,* 2: 977-997.

Bridgwater, A. V. 2004. Biomass Fast Pyrolysis. *Therm. Sci.,* 8: 21-49.

Bridgwater, A. V. 2018a. Challenges and Opportunities in Fast Pyrolysis of Biomass: Part I Introduction to the Technology, Feedstocks and Science Behind a Promising Source of Fuels And Chemicals. *Johnson Matthey Technol. Rev.*, 62(1): 118-130. www.technology.matthey.com/pdf/118-130-jmtr-jan18.pdf.

Bridgwater, A.V. 2018b. Challenges and Opportunities in Fast Pyrolysis of Biomass: Part II Upgrading Options and Promising Applications in Energy, Biofuels and Chemicals. *Johnson Matthey Technol. Rev.*, 62(2): 150-160. www.technology.matthey.com/pdf/150-160-jmtr-apr18.pdf.

Chen, J., Zhu, D., and Sun, C. 2007. Effect of Heavy Metals on the Sorption of Hydrophobic Organic Compounds to Wood Charcoal. *Environ. Sci. Technol.,* 2536-2541.

Chiaramonti, D., Oasmaa, A., and Solantausta, Y. 2007. Power Generation Using Fast Pyrolysis Liquids from Biomass. *Renew. Sustain. Energy Rev.*, 11: 1056-1086.

Cornelissen, T., Yperman, Y., Reggers, G., Schreurs, S., Carleer, R. 2008. Flash Co-pyrolysis of Biomass with Polylactic Acid. Part 1: Influence on Bio-Oil Yield and Heating Value. *Fuel* 2008, 87, 1031–1041.

Czernik, S., and Bridgwater, A. V. 2004. Applications of biomass fast pyrolysis oil. *Energy Fuel,* 18, 590-598.

Demiral, I., and Sensoz, S. 2008. The Effects of Different Catalysts on the Pyrolysis of Industrial Wastes (Olive and Hazelnut Bagasse), *Bioresour. Technol.*, 99: 8002-8007.

Demirbaş, A. 2000a. Recent Advances in Biomass Conversion Technologies. *Energy Educ. Sci. Technol.* 2000, 6, 77–83.

Demirbaş, A. 2000b. Mechanisms of Liquefaction and Pyrolysis Reactions of Biomass. *Energy Convers. Manag.,* 41: 633-646.

Demirbaş, A. 2002. Partly Chemical Analysis of Liquid Fraction of Flash Pyrolysis Products from Biomass in the Presence of Sodium Carbonate. *Energy Convers. Manag.,* 43, 1801-1809.

Demirbaş, A., and Arin, G. 2002. An Overview of Biomass Pyrolysis. *Energy Sources Part A*, 24: 471-482.

Demirbaş, A. 2004. Current Technologies for the Thermo-Conversion of Biomass into Fuels and Chemicals. *Energy Sources Part A,* 26: 715-730.

Demirbaş, A. 2007. The Influence of Temperature on the Yields of Compounds Existing in Bio-Oils Obtained from Biomass Samples Via Pyrolysis. *Fuel Proc. Technol.* 2007, 88, 591–597.

Dobele, G., Urbanovich, I., Volpert, A., Kampars, V., and Samulis, E. 2007. Fast pyrolysis – Effect of wood drying on the yield and properties of bio-oil. *Bioresources,* 2: 699-706.

Domínguez, A., Menéndez, J. A., Fernández, Y., Pis, J. J., Valente Nabais, J. M., Carrott, P. J. M., and Ribeiro Carrott, M. M. L. 2007. Conventional and Microwave Induced Pyrolysis of Coffee Hulls for the Production of a Hydrogen Rich Fuel Gas. *J. Anal. Appl. Pyrolysis,* 79: 128-135.

Fegbemi, L., Khezami, L., and Capart, R. 2001. Pyrolysis Products from Different Biomasses: Application to the Thermal Cracking f Tar. *Appl. Energy,* 69: 293-306.

Fernández, Y., and Menéndez, J. A. 2011. Influence of Feed Characteristics on the Microwave-Assisted Pyrolysis Used to Produce Syngas from Biomass Wastes. *J. Anal. Appl. Pyrolysis,* 91: 316-322.

Fisher, T., Hajaligol, M., Waymack, B., and Kellogg, D. 2002. Pyrolysis Behavior and Kinetics of Biomass Derived Feedstocks. *J. Appl. Pyrolysis,* 62: 331-349.

French, R., and Czernik, S. 2010. Catalytic Pyrolysis of Biomass for Biofuels Production. *Fuel Processing Technology,* 91(1): 25-32.

Gary, J. G., Handwerk, G. E., and Kaiser, M. J. 2007. *Crude oil Refining: Technology and Economics,* 5th Edition. CRC Press, Taylor & Francis Group, Boca Raton, Florida.

Goyal, H. B., Seal, D., Saxena, R. C. 2008. Bio-fuels from Thermochemical Conversion of Renewable Resources: A Review. *Renew. Sustain. Energy Rev.,* 12: 504-517.

Halpern, Y.; Riffer, R.; Broido, A. 1973. Levoglucosenone (1,6-anhydro-3,4-dideoxy-delta.3-beta-d-pyranosen-2-one). Major Product of the Acid-Catalyzed Pyrolysis of Cellulose and Related Carbohydrates. *Journal of Organic Chemistry;* 38(2): 204-209.

Hoekstra, E., Hogendoorn, K. J. A., Wang, X., Westerhof, R. J. M., Kersten, S. R. A., Van Swaaij, W. P. M., and Groenveld, M. J. 2009. Fast Pyrolysis of Biomass in a Fluidized Bed Reactor: In Situ Filtering of the Vapors. *Ind. Eng. Chem. Res.*, 48(10): 4744-4756.

Hsu, C. S., and Robinson, P. R. (Editors). 2017. *Handbook of Petroleum Technology.* Springer International Publishing AG, Cham, Switzerland.

Jensen, P. A., Sander, B., Dam-Johansen, K. 2001. Pretreatment of Straw for Power Production by Pyrolysis and Char Wash. *Biomass Bioenergy*, 20: 431-446.

Karaosmanoglu, F., and Tetik, E. 1999. Fuel Properties of Pyrolytic Oil of the Straw and Stalk of Rape Plant. *Renew. Energy*, 16: 1090–1093.

Lakshmanan, C. M., and Hoelscher, H. E. 1970. Production of Levoglucosan by Pyrolysis of Carbohydrates. Pyrolysis in Hot Inert Gas Stream. *Industrial & Engineering Chemistry Product Research and Development*, 9: 57-59.

Lam, S. S., Russell, A. D., and Chase, H. A. 2010. Microwave Pyrolysis: A Novel Process for Recycling Waste Automotive Engine Oil. *Energy*, 35, 2985-2991.

Lédé, J., Broust, F., Fatou-Toutie, N., and Ferrer, M. 2007. Properties of Bio-Oils Produced by Biomass Fast Pyrolysis in a Cyclone Reactor. *Fuel,* 86(12-13): 1800-1810.

Lee, S., Speight, J. G., and Loyalka, S. 2014. *Handbook of Alternative Fuel Technologies.* 2nd Edition. CRC Press, Taylor & Francis Group, Boca Raton, Florida.

Mathew, M., and Muruganandam, L. 2017. Pyrolysis of Agricultural Biomass using an Auger Reactor: A Parametric Optimization. *International Journal of Chemical Reactor Engineering January.* file:///C:/Users/James/Downloads/Melvinet.al.2017.pdf.

Mohan, D., Pittman, C. U., and Steele, P. H. 2006. Pyrolysis of Wood/Biomass for Bio-Oil: A Critical Review. *Energy Fuels*, 20: 848-889.

Parkash, S. 2003. *Refining Processes Handbook.* Gulf Professional Publishing, Elsevier, Amsterdam, Netherlands.

Pütün, E., Uzun, B. B., and Pütün, A. E. 2006. Fixed-bed Catalytic Pyrolysis of Cotton-Seed Cake: Effects of Pyrolysis Temperature, Natural Zeolite Content and Sweeping Gas Flow Rate. *Bioresour. Technol.* 2006, 97, 701-710.

Ringer, M., Putsche, V., and Scahill, J. 2006. *Large-Scale Pyrolysis Oil Production: A Technology Assessment and Economic Analysis*. Technical Report No. NREL/TP-510-37779. National Renewable Energy Laboratory, Golden, Colorado. November.

Rony, A. H., Mosiman, D., Sun, Z., Qin, D., Zheng, Y., Boman, J. H., and Fan, M. 2018. A Novel Solar Powered Biomass Pyrolysis Reactor for Producing Fuels and Chemicals. *Journal of Analytical and Applied Pyrolysis*, 132(June): 19-32.

Scott, D. S., and Piskorz, J. 1984. The Continuous Flash Pyrolysis of Biomass. Canadian *Journal of Chemical Engineering*, 62(3): 404-412.

Speight, J. G. 2008. *Synthetic Fuels Handbook: Properties, Processes, and Performance.* McGraw-Hill, New York.

Speight, J. G. (Editor). 2011. *The Biofuels Handbook. Royal Society of Chemistry,* London, United Kingdom.

Speight, J. G. 2013. *The Chemistry and Technology of Coal* 3rd Edition. CRC Press, Taylor & Francis Group, Boca Raton, Florida.

Speight, J. G. 2014a. *The Chemistry and Technology of Petroleum* 5th Edition. CRC Press, Taylor & Francis Group, Boca Raton, Florida.

Speight, J. G. 2014b. *Gasification of Unconventional Feedstocks*. Gulf Professional Publishing Company, Elsevier, Oxford, United Kingdom, 2014.

Speight, J. G. 2017. *Handbook of Petroleum Refining*. CRC Press, Taylor & Francis Group, Boca Raton, Florida.

Speight, J. G. 2019. *Handbook of Petrochemical Processes*. CRC Press, Taylor & Francis Group, Boca Raton, Florida.

Tang, L., and Huang, H. 2005. Plasma Pyrolysis of Biomass for the Production of Syngas and Carbon Adsorbent. *Energy Fuels*, 19: 1174-1178.

Tsai, W. T., Lee, M. K., and Chang, Y. M. 2007. Fast pyrolysis of rice husk: Product yields and compositions. *Bioresour. Technol.*, 98: 22-28.

Visscher, F., van der Schaaf, J., Nijhuis, X., and Schouten, J. 2013. Rotating Reactors – A Review, *Chemical Engineering Research and Design*, 91(10): 1923-1940.

Wagenaar, B. M., Prins, W., and Van Swaaij, W. P. M. 1994. Pyrolysis of Biomass in the Rotating Cone Reactor: Modelling and Experimental Justification. *Chemical Engineering Science*, 49(24 Part 2): 5109-5126.

Wang, X., Kersten, S. R. A., Prins, W., Van Swaaij, W. P. M. 2005. Biomass Pyrolysis in a Fluidized Bed Reactor. Part 2: Experimental Validation of Model Results. *Ind. Eng. Chem. Res*. 2005, 44, 8786–8795.

Yang, W., Ponzio, A., Lucas, C., and Blasiak, W. 2006. Performance Analysis of a Fixed–Bed Biomass Gasifier Using High-Temperature Air. *Fuel Proc. Technol.*, 87: 235-245.

Yang, H., Yan, R., Chen, H., Lee, D. H., and Zheng, C. 2007. Characteristics of Hemicellulose, Cellulose and Lignin Pyrolysis. *Fuel*, 86: 1781-1788.

Zhang, O., Chang, J., Wang, T., and Xu, Y. 2007. Review of Biomass Pyrolysis Oil Properties and Upgrading Research. *Energy Convers. Manag.,* 48: 87-92.

Zhao, X., Song, Z., Liu, H., Li, Z., Li, L., Ma, C. 2010. Microwave Pyrolysis of Corn Stalk Bale: A Promising Method for Direct Utilization of Large-Sized Biomass and Syngas Production. J. *Anal. Appl. Pyrolysis* 2010, 89, 87–94.

Zhao, X., Wang, M., Liu, H., Zhong, Li, L., Ma, C., and Song, Z. 2012. A Microwave Reactor for Characterization of Pyrolyzed Biomass. *Bioresource Technology*, 104: 673-678.

Chapter 6

BIOGAS BY GASIFICATION

1. INTRODUCTION

The gasification process is an extension of the pyrolysis process insofar as the thermal process is optimized to give the highest yield gas rather than to produce bio-oil and/or biochar. In any case, bio-oil and biochar can be used for the production of biogas through the agency of the gasification process. The technology used for the gasification process is a tried and true technology for coal (Speight, 2013a) and other carbonaceous feedstocks such as crude oil residua (Speight, 2014b). The majority of gasifiers are partial oxidation reactors, in which just sufficient air or oxygen is introduced to burn part of the input biomass to provide the heat for pyrolysis and gasification. In cases where the oxidant is air, the product gas is diluted by the nitrogen present, and although air is 79% v/v nitrogen, the stoichiometry of partial oxidation is such that the final product gas has approximately 50% v/v nitrogen as a diluent.

As a result the gas heating value of the fuel gas derived from air driven partial oxidation gasifiers is relatively low. The value can be higher if the feedstock is very dry, thus minimizing the heat demand for the process and the amount of oxidant required. The product gas has wide range of applications ranging from power generation to chemicals production. The power derived from the gasification of carbonaceous feedstocks followed by the combustion of the product gas(es) is considered to be a source of renewable energy of derived gaseous products (Table 6.1) that are generated from a carbonaceous source (e.g., biomass) (Speight, 2008, 2014b).

If the requirement of the process is a high heat content gas, the production of products such as methane, ethane, and propane further increases the calorific value. The use of pure oxygen as the gasification agent eliminates the nitrogen diluent and can produce medium calorific value gaseous products. An alternative strategy is to carry out the gasification process by means of indirect heat input. In this case the product stream is even higher in calorific value, as neither nitrogen nor the carbon dioxide produced from the combustion *in-situ* of the partial oxidation processes is present in the product gas stream. The challenges to achieve a clean and useable fuel gas have been addressed through gasifier design and post-gasification processing to remove tar and particulate contaminants in the gas stream.

Table 6.1. Gasification Products

Product	Properties
Low-Btu gas	150 to 300 Btu/scf
	Approximately 50% v/v nitrogen
	Smaller amounts of carbon monoxide and hydrogen
	Some carbon dioxide
	Trace amounts of methane
Medium Btu gas	300 to 550 Btu/scf
	Predominantly carbon monoxide and hydrogen
	Small amounts of methane
	Some carbon dioxide
High-Btu gas	980 to 1080 Btu/scf
	Predominantly methane - typically >85% v/v

The composition and amount of tar depends on the gasifier type and gasification temperature. Tar from the fluidized bed gasifier can be characterized as secondary and tertiary tar. For example, increasing the temperature of the gasification process typically reduces the amount of tar (the primary tar) but increases the share of higher molecular weight (and higher-boiling) constituents (the secondary tar). The effect of pressure on tar composition is variable and process dependent. However, it has been observed that an increase in pressure may drive the composition of the tar to the higher boiling constituents.

In terms of the use of biomass feedstocks and, hence, the production of biogas, the process has grown from a predominately coal conversion process used for making *town gas* for industrial lighting to an advanced process for the production of multi-product, carbon-based fuels from a variety of feedstocks such as the biomass alone, biomass mixed with coal, and biomass mixed with the viscous products from crude oil refining and any other carbonaceous feedstocks (Figure 6.1) (Kumar et al., 2009; Speight, 2013a, 2014a, 2014b; Luque and Speight, 2015).

Gasification is a key process for the thermo-chemical conversion of biomass. In the presence of a gasifying agent (such as oxygen or air), biomass is converted to a multifunctional gaseous mixture, usually called bio-synthesis gas or bio-syngas, which can be used for the production of energy (heat and/or electricity generation), chemicals (ammonia), and biofuels. Furthermore, a solid residue after biomass conversion (char) is generally found (Bridgwater, 2003; Rauch et al., 2014; Molino et al., 2016; Sikarwar et al., 2017). Syngas consists of a mixture of carbon monoxide, hydrogen, methane, carbon dioxide (the primary components) and water, hydrogen sulfide, ammonia (secondary components), as well as tarry constituents and other trace species with a composition dependent on feedstock type and characteristics, operating conditions (i.e., the gasifying agents, gasifier temperature and pressure, type of bed materials), and gasification technology (Molino et al., 2013; Asadullah, 2014; Rodríguez-Olalde et al., 2015; Ahmad et a., 2016).

Gasification is an appealing process for the utilization of relatively inexpensive feedstocks, such as biomass and waste, that might in the case of waste be sent to a landfill (where the production of methane – a so-called greenhouse gas – will be produced and often escapes into the atmosphere) or combusted which may not (depending upon the feedstock) be energy efficient. Because of the development of the process using coal as the feedstock (Speight, 2013), biomass is often considered to be an unconventional carbonaceous feedstock which also include as heavy oil, extra heavy oil, tar sand bitumen, and crude oil residua (Speight, 2014b).

A biomass gasifier needs uniform-sized and dry fuel for smooth and trouble-free operation. Most gasifier systems are designed either for woody biomass (or dense briquettes made from loose biomass) or for loose, pulverized biomass. The feedstock requirements vary depending upon the biomass fed to the gasifier, for example: (i) for woody biomass: pieces smaller than 5 to 10 cm (2 to 4 inches) in any dimension, depending on design and bulk density of wood or briquettes, and (ii) for loose biomass: pulverized biomass, depending on design, moisture content up to 15 to 25% w/w, depending on gasifier design, and a mineral matter content (manifested as product ash) below 5% w/w is preferred; with a maximum limit of 20% w/w.

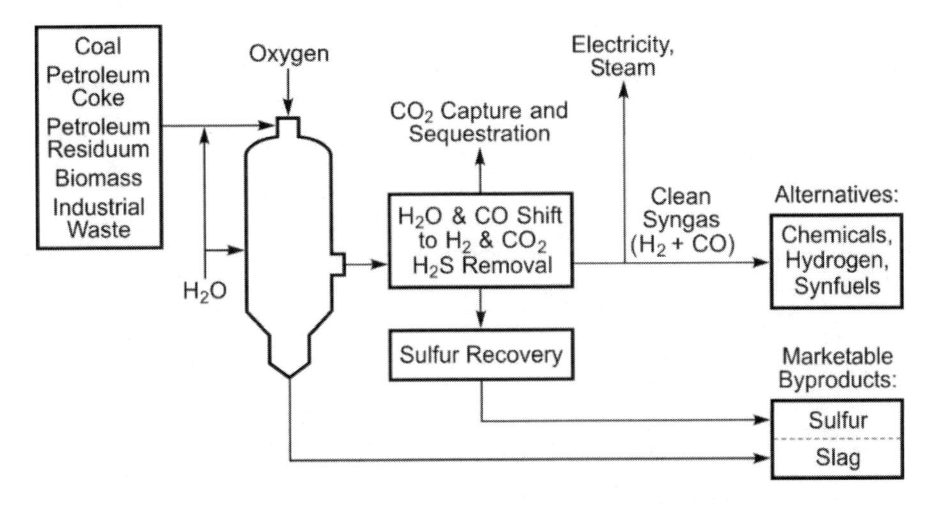

Figure 6.1. The Gasification Process Can Accommodate a Variety of Carbonaceous Feedstocks.

Overall, use of a gasification technology (Speight, 2013a, 2014b) with the necessary gas cleanup options can have a smaller environmental footprint and lesser effect on the environment than landfill operations or combustion of the biomass or organic waste. In fact, there are strong indications that gasification is a technically viable option for the waste conversion, including residual waste from separate collection of municipal solid waste. The process can meet existing emission limits and can have a significant effect on the reduction of landfill disposal using known gasification technologies to produce a biogas product than can be employed as a fuel, a petrochemical feedstock as well as for use as a feedstock to the Fischer-Tropsch process (Arena, 2012; Fabry et al., 2013; Speight, 2014b; Luque and Speight, 2015).

2. Gasification Chemistry

The gasification process consists of the conversion of the organic carbon in the biomass into various constituents that make up the biogas constituents by reacting the biomass at high temperatures (>700°C, 1290°F), without combustion using a controlled amount of oxygen and/or steam (Marano, 2003; Lee et al., 2007; Higman and Van der Burgt, 2008; Speight, 2008; Sutikno and Turini, 2012; Speight, 2013a, 2014b; Luque and Speight, 2015). The products include gaseous hydrocarbon derivatives (methane, ethane, and propane), carbon monoxide, carbon dioxide, and hydrogen. Other products may also be produced depending upon the composition of the feedstock.

Table 6.2. Reactions that Occur During Gasification of a Carbonaceous Feedstock

Reaction
$2C + O_2 \rightarrow 2CO$
$C + O2 \rightarrow CO_2$
$C + CO_2 \rightarrow 2CO$
$CO + H_2O \rightarrow CO_2 + H_2$ (shift reaction)
$C + H_2O \rightarrow CO + H_2$ (water gas reaction)
$C + 2H_2 \rightarrow CH_4$
$2H_2 + O_2 \rightarrow 2H_2O$
$CO + 2H2 \rightarrow CH_3OH$
$CO + 3H2 \rightarrow CH_4 + H_2O$ (methanation reaction)
$CO_2 + 4H_2 \rightarrow CH_4 + 2H_2O$
$C + 2H_2O \rightarrow 2H_2 + CO_2$
$2C + H_2 \rightarrow C_2H_2$
$CH_4 + 2H_2O \rightarrow CO_2 + 4H_2$

However, it is important to distinguish the gasification process from a pyrolysis process. The main difference between pyrolysis and gasification is the absence of a gasifying agent in a pyrolysis process (Chapter 5). Pyrolysis is a thermal degradation of organic compounds, at a range of temperatures on the order of 300 to 900°C (570 to

1650°F), under oxygen-deficient circumstances to produce various types of products such as gases (often referred to as biogas), a liquid product (referred to as called bio-oil), and a solid product (referred to as biochar).

On the other hand, the gasification process involves thermal cracking of solid or high boiling liquid carbonaceous material (such as bio-oil biomass, biomass char) into a combustible gas mixture, mainly composed of hydrogen, carbon monoxide (CO), carbon dioxide (CO_2), and methane (CH_4) and other gases with some byproducts (solid char or slag, oils, and water). The biogas product has a chemical composition and properties that are largely affected by (i) the type and composition of the biomass feedstock, (ii) the operational conditions throughout pyrolysis and gasification such as reactor temperature, residence time, pressure, and (iii) the reactor geometry.

In the current context, and in addition to the potential reactions that can occur in the gasifier (Table 6.2), the gasification process is often considered to involve two distinct physico-chemical stages: (i) devolatilization of the feedstock to produce volatile matter and char, (ii) followed by char gasification, which is complex and specific to the conditions of the reaction – both processes contribute to the complex kinetics of the gasification process (Sundaresan and Amundson, 1978; Luque and Speight, 2015).

2.1. General Aspects

Generally, the gasification of carbonaceous feedstocks (such as biomass, waste, heavy oil, extra heavy oil, tar sand bitumen, crude oil residua,) includes a series of reaction steps that convert the feedstock into *synthesis gas* (carbon monoxide, CO, plus hydrogen, H_2) or, if biomass is the feedstock, into bio-synthesis gas and other gaseous products (Table 6.2). The gaseous products and the proportions of these product gases (such as methane, CH_4, and other hydrocarbon derivatives, water vapor (H_2O), carbon dioxide, (CO_2), carbon monoxide (CO), hydrogen (H_2), hydrogen sulfide (H_2S), and sulfur dioxide, (SO_2) depends on: (i) the type of feedstock, (ii) the chemical composition of the feedstock, (iii) the gasifying agent or gasifying medium, as well as (iv) the thermodynamics and chemistry of the gasification reactions as controlled by the process operating parameters (Singh et al., 1980; Pepiot et al., 2010; Shabbar and Janajreh, 2013; Speight, 2013a, 2013b, 2014b; Luque and Speight, 2015). In addition, the kinetic rates, the extent of the conversion for any feedstock (irrespective of the composition), as well as the several chemical reactions that are a part of the gasification process are variable and are typically functions of (i) temperature, (ii) pressure, and (iii) reactor configuration, the (iv) gas composition of the product gases, and (v) whether or not the product gases influence the outcome of the reaction (Johnson, 1979; Penner, 1987; Müller et al., 2003; Slavinskaya et al., 2009; Speight, 2013a, 2013b, 2014b; Luque and Speight, 2015).

At a temperature in excess of 500°C (930°F) the conversion of the feedstock to char and ash and char is completed. In most of the early gasification processes, this was the desired by-product but for gas generation the char provides the necessary energy to effect further heating and – typically, the char is contacted with air or oxygen and steam to generate the product gases. Furthermore, with an increase in heating rate, feedstock particles are heated more rapidly and are burned in a higher temperature region, but the increase in heating rate may have very little effect on the mechanism (Irfan, 2009).

Most notable effects in the physical chemistry of the gasification process are those effects due to the chemical character of the feedstock as well as the physical composition of the feedstock (Speight, 2011a, 2013a, 2014a, 2014b). In more general terms of the character of the feedstock, gasification technologies generally require some initial processing of the feedstock with the type and degree of pretreatment a function of the process and/or the type of feedstock. This is especially true for biomass there the feedstock can have several constituents that strongly influence the course and outcome of the conversion process (Chapter 2).

Another factor, often presented as very general *rule of thumb*, is that optimum gas yields and gas quality are obtained at operating temperatures of approximately 595 to 650°C (1100 to 1200°F. A gaseous product with a higher heat content (BTU/ft.^{3}) can be obtained at lower system temperatures but the overall yield of gas (determined as the *fuel-to-gas ratio*) is reduced by the unburned char fraction.

With some feedstocks, the higher the amounts of volatile material produced in the early stages of the process the higher the heat content of the product gas. However, in the case of biomass, there is the potential for carbon dioxide to be an initial product of the process which will serve to lower the heat content of the product gas. Also, when the process temperature is not sufficiently high, the char oxidation reaction is suppressed and the overall heat content of the product gas is diminished. All such events serve to complicate the reaction rate and make the attempted derivation of a kinetic relationship that can be applied to all types of feedstock (and gasifier configurations) subject to serious question and doubt.

Pressure also plays a role in the type of product produced and is, as might be anticipated, dependent on the type of feedstock being processed and the gas product that is desired. In fact, some (or all) of the following processing steps will be required: (i) pretreatment of the feedstock, (ii) primary gasification of the feedstock, (iii) secondary gasification of the carbonaceous residue from the primary gasifier; (iv) removal of carbon dioxide, hydrogen sulfide, and other acid gases, (v) shift conversion for adjustment of the carbon monoxide/hydrogen mole ratio to the desired ratio, and (vi) catalytic methanation of the carbon monoxide/hydrogen mixture to form methane.

While pretreatment of the biomass feedstock for introduction into the gasifier is often considered to be a physical process in which the feedstock is prepared for gasifier – typically as pellets or finely-ground feedstock – there are chemical aspects that must also be considered (Chapter 2). Co-gasification of other feedstocks, such as coal and with the residuum or coke from crude oil refining (often referred to as petcoke) offers a bridge between the depletion of crude oil stocks when coal, residua, and/or petcoke are used (Speight, 2014b).

Table 6.3. Characteristics of the Different Types of Gasifiers

Gasifier type	Fuel properties
Fixed/moving bed	Particle size: 1 to 10 cm
	Mechanically stable fuel particles (unblocked passage of gas through the bed)
	Pellets or briquettes preferred
	Updraft configuration more tolerant to biomass moisture content (up to 40 to 50% w/w
	Drying occurs as biomass moves down the gasifier
Fluidized bed	Ash melting temperature of fuel: higher limit for operating temperature.
	Fuel particle size relatively small to ensure good contact with bed material; typically <40 mm
	Good fuel flexibility due to high thermal inertia of the bed.
Entrained bed	Fuel particle size: < 50 micrometers.
	Pulverized for high fuel conversion in short residence times
	Low moisture content
	Ash melting behavior can influence for reactor/process design.

The high reactivity of biomass and the accompanying high production of volatile products suggest that some synergetic effects might occur in simultaneous thermochemical treatment of coal, residua, or petcoke, depending on the gasification conditions such as: (i) feedstock type and origin, (ii) reactor type, and (iii) process parameters (Penrose et al., 1999; Gray and Tomlinson, 2000; McLendon et al., 2004; Lapuerta et al., 2008; Fermoso et al., 2009; Shen et al., 2012; Khosravi and Khadse, 2013; Speight, 2013a, 2014a, 2014b; Luque and Speight, 2015).

For example, carbonaceous fuels are gasified in reactors, a variety of gasifiers such as the fixed- or moving-bed, fluidized-bed, entrained-flow, and molten bath gasifiers have been developed that have differing feedstock requirements (Table 6.3) (Shen et al., 2012; Speight, 2014b; Luque and Speight, 2015). If the flow patterns are considered, the fixed-bed and fluidized-bed gasifiers intrinsically pertain to a counter-current reactor in that fuels are usually sent into the reactor from the top of the gasifier, whereas the oxidant is blown into the reactor from the bottom.

2.2. Reactions

The gasification process is not a single step process, but involves a variety of reactions (Table 6.2), and like any gasification process, the gasification of biomass (including bio-oil and biochar) is a complex thermal process that depends on the pyrolysis mechanism to generate gaseous precursors, which in the presence of reactive gases such as oxygen and steam convert the majority of the biomass into a biogas. The biogas product can in turn be further purified to synthesis gas, which is mainly composed of carbon monoxide and hydrogen, and used to produce chemicals and liquid fuels over catalysts (Speight, 2016, 2017; 2019). The distribution of weight and chemical composition of the products are also influenced by the prevailing conditions (i.e., temperature, heating rate, pressure, and residence time) and, last but by no means least, the nature of the feedstock (Speight, 2014a, 2014b).

If air is used for combustion, the product gas will have a heat content of approximately 150-300 Btu/ft^3 (depending on process design characteristics) and will contain undesirable constituents such as carbon dioxide, hydrogen sulfide, and nitrogen. The use of pure oxygen, although expensive, results in a product gas having a heat content on the order of 300 to 400 Btu/ft^3 with carbon dioxide and hydrogen sulfide as byproducts (both of which can be removed from low or medium heat-content, low-Btu or medium-Btu) gas by any of several available processes (Speight, 2013a, 2014a).

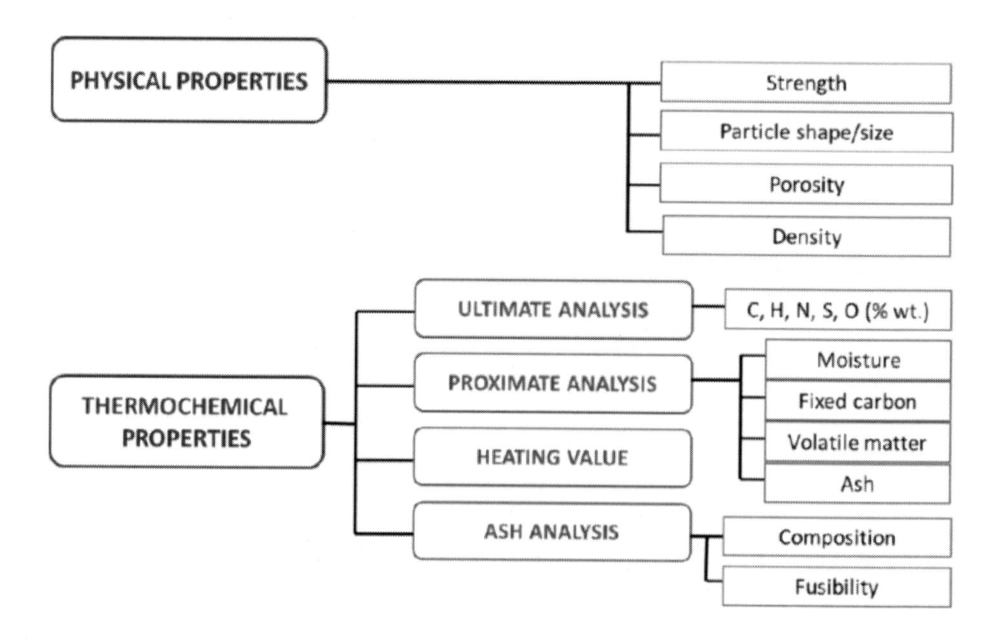

Figure 6.2. Biomass Properties that Influence the Gasification Process.

If a high heat-content (high-Btu) gas (900 to 1000 Btu/ft^3) is required, efforts must be made to increase the methane content of the gas. The reactions which generate methane are all exothermic and have negative values, but the reaction rates are relatively slow and catalysts may, therefore, be necessary to accelerate the reactions to acceptable commercial rates. Indeed, the overall reactivity of the feedstock and char may be subject to catalytic effects insofar as the mineral constituents of the feedstock (such as the mineral matter in biomass) may modify the reactivity by a direct catalytic effect (Davidson, 1983; Baker and Rodriguez, 1990; Mims, 1991; Martinez-Alonso and Tascon, 1991).

Biomass begins to rapidly decompose with heat once its temperature rises above 240°C (465°F) and the biomass breaks down into a combination of gases, liquids, and solid (char). In the process, the feedstock undergoes three steps in the conversation to gas – the first two steps, pyrolysis and combustion, occur very rapidly – all of which are highly dependent upon the properties of the biomass (Figure 6.2). However, there is also a preliminary step in which any moisture in the biomass before it enters the pyrolysis step. All of the moisture needs to be (or will be) removed from the biomass at some point (typically between 100 and 150°C, 212 and 300°F) in the higher temp processes. Moisture removal (often referred to as drying) is one of the major issues that has to be solved for successful gasification of biomass. The high moisture content of the feedstock and/or poor handling of the moisture internally, is one of the most common reasons for failure to produce clean gas in the process.

In the pyrolysis step, char is produced as the feedstock heats up and volatiles are released. In the combustion step, the volatile products and some of the char reacts with oxygen to produce various products (primarily carbon dioxide and carbon monoxide) and the heat required for subsequent gasification reactions. Finally, in the gasification step, the feedstock char reacts with steam to produce hydrogen (H_2) and carbon monoxide (CO).

Drying:

$$\text{Biomass (wet)} \rightarrow \text{biomass (dry)} + H_2O$$

Pyrolysis:

$$\text{Biomass} \rightarrow \text{volatiles} + \text{char}$$

Combustion:

$$2C_{\text{volatiles + char}} + O_2 \rightarrow 2CO + H_2O$$

Gasification:

$$C_{\text{char}} + H_2O \rightarrow H_2 + CO$$

$$CO + H_2O \rightarrow H_2 + CO_2$$

Thus, in the initial stages of gasification, the rising temperature of the feedstock initiates devolatilization and the breaking of weaker chemical bonds to yield volatile tar, volatile oil, phenol derivatives, and hydrocarbon gases. These products generally react further in the gaseous phase to form hydrogen, carbon monoxide, and carbon dioxide. The char (fixed carbon) that remains after devolatilization reacts with oxygen, steam, carbon dioxide, and hydrogen.

In essence, the direction of the gasification process is subject to the constraints of thermodynamic equilibrium and variable reaction kinetics. The combustion reactions (reaction of the feedstock or char with oxygen) essentially go to completion. The thermodynamic equilibrium of the rest of the gasification reactions are relatively well defined and collectively have a major influence on thermal efficiency of the process as well as on the gas composition. Thus, thermodynamic data are useful for estimating key design parameters for a gasification process, such as: (i) calculating of the relative amounts of oxygen and/or steam required per unit of feedstock, (ii) estimating the composition of the produced bio-synthesis gas, and (iii) optimizing process efficiency at various operating conditions.

Relative to the thermodynamic understanding of the gasification process, the kinetic behavior is much more complex. In fact, very little reliable kinetic information (for general application) on gasification reactions exists, partly because it is highly dependent on (i) the chemical nature of the feed, which varies significantly with respect to composition, mineral impurities, (ii) feedstock reactivity, and (iii) process parameters, such as temperature, pressure, and residence time. In addition, physical characteristics of the feedstock (or char) also play a role in phenomena such boundary layer diffusion, pore diffusion and ash layer diffusion which also influence the kinetic outcome. Furthermore, certain impurities, in fact, are known to have catalytic activity on some of the gasification reactions which can have further influence on the kinetic imprint of the gasification reactions.

Alkali metal salts are known to catalyze the steam gasification reaction of carbonaceous materials, including biomass – hence the need for pretreatment of biomass to remove such metals (Chapter 2). The process is based on the concept that alkali metal salts (such as potassium carbonate, sodium carbonate, potassium sulfide, sodium sulfide, and the like) will catalyze the steam gasification of feedstocks. The order of catalytic activity of alkali metals on the gasification reaction is:

Cesium (Cs) > rubidium (Rb) > potassium (K) > sodium (Na) > lithium (Li)

Ruthenium-containing catalysts are used primarily in the production of ammonia. It has been shown that ruthenium catalysts provide five to 10 times higher reactivity rates than other catalysts. However, ruthenium quickly becomes inactive due to its necessary supporting material, such as activated carbon, which is used to achieve effective reactivity. However, during the process, the carbon is consumed, thereby reducing the effect of the ruthenium catalyst.

Catalysts can also be used to favor or suppress the formation of certain components in the gaseous product by changing the chemistry of the reaction, the rate of reaction, and the thermodynamic balance of the reaction. For example, in the production of synthesis gas (mixtures of hydrogen and carbon monoxide), methane is also produced in small amounts. Catalytic gasification can be used to either promote methane formation or suppress it.

2.2.1. Water Gas Shift Reaction

If a high-Btu gas is required, the option is application of the water gas shift reaction, in which carbon monoxide is converted to hydrogen and carbon dioxide in the presence of a catalyst:

$$CO + H_2O \rightarrow CO_2 + H_2$$

This reaction maximizes the hydrogen content of the gas and the product gas is then scrubbed of particulate matter and sulfur is removed via physical absorption (Speight, 2013a, 2014a). The carbon dioxide is captured by physical absorption or a membrane and either vented or sequestered.

The water gas shift reaction in its forward direction is mildly exothermic and although all of the participating chemical species are in gaseous form, the reaction is believed to be heterogeneous insofar as the chemistry occurs at the surface of the feedstock and the reaction is actually catalyzed by carbon surfaces. In addition, the reaction can also take place homogeneously as well as heterogeneously and a generalized understanding of the water gas shift reaction is difficult to achieve. Even the published kinetic rate information is not immediately useful or applicable to a practical reactor situation.

The water gas shift reaction is one of the major reactions in the steam gasification process, where both water and carbon monoxide are present in ample amounts. Although the four chemical species involved in the water gas shift reaction are gaseous compounds at the reaction stage of most gas processing, the water gas shift reaction, in the case of steam gasification of feedstock, predominantly takes place on the solid surface of feedstock (heterogeneous reaction). If the bio-synthesis gas from a gasifier needs to be reconditioned by the water gas shift reaction, this reaction can be catalyzed by a variety of metallic catalysts.

Choice of specific kinds of catalysts has always depended on the desired outcome, the prevailing temperature conditions, composition of gas mixture, and process economics. Typical catalysts used for the reaction include catalysts containing iron, copper, zinc, nickel, chromium, and molybdenum.

2.2.2. Carbon Dioxide Gasification

The reaction of carbonaceous feedstocks with carbon dioxide produces carbon monoxide (*Boudouard reaction*) and (like the steam gasification reaction) is also an endothermic reaction:

$$C(s) + CO_2(g) \rightarrow 2CO(g)$$

The reverse reaction results in carbon deposition (carbon fouling) on many surfaces including the catalysts and results in catalyst deactivation. This reaction may be valuable when biomass is the feedstock because of the potential for the production of carbon dioxide directly from the biomass by the decomposition of carboxylic acid functions:

$$[\equiv C\text{-}CO_2H] \rightarrow CO_2 + [\equiv CH]$$

This gasification reaction is thermodynamically favored at high temperatures (>680°C, >1255°F), which is also quite similar to the steam gasification. If carried out alone, the reaction requires high temperature (for fast reaction) and high pressure (for higher reactant concentrations) for significant conversion but as a separate reaction a variety of factors come into play: (i) low conversion, (ii) slow kinetic rate, and (iii) low thermal efficiency.

Also, the rate of the carbon dioxide gasification of a feedstock is different to the rate of the carbon dioxide gasification of carbon. Generally, the carbon-carbon dioxide reaction follows a reaction order based on the partial pressure of the carbon dioxide that is approximately 1.0 (or lower) whereas the feedstock-carbon dioxide reaction follows a reaction order based on the partial pressure of the carbon dioxide that is 1.0 (or higher). The observed higher reaction order for the feedstock reaction is also based on the relative reactivity of the feedstock in the gasification system.

2.2.3. Hydrogasification

Hydrogasification is the gasification of feedstock, such as biomass, in the presence of an atmosphere of hydrogen under pressure. Not all high heat-content (high-Btu) gasification technologies depend entirely on catalytic methanation and, in fact, a number of gasification processes use hydrogasification, that is, the direct addition of hydrogen to feedstock under pressure to form methane.

$$C_{char} + 2H_2 \rightarrow CH_4$$

The hydrogen-rich gas for hydrogasification can be manufactured from steam and char from the hydrogasifier. Appreciable quantities of methane are formed directly in the primary gasifier and the heat released by methane formation is at a sufficiently high temperature to be used in the steam-carbon reaction to produce hydrogen so that less oxygen is used to produce heat for the steam-carbon reaction. Hence, less heat is lost in the low-temperature methanation step, thereby leading to higher overall process efficiency.

The hydrogen-rich gas for hydrogasification can be manufactured from steam by using the char that leaves the hydrogasifier. Appreciable quantities of methane are formed directly in the primary gasifier and the heat released by methane formation is at a sufficiently high temperature to be used in the steam-carbon reaction to produce hydrogen so that less oxygen is used to produce heat for the steam-carbon reaction. Hence, less heat is lost in the low-temperature methanation step, thereby leading to higher overall process efficiency.

The hydrogasification reaction is exothermic and is thermodynamically favored at relatively low temperature (<670°C, <1240°F), unlike the endothermic both steam gasification and carbon dioxide gasification reactions. However, at low temperatures, the reaction rate is inevitably too slow. Therefore, a high temperature is always required for kinetic reasons, which in turn requires high pressure of hydrogen, which is also preferred from equilibrium considerations. This reaction can be catalyzed by salts such as potassium carbonate (K_2CO_3), nickel chloride ($NiCl_2$), iron chloride ($FeCl_2$), and iron sulfate ($FeSO_4$). However, use of a catalyst in feedstock gasification suffers from difficulty in recovering and reusing the catalyst and the potential for the spent catalyst becoming an environmental issue.

2.2.4. Methanation

Several exothermic reactions may occur simultaneously within a methanation unit. A variety of metals have been used as catalysts for the methanation reaction; the most common, and to some extent the most effective methanation catalysts, appear to be nickel and ruthenium, with nickel being the most widely (Cusumano et al., 1978):

Ruthenium (Ru) > nickel (Ni) > cobalt (Co) > iron (Fe) > molybdenum (Mo).

Nearly all the commercially available catalysts used for this process are, however, very susceptible to sulfur poisoning and efforts must be taken to remove all hydrogen sulfide (H_2S) before the catalytic reaction starts. It is necessary to reduce the sulfur concentration in the feed gas to less than 0.5 ppm v/v in order to maintain adequate catalyst activity for a long period of time.

The gas must be desulfurized before the methanation step since sulfur compounds will rapidly deactivate (poison) the catalysts. A processing issues may arise when the concentration of carbon monoxide is excessive in the stream to be methanated since large amounts of heat must be removed from the system to prevent high temperatures and deactivation of the catalyst by sintering as well as the deposition of carbon. To eliminate this problem, temperatures should be maintained below 400°C (750°F).

The methanation reaction is used to increase the methane content of the product gas, as needed for the production of high-Btu gas.

$$4H_2 + CO_2 \rightarrow CH_4 + 2H_2O$$
$$2CO \rightarrow C + CO_2$$
$$CO + H_2O \rightarrow CO_2 + H_2$$

Among these, the most dominant chemical reaction leading to methane is the first one. Therefore, if methanation is carried out over a catalyst with a bio-synthesis gas mixture of hydrogen and carbon monoxide, the desired hydrogen-carbon monoxide ratio of the feedstock gas is on the order of 3:1. The high proportion of water (vapor) produced is removed by condensation and recirculated as process water or steam. During this process, most of the exothermic heat due to the methanation reaction is also recovered through a variety of energy integration processes.

Whereas all the reactions listed above are quite strongly exothermic except the forward water gas shift reaction, which is mildly exothermic, the heat release depends largely on the amount of carbon monoxide present in the feedstock gas. For each 1% v/v carbon monoxide in the feedstock gas, an adiabatic reaction will experience a 60°C (108°F) temperature rise, which may be termed as *adiabatic temperature rise*.

3. Gasification Processes

The typical gasification system consists of several process plants including (i) a feedstock preparation area, (ii) the type of gasifier, (iii) a gas cleaning section, and (iv) a sulfur recovery unit, as well as (v) downstream process options that are dependent on the nature of the products. By way of further explanation, the gas generated from biomass by gasification processes can be differentiated in direct (autothermal) and indirect (or allothermal) processes. For biomass applications the direct processes are typically operated with air as the gasifying agent

Typically, gasification is divided into four steps: drying (endothermic step), pyrolysis (endothermic step), oxidation (exothermic stage), and reduction (endothermic stage). In the drying step, when the biomass (which typically contains 10 to 355 w/w moisture) is heated to about 100°C (212°F), the moisture is converted into steam. After drying, as heating continues, the biomass undergoes pyrolysis which involves burning biomass completely without supplying any oxygen. As a result, the biomass is decomposed or separated into gases, liquids, and solids. During oxidation, which takes place at about 700 to 1400°C (1290 to 2550°F), the solid (carbonized) biomass, reacts with the oxygen in the air to produce carbon dioxide and heat:

$$C + O_2 \rightarrow CO_2 + \text{heat}$$

At higher temperatures and under reducing conditions, when insufficient oxygen is available, the following reactions take place forming carbon dioxide, hydrogen, and methane:

$$C + CO_2 \rightarrow 2CO$$
$$C + H_2O \rightarrow CO + H_2$$
$$CO + H_2O \rightarrow CO_2 + H_2$$
$$C + 2H_2 \rightarrow CH_4$$

Tar-reforming can also be added as a step to produce low molecular weight hydrocarbon derivatives from higher molecular weight constituents of the tarry products. Although the chemistry of biomass gasification is complex (Table 6.4), the overall process can be represented by a simple equation. Thus:

Biomass → CO +H2 + CO2 + CH4 +H2O +H2S +NH3 + CxHy + Tar + Char

Table 6.4. Predominant Reactions Occurring During Biomass Gasification

Sub-process	Reaction
Pyrolysis	Biomass → CO +H_2 + CO +CH_4 + H_2O + Tar + Char
Oxidation	Char + O_2 → CO_2 (Char Oxidation)
	2C + O_2 → 2CO (Partial Oxidation)
	$2H_2$ + $2O_2$ → $2H_2O$ (Hydrogen Oxidation)
Reduction	C + CO_2 ↔ 2CO (Boudouard Reaction)
	C +H_2O ↔ CO +H2 (Reforming of Char)
	CO +H_2O ↔ CO_2 +H_2 (Water Gas Shift (WGS) Reaction)
	C + $2H_2$ ↔ CH_4 (Methanation Reaction) CH_4 +H_2O ↔ CO + $3H_2$ (Steam Reforming of Methane)
	CH_4 + CO_2 ↔ 2CO + $2H_2$ (Dry Reforming of Methane)
Tar Reforming	Tar +H_2O → H_2 + CO_2 + CO + CxHy (Steam Reforming of Tar)

3.1. Gasifiers

A gasifier differs from a combustor in that the amount of air or oxygen available inside the gasifier is carefully controlled so that only a relatively small portion of the fuel burns completely. Thus, the gasifier contains separate sections for gasification and combustion. The gasification section consists of three parts: (i) the riser, (ii) the settling chamber, and (iii) the downcomer whereas the combustion section contains only one part, the combustor.

In the process, biomass is fed into the riser along with a small amount of superheated steam. Hot bed material (typically 925°C, 1695°F, sand or olivine of 0.2 to 0.3 mm particle size) enters the riser from the combustor through a hole in the riser (opposite of biomass feeding point). The bed material heats the biomass to 850°C (1550°F) causing the biomass particles to undergo conversion into gas. The volume created by the gas from the biomass results in the creation of a turbulent fluidization regime in the riser causing carry-over of the bed material together with the degasified biomass particles (char). The settling chamber reduces the vertical velocity of the gas causing the larger solids (bed material and char) to separate from the gas and to pass into the downcomer. The gas stream leaves the reactor from the top and is sent to the cooling and gas cleaning section in order to remove contaminants such as dust, tar, chloride and sulfur (Chapter 7) before the catalytic conversion of the gas into biomethane.

Four types of gasifier are currently available for commercial use: (i) the counter-current fixed bed, (ii) co-current fixed bed, (iii) the fluidized bed, and (iv) the entrained flow (Speight, 2008, 2013a; Luque and Speight, 2015).

In a *fixed-bed process*, the feedstock is supported by a grate and combustion gases (steam, air, oxygen, etc.) pass through the supported feedstock whereupon the hot produced gases exit from the top of the reactor. Heat is supplied internally or from an outside source, but some feedstocks cannot be used in an unmodified fixed-bed reactor. Due to the liquid-like behavior, the fluidized-beds are very well mixed, which effectively eliminates the concentration and temperature gradients inside the reactor. The process is also relatively simple and reliable to operate as the bed acts as a large thermal reservoir that resists rapid changes in temperature and operation conditions. The disadvantages of the process include the need for recirculation of the entrained solids carried out from the reactor with the fluid, and the non-uniform residence time of the solids that can cause poor conversion levels. The abrasion of the particles can also contribute to serious erosion of pipes and vessels inside the reactor (Kunii and Levenspiel 1991) – another reason why feedstock pretreatment (Chapter 2) is essential.

Fixed bed gasifiers are classified as updraft and downdraft gasifiers (Molino, 2018). In the former, biomass is supplied from the top, while the gasifying agent (air or oxygen) is supplied from the bottom (counter-current). In the latter type of gasifier, the biomass and the gasifying agent are introduced from the top (co-current). For updraft reactors, the sequence of the biomass is drying, pyrolysis, and reduction, finally arriving at the combustion zone, with syngas drawn out from the top. In the case of downdraft reactors, both biomass and oxygen or air are supplied in the drying zone, going the through pyrolysis, combustion, and reduction, with syngas drawn out from the bottom.

The *counter-current fixed bed (updraft) gasifier* consists of a fixed bed of carbonaceous fuel (such as biomass) through which the *gasification agent* (steam, oxygen and/or air) flows in counter-current configuration. An updraft gasifier has distinctly defined zones for partial combustion, reduction, pyrolysis, and drying. The gas leaves the gasifier reactor together with the products of pyrolysis from the pyrolysis zone and steam from the drying zone. The resulting combustible gas is rich in higher molecular weight hydrocarbon derivatives (tars) and, therefore, has a higher calorific value, which makes updraft gasifiers more suitable where the gas is to be used for heat production. The mineral ash is either removed dry or as a slag.

The *co-current fixed bed (downdraft) gasifier* is similar to the counter-current type, but the gasification agent gas flows in co-current configuration with the fuel (downwards, hence the name down draft gasifier). The term co-current is used because air moves in the same direction as that of fuel, downwards. A downdraft gasifier is so designed that tar product, which are produced in the pyrolysis zone, travel through the combustion zone, where the tar is decomposed into lower molecular weight products or burned. As a result, the mixture of gases in the exit stream is relatively clean. The position of the combustion zone is thus a critical element in the downdraft gasifier – the main advantage being that it produces gas with low tar content. The operating temperature varies from a minimum of 900°C (1650°F) to a maximum of 1000 to 1050°C (1830 to 1920°F). In a *cross-draft gasifier*, air enters from one side of the gasifier reactor and leaves from the other. Also, cross-draft gasifiers do not need a grate; the ash falls to the bottom and does not come in the way of normal operation.

In the *fluidized bed gasifier*, thc biomass is fluidized in oxygen (or air) and steam and the biomass is brought into an inert bed of fluidized material (such as sand or char) – the air is distributed through nozzles located at the bottom of the bed. The fluidized bed reactors are diverse, but characterized by the fluid behavior of the catalyst (Dry (2001). Typically, the feedstock is introduced into the fluidized system either above-bed or directly into the bed, depending upon the size and density of the feedstock and how it is affected by the bed velocities. During normal operation, the bed media is maintained at a temperature between 550 and 1000°C (1020 and 1830°F). When the biomass is introduced under such temperature conditions, the drying and pyrolysis steps proceed rapidly, driving off all gaseous portions of the fuel at relatively low temperature. The remaining char is oxidized within the bed to provide the heat source for the drying and devolatilizing reactions to continue. Fluidized bed gasifiers produce a uniformly high (800 to 1000°C, 1470 to 1830°F) bed temperature. A fluidized bed gasifier works as a hot bed of sand particles agitated constantly by air. Fluidized bed gasifiers are most useful for fuels that form highly corrosive ash that would damage the walls of slagging gasifiers. The ash is removed dry or as high molecular weight agglomerated materials – a disadvantage of biomass feedstocks is that they generally contain high levels of corrosive ash.

In the *entrained flow gasifier*, a dry pulverized solid, an atomized liquid fuel or a fuel slurry is gasified with oxygen (much less frequent: air) in co-current flow. The high temperatures and pressures also mean that a higher throughput can be achieved but thermal efficiency is somewhat lower as the gas must be cooled before it can be sent to a gas processing facility. All entrained flow gasifiers remove the major part of the ash as a slag as the operating temperature is well above the ash fusion temperature. The operating temperatures are on the order of 1200 to 1600°C (2190 to 2910°F) and the pressure is 300 to 1200 psi. Entrained-flow gasifiers are suitable for use with biomass feedstocks as long as the feedstock (low moisture) and has a low content of mineral matter (manifested as a low ash in the product). Due to the short residence time (0.5 to 4.0 seconds), high temperatures are required for such gasifiers and – as a result – entrained-flow gasifiers is that the gas contains very little tar.

Entrained flow gasifiers are classified into two types: (i) top-fed gasifiers and (ii) side-fed gasifiers. A top-fed gasifier is a vertical cylinder reactor where fine particles (pulverized) and the gasification agent are co-currently fed from the top in the form of a jet and the feedstock conversion is achieved by use of an inverted burner. The product gas is taken from the side of the lower section while slag is extracted from the bottom of the reactor. On the other hand, in a side-fed gasifier, the pulverized fed and the gasification agent are co-currently fed by nozzles installed in the lower reactor, resulting in an appropriate mixing of biomass and gasification agent. The gas is extracted from the top of the gasifier and the slag from the bottom. Both configurations are highly efficient, with a standard operating temperature (1300 to 1500°C, 2370 to 2730°F)] and pressure 300 to 1000 psi.

Also, with regard to the entrained-flow reactor, it may be necessary to pulverize the feedstock (such as biomass char). On the other hand, when the feedstock is sent into an entrained-flow gasifier, the biomass can be in either form of dry feed or slurry feed if, for example, bio-oil and a solid feedstock (including biomass and biomass char) are co-gasified. In general, dry-feed gasifiers have the advantage over slurry-feed gasifiers in that the former can be operated with lower oxygen consumption. Moreover, dry-feed gasifiers have an additional degree of freedom that makes it possible to optimize bio-synthesis gas production (Shen et al., 2012).

While there are many alternate uses for the biogas produced by gasification, and a combination of products/utilities can be produced in addition to power. A major benefit of the integrated gasification combined cycle concept is that power can be produced with the lowest sulfur oxide (Sox) and nitrogen oxide (NOx) emissions of any liquid/solid feed power generation technology.

Finally, and by way of explanation, an *indirect gasifier*, requires an external source of heat to be transferred to the biomass to pyrolyze the feedstock under conditions of high severity, i.e., long residence time at high temperature. In the case of a biomass feedstock, the process will provide a char stream using either slow or fast pyrolysis to temperatures of 600 to 750°C (1110 to 1380°F), in amounts representing between 12% and 25% w/w of the biomass feedstock.

3.2. Fischer-Tropsch Synthesis

Bio-synthesis gas is used as a raw material in different thermochemical processes for the production of second-generation biofuels, both liquid, (such as methanol, ethanol, dimethyl ether (DME), and Fischer-Tropsch diesel) and gaseous (such as hydrogen and synthetic natural gas (SNG). In particular, the type of biomass and its production process strongly influences their composition and heating value. The production of liquid biofuel as an energy carrier could be very cost-effective because it would take the same infrastructure, storage system, and transportation used for liquefied petroleum gas.

The synthesis reaction is dependent of a catalyst, mostly an iron or cobalt catalyst where the reaction takes place. There is either a low temperature Fischer-Tropsch process (often represented as the LTFT process) or high temperature Fischer-Tropsch process (often represented as the HTFT process), with temperatures ranging between 200 to 240°C (390 to 465°F) for the low temperature Fischer-Tropsch process and 300 to 350°C (570 to 660°F) for the high temperature Fischer-Tropsch process. The high temperature Fischer-Tropsch process uses an iron-based catalyst, and the low temperature Fischer-Tropsch process either an iron-based catalyst or a cobalt-based catalyst. The different catalysts include also nickel-based catalyst and ruthenium-based catalysts.

Finally, it is worthy of note that clean biogas can also be used (i) as chemical *building blocks* to produce a broad range of chemicals using processes well established in the chemical and petrochemical industry), (ii) as a fuel producer for highly efficient fuel cells (which run off the hydrogen made in a gasifier) or perhaps in the future, hydrogen turbines and fuel cell-turbine hybrid systems, and (iii) as a source of hydrogen that can be separated from the gas stream and used as a fuel or as a feedstock for refineries (which use the hydrogen to upgrade crude oil products) (Speight, 2016, 2017, 2019).

3.3. Feedstocks

For many decades, coal has been the primary feedstock for gasification units but recent concerns about the use of fossil fuels and the resulting environmental pollutants, irrespective of the various gas cleaning processes and gasification plant environmental cleanup efforts, there is a move to feedstocks other than coal for gasification processes (Speight, 2013a, 2014b). But more pertinent to the present text, the gasification process can also use carbonaceous feedstocks which would otherwise have been discarded and unused, such as waste biomass and other similar biodegradable wastes. In this respect, biomass can be used to the fullest potential for the production of hydrogen since the refining industry has seen fit to use viscous feedstock gasification as a source of hydrogen for the past several decades (Speight, 2016, 2014a).

The advantage of the gasification process when a carbonaceous feedstock (a feedstock containing carbon) or hydrocarbonaceous feedstock (a feedstock containing carbon and hydrogen) is employed is that the product of focus – bio-synthesis gas – is potentially more useful as an energy source and results in an overall cleaner process. However, the reactor must be selected on the basis of feedstock properties and predicted behavior of the feedstock in the process.

3.3.1. Biomass

Biomass can be considered as any renewable feedstock which is in principle be *carbon neutral* (while the plant is growing, it uses the sun's energy to absorb the same amount of carbon from the atmosphere as it releases into the atmosphere) (Speight, 2008, 2011a).

Raw materials that can be used to produce biomass derived fuels are widely available; they come from a large number of different sources and in numerous forms (Rajvanshi, 1986; Speight, 2008, 2011a). The main basic sources of biomass include: (i) wood, including bark, logs, sawdust, wood chips, wood pellets and briquettes, (ii) high yield energy crops, such as wheat, grown specifically for energy applications, (iii) agricultural crops and residues (e.g., straw), and (iv) industrial waste, such as wood pulp or paper pulp. These different forms of biomass include a wide range of materials that produce a variety of products which are dependent upon the feedstock (Balat, 2011; Demirbaş, 2011; Ramroop Singh, 2011; Speight, 2011a). In addition, the heat content of the different types of biomass widely varies and have to be taken into consideration when designing any conversion process (Jenkins and Ebeling, 1985).

Table 6.5. The Advantages and Disadvantages of Using Biomass as a Feedstock for Energy Production and Chemicals Production

Advantages
Theoretically inexhaustible fuel source.
Minimal environmental impact when processes such as fermentation and pyrolysis are used.
Alcohols and other fuels produced by biomass are efficient, viable, and relatively clean-burning.
Biomass is available on a world-wide basis.
Disadvantages
Could contribute to global climate change and particulate pollution when direct combustion is employed.
Production of biomass and the technological conversion to alcohols or other fuels can be expensive.
Life cycle assessments (LCA) should be considered to address energy input and output.
Possibly a net loss of energy when operated on a small scale – energy is required to grow the biomass.

Many forms of biomass contain a high percentage of moisture (along with carbohydrates and sugars) and mineral constituents (Chapter 2) – both of which can influence the economics and viability of a gasification process. The presence of high levels of moisture in biomass reduces the temperature inside the gasifier, which then reduces the efficiency of the process. Many biomass gasification technologies therefore require dried biomass to reduce the moisture content prior to feeding into the gasifier. In addition, biomass can come in a range of sizes. In many biomass gasification systems, biomass must be processed to a uniform size or shape to be fed into the gasifier at a consistent rate as well as to maximize gasification efficiency.

Biomass such as wood pellets, yard and crop waste and energy crops including switch grass and waste from pulp and paper mills can also be employed to produce bioethanol and synthetic diesel. Biomass is first gasified to produce bio-synthesis gas and then subsequently converted via catalytic processes to the aforementioned downstream products. Biomass can also be used to produce electricity – either blended with traditional feedstocks, such as coal or by itself (Shen et al., 2012; Khosravi1 and Khadse, 2013a; Speight, 2014b).

Finally, while biomass may seem to some observers to be the answer to the global climate change issue, advantages and disadvantages of biomass as feedstock must be considered carefully (Table 6.5) (Molino et al., 2018). Also, while taking the issues of global climate change into account, it must not be ignored that the Earth is in an inter-glacial period when warming will take place. The true extent of this warming is not known – no one was around to measure the temperature change in the last inter-glacial period – and by the same token the contribution of anthropological sources to global climate change (through the emissions of carbon dioxide in the past and in the present) cannot be measured accurately because, for example, of the mobility of carbon dioxide in ice (Speight and Islam, 2016; Radovanović and Speight, 2018).

3.3.2. Solid Waste

The principle behind waste gasification and the production of gaseous fuels is that waste contains carbon and it is this carbon that is converted to gaseous products via gasification chemistry. Thus when waste is fed to a gasifier, water, and volatile matter are released and a char residue is left to react further.

Waste may be municipal solid waste (MSW) which had minimal presorting, or refuse-derived fuel (RDF) with significant pretreatment, usually mechanical screening and shredding. Other more specific waste sources (excluding hazardous waste) and possibly including crude oil coke, may provide niche opportunities for co-utilization (Brigwater, 2003; John and Singh, 2011; Arena, 2012; Basu, 2013; Speight, 2013a, 2014b). The traditional waste to energy plant, based on mass-burn combustion on an inclined grate, has a low public acceptability despite the very low emissions achieved over the last decade with modern flue gas clean-up equipment. This has led to difficulty in obtaining planning permissions to construct needed new waste to energy plants. After much debate, various governments have allowed options for advanced waste conversion technologies (gasification, pyrolysis and anaerobic digestion), but will only give credit to the proportion of electricity generated from non-fossil waste.

Use of waste materials as co-gasification feedstocks may attract significant disposal credits (Ricketts et al., 2002). Cleaner biomass materials are renewable fuels and may attract premium prices for the electricity generated. Availability of sufficient fuel locally for an economic plant size is often a major issue, as is the reliability of the fuel supply. Use of more-predictably available feedstock alongside these fuels overcomes some of these difficulties and risks. However, the issues associated with gasification of municipal solid waste include, like the gasification of any mixed feedstock, feedstock homogeneity, for many gasifiers, feedstock heterogeneity and process scale up can lead to a number of mechanical problems, shutdowns, sintering and hot spots leading to corrosion and failure of the reactor wall (most of the processes proposed for waste gasification do not include a separation process).

Furthermore, the disposal of municipal and industrial waste has become an important problem because the traditional means of disposal, landfill, are much less environmentally acceptable than previously. Much stricter regulation of these disposal methods will make the economics of waste processing for resource recovery much more favorable. One method of processing waste streams is to convert the energy value of the combustible waste into a fuel. One type of fuel attainable from waste is a low heating value gas, usually 100 to 150 Btu/scf, which can be used to generate process steam or to generate electricity. Co-processing such waste with coal is also an option (Speight, 2008, 2013a, 2014b). However, co-gasification technology varies, being usually site specific and high feedstock dependent (Ricketts et al., 2002).

One of the major challenges to the gasification process of landfill waste is that such waste has high moisture content and is heterogeneous in nature. Particle size and the presence of a number of components in the waste, such as sulfur, chlorides or metal vary considerably. The interconnected properties of heating value and moisture content play an important role. Hence, pre-preparation must be carefully considered in any waste gasification process. There are a number of different approaches to pre-preparation. Most of these involve mechanical shredding and metals removal using magnetic and electric devices.

The potential variability of biomass feedstocks, longer-term changes in refuse and the size limitation of a power plant using only waste and/or biomass can be overcome combining biomass, refuse, and coal. It also allows benefit from a premium electricity price for electricity from biomass and the gate fee associated with waste. If the power plant is gasification-based, rather than direct combustion, further benefits may be available. These include a premium price for the electricity from waste, the range of technologies available for the gas to electricity part of the process, gas cleaning prior to the main combustion stage instead of after combustion and public image, which is currently generally better for gasification as compared to combustion. These considerations lead to current studies of co-gasification of wastes/biomass with coal (Speight, 2008).

Analyses of the composition of municipal solid waste indicate that plastics do make up measurable amounts (5 to 10% w/w or more) of solid waste streams (EPCI, 2004; Mastellone and Arena, 2007). Many of these plastics are worth recovering as energy. In fact, many plastics, particularly the poly-olefin derivatives, have high calorific values and simple chemical constitutions of primarily carbon and hydrogen. As a result, waste plastics are ideal candidates for the gasification process. Because of the myriad of sizes and shapes of plastic products size reduction is necessary to create a feed material of a size less than 2 inches in diameter. Some forms of waste plastics such as thin films may require a simple agglomeration step to produce a particle of higher bulk density to facilitate ease of feeding. A plastic, such as high-density polyethylene, processed through a gasifier is converted to carbon monoxide and hydrogen and these materials in turn may be used to form other chemicals including ethylene from which the polyethylene is produced – *closed the loop recycling*.

3.3.3. Black Liquor

Black liquor is the spent liquor from the Kraft process in which pulpwood is converted into paper pulp by removing lignin and hemicellulose constituents as well as other extractable materials from wood to free the cellulose fibers. The present day chemical pulping process uses a complex combustion system called a recovery boiler to generate process heat and electricity as well as to recover the processing chemicals in an almost closed cycle. The recovery boiler is a very complex device, which is actually operated as a gasifier-combustor. After evaporation of the majority of the water, the very high solids black liquor is sprayed onto a mass of char in the bottom of the boiler.

Black liquor comprises an aqueous solution of lignin residues, hemicellulose, and the inorganic chemical used in the process and 15% w/w solids of which 10% w/w are inorganic and 5% w/w are organic. Typically, the organic constituents in black liquor are 40 to 45% w/w soaps, 35 to 45% w/w lignin, and 10 to 15% w/w other (miscellaneous) organic materials.

The organic constituents in the black liquor are made up of water/alkali soluble degradation components from the wood. Lignin is partially degraded to shorter fragments with sulfur contents in the order of 1 to 2% w/w and sodium content at approximately 6% w/w of the dry solids. Cellulose (and hemicellulose) is degraded to aliphatic carboxylic acid soaps and hemicellulose fragments. The extractable constituents yield *tall oil soap* and crude turpentine. The tall oil soap may contain up to 20% w/w sodium. Lignin components currently serve for hydrolytic or pyrolytic conversion or combustion. Alternative, hemicellulose constituents may be used in fermentation processes.

In another aspect, lignin pyrolysis (Chapter 5) produces reducing gases and char which react with the spent pulping chemicals to produce sodium carbonate (Na_2CO_3) and sodium sulfide (Na_2S). Other minerals in the feedstock appear as non-usable chemical ash and have to be removed from the cycle. The gas product from the char bed passes to an oxidizing zone in the furnace where the gas is combusted to produce process steam (and electricity) as well as provide radiant heat back to the char bed for the reduction chemistry to take place. The product chemicals are molten, drained from the char bed to collectors, and then poured into water to produce green liquor.

Thus, the pulp and paper industry offers unique opportunities for the production of biogas insofar as an important part of many pulp and paper plants is the chemicals recovery cycle where black liquor is combusted in boilers. Substituting the boiler by a gasification plant with additional biofuel and electricity production is very attractive, especially when the old boiler has to be replaced. The equivalent spent cooking liquor in the sulfite process is usually called *brown liquor*, but the terms *red liquor*, *thick liquor*, and *sulfite liquor* are also used. Approximately seven units of black liquor are produced in the manufacture of one unit of pulp (Biermann, 1993).

3.3.4. Gasification of Biomass with Coal

Recently, co-gasification of various biomass and coal mixtures has attracted a great deal of interest from the scientific community. Feedstock combinations including Japanese cedar wood and coal (Kumabe et al., 2007), coal and sawdust (Vélez et al., 2009), coal and pine chips (Pan et al., 2000), coal and silver birch wood (Collot et al., 1999), and coal and birch wood (Brage et al., 2000) have been reported in gasification practice. Co-gasification of coal and biomass has some synergy – the process not only produces a low carbon footprint on the environment, but also improves the H_2/CO ratio in the produced gas which is required for liquid fuel synthesis (Sjöström et al., 1999; Kumabe et al., 2007). In addition, the inorganic matter present in biomass catalyzes the gasification of coal. However, co-gasification processes require custom fittings and optimized processes for the coal and region-specific wood residues.

While cogasification of coal and biomass is advantageous from a chemical viewpoint, some practical problems are present on upstream, gasification, and downstream processes. On the upstream side, the particle size of the coal and biomass is required to be uniform for optimum gasification. In addition, moisture content and pretreatment (torrefaction) are very important during up-stream processing. Also, biomass decomposition occurs at a lower temperature than coal and therefore different reactors compatible to the feedstock mixture are required (Speight, 2011; Brar et al., 2012; Speight, 2013a, 2013b). Furthermore, feedstock and gasifier type along with operating parameters not only decide product gas composition but also dictate the amount of impurities to be handled downstream.

Downstream processes need to be modified if coal is co-gasified with biomass. For example, heavy metal and impurities such as sulfur and mercury present in coal can make bio-synthesis gas difficult to use and unhealthy for the environment. Alkali metals (sodium and potassium) present in biomass can also cause corrosion problems high temperatures in downstream pipes. An alternative option to downstream gas cleaning would be to process coal to remove mercury and sulfur prior to feeding into the gasifier.

However, first and foremost, coal and biomass require drying and size reduction before they can be fed into a gasifier. Size reduction is needed to obtain appropriate particle sizes; however, drying is required to achieve moisture content suitable for gasification operations. In addition, biomass densification may be conducted to prepare pellets and improve density and material flow in the feeder areas.

It is recommended that biomass moisture content should be less than 15% w/w prior to gasification. High moisture content reduces the temperature achieved in the gasification zone, thus resulting in incomplete gasification. Forest residues or wood has a fiber saturation point at 30 to 31% moisture content (dry basis) (Brar et al., 2012). Compressive and shear strength of the wood increases with decreased moisture content below the fiber saturation point. In such a situation, water is removed from the cell wall leading to shrinkage. The long-chain molecules constituents of the cell wall move closer to each other and bind more tightly. A high level of moisture, usually injected in form of steam in the gasification zone, favors formation of a water-gas shift reaction that increases hydrogen concentration in the resulting gas.

The torrefaction process is a thermal treatment of biomass in the absence of oxygen, usually at 250 to 300°C (480 to 570°F) to drive off moisture, decompose hemicellulose completely, and partially decompose cellulose (Speight, 2011a). Torrefied biomass has reactive and unstable cellulose molecules with broken hydrogen bonds and not only retains 79 to 95% of feedstock energy but also produces a more reactive feedstock with lower atomic hydrogen-carbon and oxygen-carbon ratios to those of the original biomass. Torrefaction results in higher yields of hydrogen and carbon monoxide in the gasification process.

Biomass fuel producers, coal producers and, to a lesser extent, waste companies are enthusiastic about supplying co-gasification power plants and realize the benefits of co-gasification with alternate fuels (Speight, 2008, 2011a; Lee and Shah, 2013; Speight, 2013a, 2013b). The benefits of a co-gasification technology involving coal and biomass include the use of a reliable coal supply with gate-fee waste and biomass which allows the economies of scale from a larger plant to be supplied just with waste and biomass. In addition, the technology offers a future option of hydrogen production and fuel development in refineries. In fact, oil refineries and petrochemical plants are opportunities for gasifiers when the hydrogen is particularly valuable (Speight, 2011b, 2014).

While upstream processing is influential from a material handling point of view, the choice of gasifier operation parameters (temperature, gasifying agent, and catalysts) dictate the product gas composition and quality. Biomass decomposition occurs at a lower temperature than coal and therefore different reactors compatible to the feedstock mixture are required (Brar et al., 2012). Furthermore, feedstock and gasifier type along with operating parameters not only decide product gas composition but also dictate the amount of impurities to be handled downstream. Downstream processes need to be modified if coal is co-gasified with biomass. Heavy metals and other impurities such as sulfur-containing compounds and mercury present in coal can make synthesis gas difficult to use and unhealthy for the environment. Alkali present in biomass can also cause corrosion problems high temperatures in downstream pipes. An alternative option to downstream gas cleaning would be to process coal to remove mercury and sulfur prior to feeding into the gasifier.

Finally, the presence of mineral matter in the coal-biomass feedstock is not appropriate for fluidized bed gasification. Low melting point of ash present in woody biomass leads to agglomeration which causes defluidization of the ash and sintering, deposition as well as corrosion of the gasifier construction metal bed (Vélez et al., 2009). Biomass containing alkali oxides and salts are likely to produce clinkering/slagging problems from ash formation (McKendry, 2002). Thus, it is imperative to be aware of the melting of biomass ash, its chemistry within the gasification bed (no bed, silica/sand, or calcium bed), and the fate of alkali metals when using fluidized bed gasifiers.

3.3.5. Gasification of Other Mixed Feedstocks

Co-utilization of waste and biomass with coal may provide economies of scale that help achieve the above identified policy objectives at an affordable cost. In some countries, governments propose cogasification processes as being *well suited for community-sized developments* suggesting that waste should be dealt with in smaller plants serving towns and cities, rather than moved to large, central plants (satisfying the so-called *proximity principal*).

Co-gasification technology varies, being usually site specific and high feedstock dependent. At the largest scale, the plant may include the well proven fixed bed and entrained flow gasification processes. At smaller scales, emphasis is placed on technologies which appear closest to commercial operation. Pyrolysis and other advanced thermal conversion processes are included where power generation is practical using the on-site feedstock produced. However, the needs to be addressed are (i) core fuel handling and gasification/pyrolysis technologies, (ii) fuel gas clean-up, and (iii) conversion of fuel gas to electric power (Ricketts et al., 2002).

The use of waste as gasification feedstock or as a co-gasification feedstock can mitigate an important problem. The disposal of municipal and industrial wastes has become an important problem because the traditional means of disposal, landfill, has become environmentally much less acceptable than previously. New, much stricter regulation of these disposal methods will make the economics of waste processing for resource recovery much more favorable. One method of processing waste streams is to convert the energy value of the combustible waste into a fuel. One type of fuel attainable from wastes is a low heating value gas heat content on the order of 100 to 150 Btu/scf), which can be used to generate process steam or to generate electricity.

The ability of a refinery to efficiently accommodate heavy feedstock streams (such as heavy oil, extra heavy oil, residua, deasphalter bottoms, visbreaker bottoms, and tar sand bitumen) enhances the economic potential of the refinery and the development of heavy oil and tar sand resources. A refinery with the flexibility to meet the increasing product specifications for fuels through the ability to upgrade heavy feedstocks is an increasingly attractive means of extracting maximum value from each barrel of oil produced. Upgrading can convert marginal heavy crude oil into light, higher value crude, and can convert heavy, sour refinery bottoms into valuable transportation fuels. On the downside, most upgrading processes also produce an even heavier residue whose disposition costs may approach the value of the upgrade itself.

4. Gas Production and Other Products

The gasification of a carbonaceous feedstock (i.e., char produced from the feedstock such as char from the pyrolysis of biomass) is the conversion of the feedstock (by any one of a variety of processes) to produce gaseous products that are combustible as well as a wide range of chemical products from synthesis gas (Figure 6.3). The products from the gasification of the process may be of low, medium, or high heat-content (high-Btu) content as dictated by the process as well as by the ultimate use for the gas (Baker and Rodriguez, 1990; Probstein and Hicks, 1990; Lahaye and Ehrburger, 1991; Matsukata et al., 1992; Speight, 2013a).

4.1. Gaseous Products

The products of gasification are varied insofar as the gas composition varies with the system employed (Speight, 2013a). It cannot be over-emphasized that the gas product must be first freed from any pollutants such as particulate matter and sulfur compounds before further use, particularly when the intended use is a water gas shift or methanation (Cusumano et al., 1978; Probstein and Hicks, 1990).

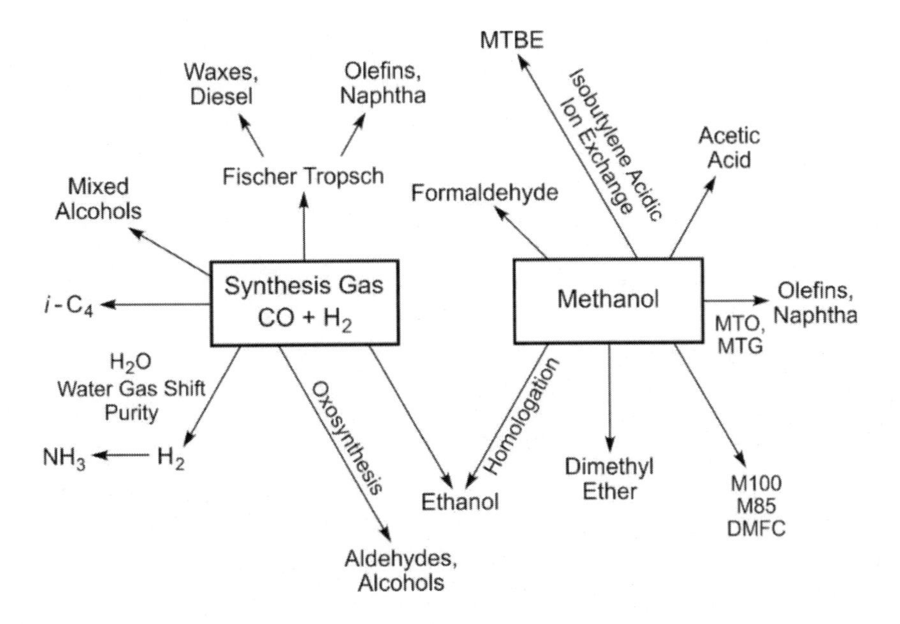

Figure 6.3. Potential Products from the Gasification Process.

In all cases, the gas exiting the gasifier contains impurities that need to be removed for the advanced use of biomass gasification gas. For example, if the gas is to be used a Fischer-Tropsch process, the gas must be ultra-clean. Any particulate matter in the gas exiting the gasifier can be removed by a hot gas filter, whereas the tar constituents and low-boiling hydrocarbon derivatives can be converted by reforming to syngas. For tar removal, another possibility is scrubbing by an organic solvent.

However, the gas stream will also, more than likely, contain other impurities, such as hydrogen sulfide, carbonyl sulfide, ammonia, and alkali metals. The sulfur constitunets will further add complications to a gas reforming process by poisoning the catalyst. In addition to hydrocarbons, ammonia is partly converted to nitrogen in the reformer. The alkali metals, in turn, will condense in hot gas filtration taking place at around 500°C (930°F). The final conditioning of gas for synthesis purposes can be realized by conventional commercial technologies for acid gas removal, such as the Rectisol process, and the H_2/CO ratio can be adjusted by a water gas shift reactor.

4.1.1. Synthesis Gas

Bio-syngas is comparable in its combustion efficiency to natural gas (Speight, 2008; Chadeesingh, 2011) which reduces the emissions of sulfur, nitrogen oxides, and mercury, resulting in a much cleaner fuel (Nordstrand et al., 2008; Lee et al., 2006; Sondreal et al., 2004, 2006; Yang et al., 2007; Wang et al., 2008). The resulting hydrogen gas can be used for electricity generation or as a transport fuel or in any aspects of refining that requires hydrogen (Speight, 2016). The gasification process also facilitates capture of carbon dioxide emissions from the combustion effluent.

Although bio-syngas can be used as a standalone fuel, the energy density of synthesis gas is approximately half that of natural gas and is therefore mostly suited for the production of transportation fuels and other chemical products. Synthesis gas is mainly used as an intermediary building block for the final production (synthesis) of various fuels such as synthetic natural gas, methanol and synthetic crude oil fuel (dimethyl ether – synthesized gasoline and diesel fuel) (Chadeesingh, 2011; Speight, 2013a). At this point, and in order to dismiss any confusion that may arise, synthesis gas, as generated from biomass, is not the same as biogas. Biogas is a clean and renewable form of energy generated from biomass that could very well substitute for conventional sources of energy. The gas is generally composed of methane (55 to 65% v/v), carbon dioxide (35 to 45% v/v), nitrogen (0 to 3% v/v), hydrogen (0 to 1% v/v), and hydrogen sulfide (0 to 1% v/v).

The use of bio-synthesis gas offers the opportunity to furnish a broad range of environmentally clean fuels and chemicals and there has been steady growth in the traditional uses of synthesis gas. Almost all hydrogen gas is manufactured from synthesis gas and there has been an increase in the demand for this basic chemical. In fact, the major use of synthesis gas is in the manufacture of hydrogen for a growing number of purposes, especially in crude oil refineries (Speight, 2014a, 2016). Methanol not only remains the second largest consumer of synthesis gas but has shown remarkable growth as part of the methyl ethers used as octane enhancers in automotive fuels.

The Fischer-Tropsch synthesis remains the third largest consumer of synthesis gas, mostly for transportation fuels but also as a growing feedstock source for the manufacture of chemicals, including polymers. The hydroformylation of olefin derivatives (the Oxo reaction), a completely chemical use of synthesis gas, is the fourth largest use of carbon monoxide and hydrogen mixtures. A direct application of synthesis gas as fuel (and eventually also for chemicals) that promises to increase is its use for *integrated gasification combined cycle* (IGCC) units for the generation of electricity (and also chemicals), crude oil coke or viscous feedstocks (Holt, 2001). Finally, synthesis gas is the principal source of carbon monoxide, which is used in an expanding list of carbonylation reactions, which are of major industrial interest.

4.1.2. Low-Btu Gas

During the production of gas by oxidation with air, the oxygen is not separated from the air and, as a result, the gas product invariably has a low Btu content (low heat-content, 150 to 300 Btu/ft^3). Several important chemical reactions and a host of side reactions, are involved in the manufacture of low heat-content gas under the high temperature conditions employed (Speight, 2013a). Low heat-content gas contains several components, four of which are always major components present at levels of at least several percent; a fifth component, methane, is marginally a major component.

The nitrogen content of low heat-content gas ranges from somewhat less than 33% v/v to slightly more than 50% v/v and cannot be removed by any reasonable means; the presence of nitrogen at these levels makes the product gas *low heat-content* by definition. The nitrogen also strongly limits the applicability of the gas to chemical synthesis. Two other noncombustible components (water, H_2O, and carbon dioxide, CO_2) further lower the heating value of the gas; water can be removed by condensation and carbon dioxide by relatively straightforward chemical means.

The two major combustible components are hydrogen and carbon monoxide; the H_2/CO ratio varies from approximately 2:3 to approximately 3:2. Methane may also make an appreciable contribution to the heat content of the gas. Of the minor components hydrogen sulfide is the most significant and the amount produced is, in fact, proportional to the sulfur content of the feedstock. Any hydrogen sulfide present must be removed by one, or more, of several procedures (Mokhatab et al., 2006; Speight, 2007, 2014).

4.1.3. Medium-Btu Gas

Medium Btu gas (medium heat content gas) has a heating value in the range 300 to 550 Btu/ft^3) and the composition is much like that of low heat-content gas, except that there is virtually no nitrogen. The primary combustible gases in medium heat-content gas are hydrogen and carbon monoxide. Medium heat-content gas is considerably more versatile than low heat-content gas; like low heat-content gas, medium heat-content gas may be used directly as a fuel to raise steam, or used through a combined power cycle to drive a gas turbine, with the hot exhaust gases employed to raise steam, but medium heat-content gas is especially amenable to synthesize methane (by methanation), higher hydrocarbon derivatives (by Fischer-Tropsch synthesis), methanol, and a variety of synthetic chemicals.

The reactions used to produce medium heat-content gas are the same as those employed for low heat-content gas synthesis, the major difference being the application of a nitrogen barrier (such as the use of pure oxygen) to keep diluent nitrogen out of the system.

In medium heat-content gas, the H_2/CO ratio varies from 2:3 C to 3:1 and the increased heating value correlates with higher methane and hydrogen contents as well as with lower carbon dioxide contents. Furthermore, the very nature of the gasification process used to produce the medium heat-content gas has a marked effect upon the ease of subsequent processing. For example, the CO_2-acceptor product is quite amenable to use for methane production because it has (i) the desired H_2/CO ratio just exceeding 3:1, (ii) an initially high methane content, and (iii) relatively low water and carbon dioxide contents. Other gases may require appreciable shift reaction and removal of large quantities of water and carbon dioxide prior to methanation.

4.1.4. High-Btu Gas

High Btu) gas (high heat-content gas) is essentially pure methane and often referred to as *synthetic natural gas* or *substitute natural gas* (SNG) (Speight, 1990, 2013a). In the context of this book, substitute natural gas (SNG) that is produced from biomass can be considered a type of biogas (Chapter 1). Synthetic natural gas contains mainly methane and can be produced from a product gas stream that contains significant amounts of methane. This generally means that gasification should take place non-catalytically at approximately 900°C (1650°F). At these temperatures, methane destruction generally is limited and the fluidized bed gasification process is one of the obvious options.

However, to qualify as substitute natural gas, a product must contain at least 95% v/v methane, giving an energy content (heat content) of synthetic natural gas on the order of 980 to 1080 Btu/ft^3). The commonly accepted approach to the synthesis of high heat-content gas is the catalytic reaction of hydrogen and carbon monoxide:

$$3H_2 + CO \rightarrow CH_4 + H_2O$$

To avoid catalyst poisoning, the feed gases for this reaction must be quite pure and, therefore, impurities in the product are rare. The large quantities of water produced are removed by condensation and recirculated as very pure water through the gasification system. The hydrogen is usually present in slight excess to ensure that the toxic carbon monoxide is reacted; this small quantity of hydrogen will lower the heat content to a small degree.

The carbon monoxide/hydrogen reaction is somewhat inefficient as a means of producing methane because the reaction liberates large quantities of heat. In addition, the methanation catalyst is troublesome and prone to poisoning by sulfur compounds and the decomposition of metals can destroy the catalyst. Hydrogasification may be thus employed to minimize the need for methanation:

$$[C]_{feedstock} + 2H2 \rightarrow CH_4$$

The product of hydrogasification is far from pure methane and additional methanation is required after hydrogen sulfide and other impurities are removed.

Besides the main components of carbon monoxide, hydrogen, carbon dioxide, methane and nitrogen, several trace substances are also present in the product gas. These include particulate matter) such as entrained ash, and bed material), sulfur compounds (hydrogen sulfide, carbonyl sulfide), alkali compounds (mainly chlorine compounds) and tars or higher hydrocarbons that are prone to condensation at temperatures on the order of 300 to 350°C (570 to 660°F) and can, therefore, foul pipes and equipment. Removal of these substances is of particular importance in the production of synthetic natural gas since the methanation reaction uses catalysts that are highly sensitive to impurities in the gas. A number of different gas cleaning technologies exist (Chapter 7).

4.2. Liquid Products

The production of liquid fuels from a carbonaceous feedstock via gasification is often referred to as the *indirect liquefaction* of the feedstock (Speight, 2013a, 2014). In these processes, the feedstock is not converted directly into liquid products but involves a two-stage conversion operation in which the feedstock is first converted (by reaction with steam and oxygen) to produce a gaseous mixture that is composed primarily of carbon monoxide and hydrogen (synthesis gas). The gas stream is subsequently purified (to remove sulfur, nitrogen, and any particulate matter) after which it is catalytically converted to a mixture of liquid hydrocarbon products.

The synthesis of hydrocarbon derivatives from carbon monoxide and hydrogen (synthesis gas) (the Fischer-Tropsch synthesis) is a procedure for the indirect liquefaction of various carbonaceous feedstocks (Speight, 2011a, 2011b). This process is the only liquefaction scheme currently in use on a relatively large commercial scale (for the production of liquid fuels from coal using the Fischer-Tropsch process (Singh, 1981).

Thus, the feedstock is converted to gaseous products at temperatures in excess of 800°C (1470°F), and at moderate pressures, to produce synthesis gas:

$$[C]_{feedstock} + H_2O \rightarrow CO + H_2$$

In practice, the Fischer-Tropsch reaction is carried out at temperatures of 200 to 350°C (390 to 660°F) and at pressures of 75 to 4000 psi. The hydrogen/carbon monoxide ratio is typically on the order of 2/2:1 or 2/5:1, since up to three volumes of hydrogen may be required to achieve the next stage of the liquids production, the synthesis gas must then be converted by means of the water-gas shift reaction) to the desired level of hydrogen:

$$CO + H_2O \rightarrow CO_2 + H_2$$

After this, the gaseous mix is purified and converted to a wide variety of hydrocarbon derivatives:

$$nCO + (2n + 1)H_2 \rightarrow C_nH_{2n+2} + nH_2O$$

These reactions result primarily in low- and medium-boiling aliphatic compounds suitable for gasoline and diesel fuel.

4.3. Solid Products

The solid product (solid waste) of a gasification process is typically ash which is the oxides of metals-containing constituents of the feedstock. The amount and type of solid waste produced is very much feedstock dependent. The waste is a significant environmental issue due to the large quantities produced, chiefly fly ash if coal is the feedstock or a co-feedstock, and the potential for leaching of toxic substances (such as heavy metals such as lead and arsenic) into the soil and groundwater at disposal sites.

At the high temperature of the gasifier, most of the mineral matter of the feedstock is transformed and melted into slag, an inert glass-like material and, under such conditions, non-volatile metals and mineral compounds are bound together in molten form until the slag is cooled in a water bath at the bottom of the gasifier, or by natural heat loss at the bottom of an entrained-bed gasifier. Slag production is a function of mineral matter content of the feedstock – coal produces much more slag per unit weight than crude oil coke. Furthermore, as long as the operating temperature is above the fusion temperature of the ash, slag will be produced. The physical structure of the slag is sensitive to changes in operating temperature and pressure of the gasifier and a quick physical examination of the appearance of the slag can often be an indication of the efficiency of the conversion of feedstock carbon to gaseous product in the process.

Slag is comprised of black, glassy, silica-based materials and is also known as *frit*, which is a high density, vitreous, and abrasive material low in carbon and formed in various shapes from jagged and irregular pieces to rod and needle-like forms. Depending upon the gasifier process parameters and the feedstock properties, there may also be residual carbon char. Vitreous slag is much preferable to ash, because of its habit of encapsulating toxic constituents (such as heavy metals) into a stable, non-leachable material. Leachability data obtained from different gasifiers unequivocally shows that gasifier slag is highly non-leachable, and can be classified as non-hazardous. Because of its particular properties and non-hazardous, non-toxic nature, slag is relatively easily marketed as a by-product for multiple advantageous uses, which may negate the need for its long-term disposal.

The physical and chemical properties of gasification slag are related to (i) the composition of the feedstock, (ii) the method of recovering the molten ash from the gasifier, and (iii) the proportion of devolatilized carbon particles (char) discharged with the slag. The rapid water-quench method of cooling the molten slag inhibits recrystallization, and results in the formation of a granular, amorphous material. Some of the differences in the properties of the slag may be attributed to the specific design and operating conditions prevailing in the gasifiers.

Char is the finer component of the gasifier solid residuals, composed of unreacted carbon with various amounts of siliceous ash. Char can be recycled back into the gasifier to increase carbon usage and has been used as a supplemental fuel source for use in a combustor. The irregularly shaped particles have a well-defined pore structure and have excellent potential as an adsorbent and precursor to activated carbon. In terms of recycling char to the gasifier, a property that is important to fluidization is the effective particle density. If the char has a large internal void space, the density will be much less than that of the feedstock (especially coal) or char from slow carbonization of a carbonaceous feedstock.

5. The Future

The future of biogas production by gasification of biomass depends very much on the effect of gasification processes on the surrounding environment. It is these environmental effects and issues that will direct the success of gasification. In fact, there is the distinct possibility that within the foreseeable future the gasification process will increase in popularity in crude oil refineries – some refineries may even be known as gasification refineries (Speight, 2011b). A gasification refinery would have, as the center-piece of the refinery, gasification technology as is the case of the Sasol refinery in South Africa (Couvaras, 1997). The refinery would produce synthesis gas (from the carbonaceous feedstock) from which liquid fuels would be manufactured using the Fischer-Tropsch synthesis technology.

In fact, gasification to produce synthesis gas can proceed from any carbonaceous material, including biomass. Inorganic components of the feedstock, such as metals and minerals, are trapped in an inert and environmentally safe form as char, which may have use as a fertilizer. Biomass gasification is therefore one of the most technically and economically convincing energy possibilities for a potentially carbon neutral economy.

The manufacture of gas mixtures of carbon monoxide and hydrogen has been an important part of chemical technology for about a century. Originally, such mixtures were obtained by the reaction of steam with incandescent coke and were known as *water gas*. Eventually, steam reforming processes, in which steam is reacted with natural gas (methane) or crude oil naphtha over a nickel catalyst, found wide application for the production of synthesis gas.

A modified version of steam reforming known as autothermal reforming, which is a combination of partial oxidation near the reactor inlet with conventional steam reforming further along the reactor, improves the overall reactor efficiency and increases the flexibility of the process. Partial oxidation processes using oxygen instead of steam also found wide application for synthesis gas manufacture, with the special feature that they could utilize low-value feedstocks such as viscous crude oil residues. In recent years, catalytic partial oxidation employing very short reaction times (milliseconds) at high temperatures (850 to 1000°C, 1560 to 1830°F) is providing still another approach to synthesis gas manufacture (Hickman and Schmidt, 1993).

In a gasifier, the carbonaceous material undergoes several different processes: (i) pyrolysis of carbonaceous fuels, (ii) combustion, and (iii) gasification of the remaining char. The process is very dependent on the properties of the carbonaceous material and determines the structure and composition of the char, which will then undergo gasification reactions.

The conversion of the gaseous products of gasification processes to synthesis gas, a mixture of hydrogen (H_2) and carbon monoxide (CO), in a ratio appropriate to the application, needs additional steps, after purification. The product gases - carbon monoxide, carbon dioxide, hydrogen, methane, and nitrogen - can be used as fuels or as raw materials for chemical or fertilizer manufacture.

Gasification by means other than the conventional methods has also received some attention and has provided rationale for future processes (Rabovitser et al., 2010). In the process, a carbonaceous material and at least one oxygen carrier are introduced into a non-thermal plasma reactor at a temperature in the range of approximately 300°C to approximately 700°C (570 to approximately 1290°F) and a pressure in a range from atmospheric pressure to approximate 1030 psi and a non-thermal plasma discharge is generated within the non-thermal plasma reactor. Plasma gasification technology has been shown to be an effective and environmentally friendly method for solid waste treatment and energy utilization.

Plasma gasification uses an external heat source to gasify the waste, resulting in very little combustion. Almost all of the carbon is converted to fuel synthesis gas. The high operating temperatures (above 1800°C, 3270°F) allow for the breaking down of all tars, char and dioxins. The exit gas from the reactor is cleaner and there is no ash at the bottom of the reactor. The waste feed sub-system is used for the treatment of each type of waste in order to meet the inlet requirements of the plasma furnace. For example, for a waste material with high moisture content, a drier will be required. However a typical feed system consists of a shredder for solid waste size reduction prior to entering the plasma furnace.

In the plasma reactor two torches are used together with a gas (such as oxygen, helium or air) to generate the plasma. The torches which extend into the plasma furnace are fitted with graphite electrodes. An electric current is passed through the electrodes and an electric arc is generated between the tip of the electrodes and the conducting receiver which is the slag at the bottom of the furnace. Because of the electrical resistivity across the system significant heat is required to strip away electrons from the molecules of the gas introduced resulting in an ionized superheated gas stream or plasma. This gas exits the torch at temperatures up to 10000°C (18000°F). Such high temperatures are used to break down the waste primarily into elemental gas and solid waste (slag) as well as reduce dioxins, sulfur oxides and carbon dioxide emissions. Unlike waste incineration plasma gasification waste to energy technology has proven to have a benign impact on the environment.

The plasma furnace is the central component of the system where gasification and vitrification takes place. Plasma torches are mounted at the bottom of the reactor, they provide high temperature air (almost three times higher than traditional combustion temperatures) which allow for the gasification of the waste materials. It is a non-incineration thermal process that uses extremely high temperatures in a partial oxygen environment to decompose completely the input waste material into very simple molecules. In the process, the carbonaceous feedstock and the oxygen carrier are exposed to the non-thermal plasma discharge, resulting in the formation of a product gas which comprises substantial amounts of hydrocarbon derivatives, such as methane, hydrogen and/or carbon monoxide. The products of the process are a fuel or gas known as synthesis gas and an inert vitreous material known as slag.

Furthermore, gasification and conversion of carbonaceous solid fuels to synthesis gas for application of power, liquid fuels and chemicals is practiced worldwide. Crude oil coke, coal, biomass, and refinery waste are major feedstocks for an on-site refinery gasification unit. The concept of blending of a variety of carbonaceous feedstocks (such as coal, biomass, or refinery waste) with a viscous feedstock of the coke from the thermal processing of the viscous feedstock is advantageous in order to obtain the highest value of products as compared to gasification of crude oil coke alone. Furthermore, based on gasifier type, co-gasification of carbonaceous feedstocks can be an advantageous and efficient process. In addition, the variety of upgrading and delivery options that are available for application to synthesis gas enable the establishment of an integrated energy supply system whereby synthesis gases can be upgraded, integrated, and delivered to a distributed network of energy conversion facilities, including power, combined heat and power, and combined cooling, heating and power (sometime referred to as *tri-generation*) as well as used as fuels for transportation applications.

As a final note, the production of chemicals from biomass is based on thermochemical conversion routes which are, in turn, based on biomass gasification (Roddy and Manson-Whitton, 2012). The products are (i) a gas, which is the desired product and (ii) a solid ash residue whose composition depends on the type of biomass. Continuous gasification processes for various feedstocks have been under development since the early 1930s. Ideally, the gas produced would be a mixture of hydrogen and carbon monoxide, but, in practice, it also contains methane, carbon dioxide, and a range of contaminants.

A variety of gasification technologies is available across a range of sizes, from small updraft and downdraft gasifiers through a range of fluidized bed gasifiers at an intermediate scale and on to larger entrained flow and plasma gasifiers (Bridgwater, 2003; Roddy and Manson-Whitton, 2012). In an updraft gasifier, the oxidant is blown up through the fixed gasifier bed with the syngas exiting at the top whereas in a downdraft gasifier, the oxidant is blown through the reactor in a downward direction with the synthesis gas exiting at the bottom.

Gasification processes tend to operate either above the ash melting temperature (typically 1200°C, 2190°F) or below the ash melting temperature (typically >1000°C, >1830°F). In the higher temperature processes, there is little methane or tar formation. The question of which gasification technology is the most appropriate depends on whether the priority is to (i) produce a very pure bio-synthesis gas, (ii) accommodate a wide range of feedstock types, (iii) avoid pre-processing of biomass, or (iv) operate at a large scale, and produce chemical products from a variety of feedstocks.

In summary, a refinery that is equipped with a gasifier is a suitable refinery for the complete conversion of heavy feedstocks and (including petroleum coke) to valuable products (including petrochemicals). In fact, integration between bottoms processing units and gasification, presents some unique synergies including the production of feedstocks for a petrochemical complex.

REFERENCES

Abadie, L. M., Chamorro, J. M. 2009. The Economics of Gasification: A Market-Based Approach. *Energies,* 2: 662-694.

Adhikari, U., Eikeland, M. S., and Halvorsen, B. M. 2015. Gasification of Biomass for Production of Syngas for Biofuel. *Proceedings. 56th SIMS,* Linköping, Sweden. October 7-9. Page 255-260. http://www.ep.liu.se/ecp/119/025/ecp15119025.pdf.

Ahmad, A. A., Zawawi, N. A., Kasim, F. H., Inayat, A., and Khasri, A. 2016. Assessing the gasification performance of biomass: A Review on Biomass Gasification Process Conditions, Optimization and Economic Evaluation. *Renew. Sustain. Energy Rev.,* 53: 1333-1347.

Arena, U. 2012. Process and Technological Aspects of Municipal Solid Waste Gasification. A Review. *Waste Management*, 32: 625-639.

Asadullah, M. 2014. Barriers of Commercial Power Generation Using Biomass Gasification Gas: A Review. *Renew. Sustain. Energy Rev.*, 29: 201-215.

ASTM D388. 2019. Standard Classification of Coal by Rank. *Annual Book of Standards*, ASTM International, West Conshohocken, Pennsylvania.

Baker, R. T. K., and Rodriguez, N. M. 1990. In: *Fuel Science and Technology Handbook*. Marcel Dekker Inc., New York. Chapter 22.

Balat, M. 2011. Fuels from Biomass – An Overview. In: *The Biofuels Handbook*. J. G. Speight (Editor). Royal Society of Chemistry, London, United Kingdom. Part 1, Chapter 3.

Bernetti, A., De Franchis, M., Moretta, J. C., and Shah, P. M. 2000. Solvent Deasphalting and Gasification: A Synergy. *Petroleum Technology Quarterly*, Q4: 1-7.

Bhattacharya, S., Md. Mizanur Rahman Siddique, A. H., and Pham, H-L. 1999. A Study in Wood Gasification on Low Tar Production. *Energy*, 24: 285-296.

Biermann, C. J. 1993. *Essentials of Pulping and Papermaking*. Academic Press Inc., New York.

Boateng, A. A., Walawender, W. P., Fan, L. T., and Chee, C. S. 1992. Fluidized-Bed Steam Gasification of Rice Hull. *Bioresource Technology*, 40(3): 235-239.

Brage, C., Yu, Q., Chen, G., and Sjöström, K. 2000. Tar Evolution Profiles Obtained from Gasification of Biomass and Coal. *Biomass and Bioenergy*, 18(1): 87-91.

Brar, J. S., Singh, K., Wang, J., and Kumar, S. 2012. Cogasification of Coal and Biomass: A Review. *International Journal of Forestry Research*, 2012: 1-10.

Bridgwater, A. V. 2003. Renewable Fuels and Chemicals by Thermal Processing of Biomass. *Chem. Eng. Journal*, 91: 87-102.

Chadeesingh, R. 2011. The Fischer-Tropsch Process. In: *The Biofuels Handbook*. J. G. Speight (Editor). The Royal Society of Chemistry, London, United Kingdom. Part 3, Chapter 5, Page 476-517.

Chen, G., Sjöström, K. and Bjornbom, E. 1992. Pyrolysis/Gasification of Wood in a Pressurized Fluidized Bed Reactor. *Ind. Eng. Chem. Research*, 31(12): 2764-2768.

Collot, A. G., Zhuo, Y., Dugwell, D. R., and Kandiyoti, R. 1999. Co-Pyrolysis and Cogasification of Coal and Biomass in Bench-Scale Fixed-Bed and Fluidized Bed Reactors. *Fuel*, 78: 667-679.

Couvaras, G. 1997. Sasol's Slurry Phase Distillate Process and Future Applications. *Proceedings. Monetizing Stranded Gas Reserves Conference*, Houston. December 1997.

Cusumano, J. A., Dalla Betta, R. A., and Levy, R. B. 1978. *Catalysis in Coal Conversion*. Academic Press Inc., New York.

Davidson, R. M. 1983. *Mineral Effects in Coal Conversion*. Report No. ICTIS/TR22, International Energy Agency, London, United Kingdom.

Demirbaş, A. 2011. Production of Fuels from Crops. In: *The Biofuels Handbook.* J.G. Speight (Editor). Royal Society of Chemistry, London, United Kingdom. Part 2, Chapter 1.

Dutcher, J. S., Royer, R. E., Mitchell, C. E., and Dahl, A. R. 1983. In: *Advanced Techniques in Synthetic Fuels Analysis.* C. W. Wright, W. C. Weimer, and W. D. Felic (Editors). Technical Information Center, United States Department of Energy, Washington, DC. Page 12.

EIA. 2007. *Net Generation by Energy Source by Type of Producer.* Energy Information Administration, United States Department of Energy, Washington, DC. http://www.eia.doe.gov/cneaf/electricity/epm/table1_1.html.

Ergudenler, A., and Ghaly, A. E. 1993. Agglomeration of Alumina Sand in a Fluidized Bed Straw Gasifier at Elevated Temperatures. *Bioresource Technology*, 43(3): 259-268.

Fabry, F., Rehmet, C., Rohani, V. J., and Fulcheri, L. 2013. Waste Gasification by Thermal Plasma: A Review. *Waste and Biomass Valorization,* 4(3): 421-439.

Fermoso, J., Plaza, M. G., Arias, B., Pevida, C., Rubiera, F., and Pis, J. J. 2009. Co-Gasification of Coal with Biomass and Petcoke in a High-Pressure Gasifier for Syngas Production. *Proceedings. 1st Spanish National Conference on Advances in Materials Recycling and Eco-Energy.* Madrid, Spain. November 12-13.

Furimsky, E. 1999. Gasification in a Petroleum Refinery of the 21st Century. Oil & Gas Science and Technology, *Revue Institut Français du Pétrole*, 54(5): 597-618.

Gabra, M., Pettersson, E., Backman, R., and Kjellström, B. 2001. Evaluation of Cyclone Gasifier Performance for Gasification of Sugar Cane Residue – Part 1: Gasification of Bagasse. *Biomass and Bioenergy*, 21(5): 351-369.

Gary, J. G., Handwerk, G. E., and Kaiser, M. J. 2007. *Crude oil Refining: Technology and Economics*, 5th Edition. CRC Press, Taylor & Francis Group, Boca Raton, Florida.

Gray, D., and Tomlinson, G. 2000. Opportunities for Petroleum Coke Gasification under Tighter Sulfur Limits for Transportation Fuels. Proceedings. 2000 Gasification Technologies Conference, San Francisco, California. October 8-11.

Hanaoka, T., Inoue, S., Uno, S., Ogi, T., and Minowa, T. 2005. Effect of Woody Biomass Components on Air-Steam Gasification. *Biomass and Bioenergy*, 28(1): 69-76.

Hickman, D. A., and Schmidt, L. D. 1993. Syngas Formation by Direct Catalytic Oxidation of Methane. *Science,* 259: 343-346.

Higman, C., and Van der Burgt, M. 2008. *Gasification* 2nd Edition. Gulf Professional Publishing, Elsevier, Amsterdam, Netherlands.

Holt, N. A. H. 2001. Integrated Gasification Combined Cycle Power Plants. *Encyclopedia of Physical Science and Technology.* 3rd Edition. Academic Press Inc., New York.

Hotchkiss, R. 2003. Coal Gasification Technologies. *Proceedings. Institute of Mechanical Engineers Part A*, 217(1): 27-33.

Hsu, C. S., and Robinson, P. R. (Editors). 2017. *Handbook of Petroleum Technology.* Springer International Publishing AG, Cham, Switzerland.

Irfan, M. F. 2009. Research Report: Pulverized Coal Pyrolysis & Gasification in $N_2/O_2/CO_2$ Mixtures by Thermo-Gravimetric Analysis. *Novel Carbon Resource Sciences Newsletter*, Kyushu University, Fukuoka, Japan. Volume 2: 27-33.

Jenkins, B. M., and Ebeling, J. M. 1985. *Thermochemical Properties of Biomass Fuels.* California Agriculture (May-June), Page 14-18.

John, E., and Singh, K. 2011. Properties of Fuels from Domestic and Industrial Waste. In: *The Biofuels Handbook.* J. G. Speight (Editor). The Royal Society of Chemistry, London, United Kingdom. Part 3, Chapter 2, Page 377-407.

Johnson J. L. 1979. *Kinetics of Coal Gasification.* John Wiley and Sons Inc., Hoboken, New Jersey.

Khosravi1, M., and Khadse, A., 2013. Gasification of Petcoke and Coal/Biomass Blend: A Review. *International Journal of Emerging Technology and Advanced Engineering,* 3(12): 167-173.

Ko, M. K., Lee, W. Y., Kim, S. B., Lee, K. W., and Chun, H. S. 2001. Gasification of Food Waste with Steam in Fluidized Bed. *Korean Journal of Chemical Engineering*, 18(6): 961-964.

Kumabe, K., Hanaoka, T., Fujimoto, S., Minowa, T., and Sakanishi, K. 2007. Cogasification of Woody Biomass and Coal with Air and Steam. *Fuel,* 86: 684-689.

Kumar, A., Jones, D. D., and Hanna, M. A. 2009. Thermochemical Biomass Gasification: A Review of the Current Status of the Technology. *Energies,* 2: 556-581.

Kunii, D., and Levenspiel, O. 2013. *Fluidization Engineering* 2nd Edition. Butterworth-Heinemann, Elsevier, Amsterdam, Netherlands.

Lahaye, J., and Ehrburger, P. (Editors). 1991. *Fundamental Issues in Control of Carbon Gasification Reactivity*. Kluwer Academic Publishers, Dordrecht, Netherlands.

Lapuerta M., Hernández J. J., Pazo A., and López, J. 2008. Gasification and co-gasification of biomass wastes: Effect of the biomass origin and the gasifier operating conditions. *Fuel Processing Technology*, 89(9): 828-837.

Lee, S. 2007. Gasification of Coal. In: *Handbook of Alternative Fuel Technologies*. S. Lee, J. G. Speight, and S. Loyalka (Editors). CRC Press, Taylor & Francis Group, Boca Raton, Florida. 2007.

Lee, S., Speight, J. G., and Loyalka, S. 2007. *Handbook of Alternative Fuel Technologies.* CRC-Taylor & Francis Group, Boca Raton, Florida.

Lee, S., and Shah, Y. T. 2013. *Biofuels and Bioenergy*. CRC Press, Taylor & Francis Group, Boca Raton, Florida.

Luque, R., and Speight, J. G. (Editors). 2015. *Gasification for Synthetic Fuel Production: Fundamentals, Processes, and Applications*. Woodhead Publishing, Elsevier, Cambridge, United Kingdom.

Lv, P. M., Xiong, Z. H., Chang, J., Wu, C. Z., Chen, Y., and Zhu, J. X. 2004. An Experimental Study on Biomass Air-Steam Gasification in a Fluidized Bed. *Bioresource Technology,* 95(1): 95-101.

Marano, J. J. 2003. *Refinery Technology Profiles: Gasification and Supporting Technologies*. Report prepared for the United States Department of Energy, National Energy Technology Laboratory. United States Energy Information Administration, Washington, DC. June.

Martinez-Alonso, A., and Tascon, J. M. D. 1991. In: *Fundamental Issues in Control of Carbon Gasification Reactivity*. Lahaye, J., and Ehrburger, P. (Editors). Kluwer Academic Publishers, Dordrecht, Netherlands.

Matsukata, M., Kikuchi, E., and Morita, Y. 1992. A New Classification of Alkali and Alkaline Earth Catalysts for Gasification of Carbon. *Fuel*. 71: 819-823.

McKendry, P. 2002. Energy Production from Biomass Part 3: Gasification Technologies. *Bioresource Technology*, 83(1): 55-63.

McLendon T. R., Lui A. P., Pineault R. L., Beer S. K., and Richardson S. W. 2004. High-Pressure Co-Gasification of Coal and Biomass in a Fluidized Bed. *Biomass and Bioenergy*, 26(4): 377-388.

Mims, C. A. 1991. In *Fundamental Issues in Control of Carbon Gasification Reactivity*. Lahaye, J., and Ehrburger, P. (Editors). Kluwer Academic Publishers, Dordrecht, Netherlands. Page 383.

Mokhatab, S., Poe, W. A., and Speight, J. G. 2006. *Handbook of Natural Gas Transmission and Processing*. Elsevier, Amsterdam, Netherlands.

Molino, A., Iovane, P., Donatelli, A., Braccio, G., Chianese, S., Musmarra, D. 2013. Steam Gasification of Refuse-Derived Fuel in a Rotary Kiln Pilot Plant: Experimental Tests. *Chem. Eng. Trans.,* 32: 337-342.

Molino, A., Chianese, S., and Musmarra, D. 2016. Biomass Gasification Technology: The State of the Art Overview. *J. Energy Chem.,* 25: 10-25.

Nordstrand D., Duong D. N. B., Miller, B. G. 2008. *Combustion Engineering Issues for Solid Fuel Systems-Chapter 9 Post-combustion Emissions Control*. B. G. Miller and D. Tillman (Editors). Elsevier, London, United Kingdom.

Pakdel, H., and Roy, C. 1991. Hydrocarbon Content of Liquid Products and Tar from Pyrolysis and Gasification of Wood. *Energy & Fuels*, 5: 427-436.

Pan, Y. G., Velo, E., Roca, X., Manyà, J. J., and Puigjaner, L. 2000. Fluidized-Bed Cogasification of Residual Biomass/Poor Coal Blends for Fuel Gas Production. *Fuel,* 79: 1317-1326.

Parkash, S. 2003. *Refining Processes Handbook.* Gulf Professional Publishing, Elsevier, Amsterdam, Netherlands.

Penrose, C. F., Wallace, P. S., Kasbaum, J. L., Anderson, M. K., and Preston, W. E. 1999. Enhancing Refinery Profitability by Gasification, Hydroprocessing and Power Generation. *Proceedings. Gasification Technologies Conference*. San Francisco, California. October. https://www.globalsyngas.org/uploads/eventLibrary/GTC99270.pdf.

Pepiot, P., Dibble, and Foust, C. G. 2010. In: Computational Fluid Dynamics Modeling of Biomass Gasification and Pyrolysis. In: *Computational Modeling in Lignocellulosic Biofuel Production.* M. R. Nimlos and M. F. Crowley (Editors). ACS Symposium Series; American Chemical Society, Washington, DC.

Probstein, R. F., and Hicks, R. E. 1990. *Synthetic Fuels.* pH Press, Cambridge, Massachusetts. Chapter 4.

Rabovitser, I. K., Nester, S., and Bryan, B. 2010. *Plasma Assisted Conversion of Carbonaceous Materials into a Gas.* United States Patent 7,736,400. June 25.

Radovanović, L., and Speight, 2018. J. G. *Global Warming – Truth and Myths. Petroleum and Chemical Industry International*, 1(1). https://www.opastonline.com/wp-content/uploads/2018/10/Globalwarming-truthandmyths-pcii-18.pdf.

Rajvanshi, A. K. 1986. Biomass Gasification. In: *Alternative Energy in Agriculture*, Vol. II. D. Y. Goswami (Editor). CRC Press, Boca Raton, Florida. Page 83-102.

Ramroop Singh, N. 2011. Biofuel. In: *The Biofuels Handbook.* J. G. Speight (Editor). Royal Society of Chemistry, London, United Kingdom. Part 1, Chapter 5.

Rapagnà, N. J., and Latif, A. 1997. Steam Gasification of Almond Shells in a Fluidized Bed Reactor: The Influence of Temperature and Particle Size on Product Yield and Distribution. *Biomass and Bioenergy*, 12(4): 281-288.

Rapagnà, N. J., and, A. Kiennemann, A., and Foscolo, P. U. 2000. Steam-Gasification of Biomass in a Fluidized-Bed of Olivine Particles. *Biomass and Bioenergy*, 19(3): 187-197.

Rauch, R., Hrbek, J., Hofbauer, H. 2014. Biomass Gasification for Synthesis Gas Production and Applications of the Syngas. Wiley Interdiscip. *Rev. Energy Environ.*, 3: 343-362.

Ricketts, B., Hotchkiss, R., Livingston, W., and Hall, M. 2002. Technology Status Review of Waste/Biomass Co-Gasification with Coal. Proceedings. *Inst. Chem. Eng. Fifth European Gasification Conference.* Noordwijk, Netherlands. April 8-10.

Roddy, D. J., and Manson-Whitton, C. 2012. Biomass Gasification and Pyrolysis. In: *Comprehensive Renewable Energy.* Volume 5: Biomass and Biofuels. D. J. Roddy (Editor). Elsevier, Amsterdam, Netherlands.

Rodríguez-Olalde, N.E., Mendoza-Chávez, E., Castro-Montoya, A. J., Saucedo-Luna, J., Maya-Yescas, R., Rutiaga-Quiñones, J. G., and Ponce Ortega, J. M. 2015. Simulation of Syngas Production from Lignin Using Guaiacol as a Model Compound. *Energies,* 8: 6705-6714.

Sha, X. 2005. Coal Gasification. In: *Coal, Oil Shale, Natural Bitumen, Heavy Oil and Peat. Encyclopedia of Life Support Systems (EOLSS),* Developed under the Auspices of the UNESCO, EOLSS Publishers, Oxford, UK, [http://www.eolss.net].

Shabbar, S., and Janajreh, I. 2013. Thermodynamic Equilibrium Analysis of Coal Gasification Using Gibbs Energy Minimization Method. *Energy Conversion and Management,* 65: 755-763.

Shen, C. H., Chen, W. H., Hsu, H. W., Sheu, J. Y., and Hsieh, T. H. 2012. Co-Gasification Performance of Coal and Petroleum Coke Blends in A Pilot-Scale Pressurized Entrained-Flow Gasifier. *Int. J. Energy Res.*, 36: 499-508.

Sikarwar, V. S., Zhao, M., Fennell, P. S., Shah, N., Anthony, E. J. 2017. Progress in Biofuel Production from Gasification. *Prog. Energy Combust. Sci.*, 61: 189-248.

Singh, S. P., Weil, S. A., and Babu, S. P. 1980. Thermodynamic Analysis of Coal Gasification Processes. *Energy*, 5(8-9): 905–914.

Sjöström, K., Chen, G., Yu, Q., Brage, C., and Rosén, C. 1999. Promoted Reactivity of Char in Cogasification of Biomass and Coal: Synergies in the Thermochemical Process. *Fuel,* 78: 1189-1194.

Sondreal, E. A., Benson, S. A., Pavlish, J. H., and Ralston, N. V. C. 2004. An Overview of Air Quality III: Mercury, Trace Elements, and Particulate Matter. *Fuel Processing Technology.* 85: 425-440.

Speight, J. G. 1990. In: *Fuel Science and Technology Handbook.* J. G. Speight (Editor). Marcel Dekker Inc., New York. Chapter 33.

Speight, J. G. 2007. *Natural Gas: A Basic Handbook*. GPC Books, Gulf Publishing Company, Houston, Texas.

Speight, J. G. 2008. *Synthetic Fuels Handbook: Properties, Processes, and Performance*. McGraw-Hill, New York.

Speight, J. G. (Editor). 2011a. *Biofuels Handbook.* Royal Society of Chemistry, London, United Kingdom.

Speight, J. G. 2011b. *The Refinery of the Future*. Gulf Professional Publishing, Elsevier, Oxford, United Kingdom.

Speight, J. G. 2013a. *The Chemistry and Technology of Coal* 3rd Edition. CRC Press, Taylor & Francis Group, Boca Raton, Florida.

Speight, J. G. 2013b. *Coal-Fired Power Generation Handbook.* Scrivener Publishing, Salem, Massachusetts.

Speight, J. G. 2014a. *The Chemistry and Technology of Petroleum* 5th Edition. CRC Press, Taylor & Francis Group, Boca Raton, Florida.

Speight, J. G. 2014b. *Gasification of Unconventional Feedstocks*. Gulf Professional Publishing, Elsevier, Oxford, United Kingdom.

Speight. J. G. 2016. Hydrogen in Refineries. In: *Hydrogen Science and Engineering – Materials, Processes, Systems, and Technology*. D. Stolten and B. Emonts (Editors). Wiley-VCH Verlag GmbH & Co., Weinheim, Germany. Chapter 1. Page 3-18.

Speight, J. G., and Islam, M. R. 2016. *Peak Energy – Myth or Reality*. Scrivener Publishing, Beverly, Massachusetts.

Speight, J. G. 2017. *Handbook of Petroleum Refining*. CRC Press, Taylor & Francis Group, Boca Raton, Florida.

Speight, J. G. 2019. *Handbook of Petrochemical Processes*. CRC Press, Taylor & Francis Group, Boca Raton, Florida.

Sundaresan, S., and Amundson, N. R. 1978. Studies in Char Gasification – I: A lumped Model. *Chemical Engineering Science,* 34: 345-354.

Sutikno, T., and Turini, K. 2012. Gasifying Coke to Produce Hydrogen in Refineries. *Petroleum Technology Quarterly*, Q3: 105.

Van Heek, K. H., Muhlen, H-J. 1991. In: *Fundamental Issues in Control of Carbon Gasification Reactivity.* J. Lahaye and P. Ehrburger (editors). Kluwer Academic Publishers Inc., Dordrecht, Netherlands. Page 1.

Vélez, F. F., Chejne, F., Valdés, C. F., Emery, E. J., and Londoño, C. A. 2009. Cogasification of Colombian Coal and Biomass in a Fluidized Bed: An Experimental Study. *Fuel,* 88(3): 424-430.

Wallace, P. S., Anderson, M. K., Rodarte, A. I., and Preston, W. E. 1998. Heavy Oil Upgrading by the Separation and Gasification of Asphaltenes. *Proceedings. Presented at the Gasification Technologies Conference.* San Francisco, California. October. https://www.globalsyngas.org/uploads/eventLibrary/gtc9817p.pdf.

Wang, Y., Duan Y., Yang L., Jiang Y., Wu C., Wang Q., and Yang X. 2008. Comparison of Mercury Removal Characteristic between Fabric Filter and Electrostatic Precipitators of Coal-Fired Power Plants. *Journal of Fuel Chemistry and Technology,* 36(1): 23-29.

Wolff, J., and Vliegenthart, E. 2011. Gasification of Heavy Ends. *Petroleum Technology* Quarterly, Q2: 1-5.

Yang, H., Xua, Z., Fan, M., Bland, A. E., and Judkins, R. R. 2007. Adsorbents for Capturing Mercury in Coal-Fired Boiler Flue Gas. *Journal of Hazardous Materials.* 146:1-11.

Yuksel Orhan, İs, G., Alper, E., McApline, K., Daly, S., Sycz, M., and Elkamel. A. 2014. Gasification of Oil Refinery Waste for Power and Hydrogen Production. *Proceedings. 2014 International Conference on Industrial Engineering and Operations Management.* Bali, Indonesia, January 7-9.

Chapter 7

BIOGAS CLEANING

1. INTRODUCTION

Biogas is experiencing a period of rapid development and biogas upgrading is attracting increasing attention (Ryckebosch et al., 2011; Niesner et al., 2013; Andriani et al., 2014). Currently, biogas is mostly used as a fuel in power generators and boilers. For these uses, the hydrogen sulfide content in biogas should be less than 200 parts per million (ppm v/v) to ensure a long life for the power and heat generators. Consequently, the market for biogas upgrading is facing significant challenges in terms of energy consumption and operating costs. Selection of upgrading technology is site-specific, case-sensitive and dependent on the biogas utilization requirements and local circumstances (Ramírez et al., 2015). But first it is important to recognize the composition of the gas so that the appropriate technologies can be selected for use to remove the specific contaminants. Thus, when any biomass-derived gas (either biogas or bio-synthesis gas) is used for biofuel production, the cleaning of the raw gas is needed strictly in order to remove contaminants and potential catalyst poisons as well as to achieve the qualitative composition required before the gas is sold for use.

Biogas, as produced by any of the methods described in earlier chapters (i.e., by anaerobic digestion, pyrolysis, or gasification) is typically saturated with water vapor and contains, in addition to methane (CH_4) and carbon dioxide (CO_2), various amounts of hydrogen sulfide (H_2S) as well as other contaminants that vary in amounts depending upon the composition of the biomass feedstocks and the process parameters employed for production of the gas. The properties of the contaminants are such that failure to remove them will make the gas unusable and even poisonous to the user. For example, hydrogen sulfide is a toxic gas, with a specific, unpleasant odor, similar to rotten eggs, forming acidic products in combination with the water vapors in the biogas which can also result in corrosion of equipment. To prevent this, and similar example of the adverse effects of other contaminants, the biogas must be dried and any contaminants removed before the biogas is sent to one or more conversion units or for sales.

Typically, in the gas processing industry (also called the *gas cleaning industry* or the *gas refining industry*), the feedstock is not used directly as fuel because of the complex nature of the feedstock and the presence of one of more of the aforementioned impurities that are corrosive or poisonous. It is, therefore, essential that any gas should be contaminant free when it enters any one of the various reactors (Parkash, 2003; Gary et al., 2007; Speight, 2014; Hsu and Robinson, 2017; Speight, 2017). Typically, the primary raw biogas has been subjected to chemical and/or physical changes (refining) after being produced. Thus for many applications (Chapter 8) the quality of biogas has to be improved. In addition, landfill gas (Chapter 1, Chapter 4) often contains significant amounts of halogenated compounds which need to be removed prior to use. Occasionally the oxygen content is high when air is allowed to invade the gas producer during collection of the landfill gas or when the biomass from which the gas is produced contains an abundance of oxygen-containing organic constituents.

The main contaminants (but not necessarily the only contaminants) that may require removal in gas cleaning systems are hydrogen sulfide, water, carbon dioxide, and halogenated compounds. Desulfurisation to prevent corrosion and avoid concentrations of the toxic hydrogen sulfide for safety in use and in the workplace. Also, if hydrogen sulfide and other sulfur-containing species such as thiol derivatives (RSH, also called mercaptan derivative) and carbonyl sulfide (COS) are not removed, combustion of the gas produces sulfur dioxide and sulfur trioxide which is even more poisonous than hydrogen sulfide. At the same time, the presence of sulfur dioxide in the gas stream lowers the dew point (the temperature to which the gas must be cooled to become saturated with water vapor) of in the stack gas. The sulfurous acid formed (H_2SO_3) by reaction of the sulfur dioxide the and the water vapor as well as reaction of carbon dioxide and the water vapor to form highly corrosive acidic species:

$$SO_2 + H_2O \rightarrow H_2SO_3$$
$$CO_2 + H_2O \rightarrow H_2CO_3$$

Thus, water removal from the biogas is also essential, not only because of the potential for the reaction of water with contaminants in the gas stream but also because of potential accumulation of water vapor condensing in the gas line, the formation of a corrosive acidic solution when hydrogen sulfide is dissolved or to achieve low dew points when biogas is stored under elevated pressures in order to avoid condensation and freezing. Also, if the non-combustible carbon dioxide is not removed the energy content of the biogas is diluted and there may also be an environmental impact due to the presence and emission of this contaminant.

Generally, the contaminants are categorized as (i) particulate matter, (ii) condensable hydrocarbon derivatives, including tar products, (iii) alkali metals, such as sodium and potassium, (iv) nitrogen-containing derivatives, including ammonia and hydrogen cyanide, (v) sulfur containing derivatives, such as hydrogen sulfide, carbonyl sulfide, and carbon disulfide, and halogen containing derivatives, such as hydrogen chloride and hydrogen bromide, and hydrogen fluoride.

In fact, the constituents that make up the tar products are formed during gasification when biomass decomposes in pyrolysis and gasification reactions. Primary tar compounds are mostly oxygenated compounds that are decomposition products of biomass. These compounds react further and form secondary and tertiary tar compounds, which consists of compounds that do not exist in the source biomass. The secondary tar typically consists of alkylated one-ring and two-ring aromatic compounds (including heterocyclic compounds) whereas tertiary tar consists of aromatic hydrocarbon derivatives such as benzene, naphthalene, and various polycyclic aromatic hydrocarbon derivatives (PAHs, also called polynuclear aromatic compounds, PNAs). Tar constituents are typically classified on the basis of the number of rings in the constituents or by boiling point distribution or physical properties. More generally, tar is defined as aromatic compounds that are higher-boiling than benzene and, in addition, an operational definition for tar depending on the end-use application has been used.

Process selectivity indicates the preference with which the process removes one acid gas component relative to (or in preference to) another. For example, some processes remove both hydrogen sulfide and carbon dioxide; other processes are designed to remove hydrogen sulfide only. It is very important to consider the process selectivity for, say, hydrogen sulfide removal compared to carbon dioxide removal that ensures minimal concentrations of these components in the product, thus the need for consideration of the carbon dioxide to hydrogen sulfide in the gas stream.

To include a description of all of the possible process variations for gas cleaning is beyond the scope of this book. Therefore, the focus of this chapter is a selection of the processes that are an integral part within the concept of production of a specification-grade product (methane) for sale to the consumer.

Furthermore, it is the purpose of this chapter to present the methods by which various gaseous and liquid feedstocks can be processed and prepared for petrochemical production. This requires removal of impurities that would otherwise by deleterious to petrochemical production and analytical assurance that that feedstocks are, indeed, free of deleterious contaminants (Speight, 2015, 2018).

2. Biogas Cleaning

Whatever the source of the biogas, methane (in varying amounts) is the predominant hydrocarbon derivative in the gas stream methane. However, there are other hydrocarbon derivatives in addition to water vapor, hydrogen sulfide (H_2S), carbon dioxide, nitrogen, and other compounds.

Treated biogas consists mainly of methane (often referred to as biomethane). However, biogas is not pure methane, and its properties are modified by the presence of impurities, such as nitrogen, carbon dioxide, and small amounts of unrecovered higher boiling (non-gaseous at STP) hydrocarbon derivatives. An important property of any gas stream is its heating value – relatively high amounts of nitrogen and/or carbon dioxide reduce the heating value of the gas. Biogas, if it was pure methane, would have a heating value of 1,010 Btu/ft^3. This value is reduced to approximately 900 Btu/ft^3 if the gas contains approximately 10% v/v nitrogen and carbon dioxide – the heating value of carbon dioxide is zero).

In order to remove the unwanted byproducts and contaminants in the product gas, primary treatments, such as optimization of the properties of biomass feedstock, design of the gasifier, and the operational parameters of the gasifier are first implemented. Secondary treatment, such as a downstream cleaning system based on physical (scrubbers, filters) or catalytic strategies, have to be incorporated for the hot gas cleaning to achieve a more satisfactory reduction for end-applications and also to meet sales specification, which include meeting environmental regulations.

Thus, to reach the condition of acceptability, the biogas must be the end-product of treated by a series of processes that have successfully removed the contaminants that are present in the raw (untreated) gas. Gas processing consists of separating all of the non-methane derivatives (both the combustible and non-combustible environmentally-unfriendly and performance-unfriendly constituents) and fluids from the gas (Kidnay and Parrish, 2006; Mokhatab et al., 2006; Speight, 2014, 2019). While often assumed to be hydrocarbon derivatives in nature, there are also components of the gaseous products that must be removed prior to release of the gases to the atmosphere or prior to use of the gas, i.e., as a fuel gas or as a process feedstock.

2.1. Composition

Trace quantities of contaminants in hydrocarbon products can be harmful to many catalytic chemical processes in which these products are used. In general, the raw product gas of biomass gasification contains a range of minor species and contaminants, including particles, tar, alkali metals, chlorine compounds such as hydrogen chloride (HCl), nitrogen compounds such as ammonia (NH_3) and hydrogen cyanide (HCN), sulfur compounds such as hydrogen sulfide (H_2S) and carbonyl sulfide (COS), as well other species.

The maximum permissible levels of total contaminants are normally included in specifications for such hydrocarbon derivatives. It is recommended that this test method be used to provide a basis for agreement between two laboratories when the determination of sulfur in hydrocarbon gases is important. In the case of liquefied petroleum gas, total volatile sulfur is measured on an injected gas sample. One test method (ASTM D3246) describes a procedure for the determination of sulfur in the range from 1.5 to 100 mg/kg (ppm w/w) in hydrocarbon products that are gaseous at normal room temperature and pressure. There is also a variety of other standard test methods that can be applied to the determination of the properties of biogas and, hence the suitability of the gas for use (ASTM, 2018; Speight, 2018).

Acidic constituents such as carbon dioxide and hydrogen sulfide as well as thiol derivatives (RSH, also called mercaptan derivatives, RSH) can contribute to corrosion of refining equipment, harm catalysts, pollute the atmosphere, and prevent the use of hydrocarbon components in petrochemical manufacture (Mokhatab et al., 2006; Speight, 2014, 2019). When the amount of hydrogen sulfide is high, it may be removed from a gas stream and converted to sulfur or sulfuric acid; a recent option for hydrogen sulfide removal is the use of chemical scavengers (Kenreck, 2014). Some gases contain sufficient carbon dioxide to warrant recovery as dry ice (Bartoo, 1985).

Biogas is not always truly hydrocarbon in nature and may contain contaminants, such as carbon oxides (CO_x, where x = 1 and/or 2), sulfur oxides (SO_x, where x = 2 and/or 3), as well as ammonia (NH_3), mercaptan derivatives (RSH), carbonyl sulfide (COS), and mercaptan derivatives (RSH). The presence of these impurities may eliminate some of the sweetening processes from use since some of these processes remove considerable amounts of acid gas but not to a sufficiently low concentration. On the other hand, there are those processes not designed to remove (or incapable of removing) large amounts of acid gases whereas they are capable of removing the acid gas impurities to very low levels when the acid gases are present only in low-to-medium concentration in the gas (Katz, 1959; Mokhatab et al., 2006; Speight, 2014, 2019).

The sources of the various biogas stream are varied but, in terms of gas cleaning (i.e., removal of the contaminants before petrochemical production), the processes are largely the same but it is a question of degree. For example, gas streams for some sources may produce gases may contain higher amounts of carbon dioxide and/or hydrogen sulfide and the processes will have to be selected accordingly.

In addition to its primary importance as a fuel, biogas is also a source of hydrocarbon derivatives for petrochemical feed stocks. While the major hydrocarbon constituent of biogas is methane, there are components such as carbon dioxide (CO_2), hydrogen sulfide (H_2S), and mercaptan derivatives (thiols, also called mercaptans, RSH), as well as trace amounts of sundry other emissions such as carbonyl sulfide (COS). The fact that methane has a foreseen and valuable end-use makes it a desirable product, but in several other situations it is considered a pollutant, having been identified a greenhouse gas.

In practice, heaters and scrubbers are usually installed at or near to the source. The scrubbers serve primarily to remove sand and other large-particle impurities and the heaters ensure that the temperature of the gas does not drop too low. With gas that contains even low quantities of water, gas hydrates ($C_nH_{2n+2}.xH_2O$) tend to form when temperatures drop. These hydrates are solid or semi-solid compounds, resembling ice like crystals. If the hydrates accumulate, they can impede the passage of gas through valves and gathering systems (Zhang et al., 2007). To reduce the occurrence of hydrates small gas-fired heating units are typically installed along the gathering pipe wherever it is likely that hydrates may form.

2.2. Process Types

Biogas can also be upgraded to pipeline-quality gas and this upgraded gas may be used for residential heating and as vehicle fuel. When distributing the biogas using pipelines, gas pipeline standards will, more than likely, become applicable. Removing water vapor is easier than removing carbon dioxide and hydrogen sulfide from biogas (Table 7.1). A condensate trap at a proper location on the gas pipeline can remove water vapor as warm biogas cools by itself after leaving the digester or the gas-generating equipment.

Gas processing involves the use of several different types of processes to remove contaminants from gas streams but there is always overlap between the various processing concepts. In addition, the terminology used for gas processing can often be confusing and/or misleading because of the overlap (Curry, 1981; Maddox, 1982). Gas processing is necessary to ensure that the gas prepared for transportation (usually by pipeline) and for sales must be as clean and pure as the specifications dictate. Thus, biogas, as it is used by consumers, is much different from the gas that is brought from, for example, the anaerobic digester (Chapter 3). Moreover, although gas produced in the digester may be composed primarily of methane, is by no means as pure.

The processes that have been developed to accomplish gas purification vary from a simple once-through wash operation to complex multi-step recycling systems (Abatzoglou and Boivin, 2009; Mokhatab et al., 2006; Speight, 2014, 2019). In many cases, the process complexities arise because of the need for recovery of the materials used to remove the contaminants or even recovery of the contaminants in the original, or altered, form (Katz, 1959; Kohl and Riesenfeld, 1985; Newman, 1985; Mokhatab et al., 2006; Speight, 2014, 2015, 2019). In addition to the corrosion of equipment by acid gases (Speight, 2014) the escape into the atmosphere of sulfur-containing gases can eventually lead to the formation of the constituents of acid rain, i.e., the oxides of sulfur (sulfur dioxide, SO_2, and sulfur trioxide, SO_3).

Table 7.1. Common Methods for the Removal of Carbon Dioxide and Hydrogen Sulfide from Gas Streams

Carbon dioxide removal	
Water scrubbing	Uses principle of the higher solubility of carbon dioxide in water to separate the carbon dioxide from biogas. The process uses high pressure and removes hydrogen sulfide as well as carbon dioxide. The main disadvantage of this process is that it requires a large volume of water that must be purified and recycled.
Polyethylene glycol scrubbing	This process is more efficient than water scrubbing. It also requires the regeneration of a large volume of polyethylene glycol.
Chemical absorption	Chemical reaction between carbon dioxide and amine-based solvents (olamines)
Carbon molecular sieves	Uses differential adsorption characteristics to separate methane and carbon dioxide, which is carried out at high pressure (pressure swing adsorption). Hydrogen sulfide should be removed before the adsorption process.
Membrane separation	Selectively separates hydrogen sulfide from carbon dioxide from methane. The carbon dioxide and the hydrogen sulfide dissolve while the methane is collected for use.
Cryogenic separation	Cooling until condensation or sublimation of the carbon dioxide.
Hydrogen sulfide removal	
Biological desulfurization	Natural bacteria can convert hydrogen sulfide into elemental sulfur in the presence of oxygen and iron. This can be done by introducing a small amount (two to five per cent) of air into the head space of the digester. As a result, deposits of elemental sulfur will be formed in the digester. This process may be optimized by a more sophisticated design where air is bubbled through the digester feed material. It is critical that the introduction of the air be carefully controlled to avoid reducing the amount of biogas that is produced.
Iron/iron oxide reaction	Hydrogen sulfide reacts readily with either iron oxide or iron chloride to form insoluble iron sulfide. The reaction can be exploited by adding the iron chloride to the digester feed material or passing the biogas through a bed of iron oxide-containing material. The iron oxide media needs to be replaced periodically. The regeneration process is highly exothermic and must be controlled to avoid problems.
Activated carbon	Activated carbon impregnated with potassium iodide can catalytically react with oxygen and hydrogen sulfide to form water and sulfur. The activated carbon beds need regeneration or replacement when saturated.
Scrubbing/membrane separation	The carbon dioxide and hydrogen sulfide can be removed by washing with water, glycol solutions, or separated using the membrane technique.

Similarly, the nitrogen-containing gases can also lead to nitrous and nitric acids (through the formation of the oxides NO_x, where x = 1 or 2) which are the other major contributors to acid rain. The release of carbon dioxide and hydrocarbon derivatives as constituents of refinery effluents can also influence the behavior and integrity of the ozone layer.

However, the precise area of application of a given process is difficult to predict define and required careful consideration of several process factors before process selection. These factors are: (i) the types of contaminants in the gas, (ii) the concentrations of contaminants in the gas, (iii) the degree of contaminant removal desired, (iv) the selectivity of acid gas removal required, (v) the temperature of the gas to be processed, (vi) the pressure of the gas to be processed, (vii) the volume of the gas to be processed, (viii) the composition of the gas to be processed, (ix) the ratio of carbon dioxide to hydrogen sulfide ratio in the gas feedstock, and (x) the desirability of sulfur recovery due to environmental issues or economic issues.

Finally, throughout the various cleaning processes, it is important to minimize the loss of methane in order to achieve an economical viable cleaning operation. However, it is also important to minimize the methane slip since methane is a strong greenhouse gas. Thus, the release of methane to the atmosphere should be minimized by treating the off-gas, air or water streams leaving the plant even though the methane cannot be utilized. Methane can be present in the off-gas leaving a pressure swing adsorption-column, in air from a water scrubber with water recirculation or in water in a water scrubber without water recirculation. The off-gas from an upgrading plant seldom contains a high enough methane concentration to maintain a flame without addition of natural gas or biogas. One way of limiting the methane slip is to mix the off-gas with air that is used for combustion. Alternatively the methane can be oxidized by thermal or catalytic oxidation. Thus:

$$CH_4 + O_2 \rightarrow CO_2 + H_2O$$

In the catalytic process, the oxidation takes place at the surface of the catalyst which lowers the energy needed to oxidize the methane, thus enabling oxidation at a lower temperature. The active component of the catalyst is platinum, palladium or cobalt.

3. Water Removal

The relative humidity of biogas inside the digester is 100%, so the gas is saturated with water vapors. To protect the energy conversion equipment from wear and from eventual damage, water must be removed from the produced biogas. The quantity of water contained by biogas depends on temperature. A part of the water vapor can be condensed by cooling of the gas. This is frequently done in the gas pipelines transporting biogas from digester to

CHP unit. The water condensates on the walls of the sloping pipes and can be collected in a condensation separator, at the lowest point of the pipeline.

Water in the liquid phase causes corrosion or erosion problems in pipelines and equipment, particularly when carbon dioxide and hydrogen sulfide are present in the gas. The simplest method of water removal (refrigeration or cryogenic separation) is to cool the gas to a temperature at least equal to or (preferentially) below the dew point (Mokhatab et al., 2006; Speight, 2014, 2019).

In addition to separating any condensate from the wet gas stream, it is necessary to remove most of the associated water. Most of the liquid, free water associated with the extracted gas stream is removed by simple separation methods but, however, the removal of the water vapor that exists in solution in a gas stream requires a more complex treatment. This treatment consists of dehydrating the gas stream, which usually involves one of two processes: either absorption, or adsorption.

Moisture may be removed from hydrocarbon gases at the same time as hydrogen sulfide is removed. Moisture removal is necessary to prevent harm to anhydrous catalysts and to prevent the formation of hydrocarbon hydrates (e.g., $C_3H_8.18H_2O$) at low temperatures. A widely used dehydration and desulfurization process is the glycolamine process, in which the treatment solution is a mixture of ethanolamine and a large amount of glycol. The mixture is circulated through an absorber and a reactivator in the same way as ethanolamine is circulated in the Girbotol process. The glycol absorbs moisture from the hydrocarbon gas passing up the absorber; the ethanolamine absorbs hydrogen sulfide and carbon dioxide. The treated gas leaves the top of the absorber; the spent ethanolamine-glycol mixture enters the reactivator tower, where heat drives off the absorbed acid gases and water.

Absorption occurs when the water vapor is taken out by a dehydrating agent. Adsorption occurs when the water vapor is condensed and collected on the surface. In a majority of cases, cooling alone is insufficient and, for the most part, impractical for use in field operations. Other more convenient water removal options use (i) *hygroscopic* liquids (e.g., diethylene glycol or triethylene glycol) and (ii) solid adsorbents or desiccants (e.g., alumina, silica gel, and molecular sieves). Ethylene glycol can be directly injected into the gas stream in refrigeration plants.

3.1. Absorption

An example of absorption dehydration is known as *glycol dehydration* – the principal agent in this process is diethylene glycol which has a chemical affinity for water (Mokhatab et al., 2006; Abdel-Aal et al., 2016; Speight, 2019). Glycol dehydration involves using a solution of a glycol such as diethylene glycol (DEG) or triethylene glycol (TEG), which is brought into contact with the wet gas stream in a *contactor*. In practice, absorption systems recover 90 to 99% by volume of methane that would otherwise be flared into the atmosphere.

In the process, a liquid desiccant dehydrator serves to absorb water vapor from the gas stream. The glycol solution absorbs water from the wet gas and, once absorbed, the glycol particles become heavier and sink to the bottom of the contactor where they are removed. The dry gas stream is then transported out of the dehydrator. The glycol solution, bearing all of the water stripped from the gas stream, is recycled through a specialized boiler designed to vaporize only the water out of the solution. The boiling point differential between water (100°C, 212°F) and glycol (204°C, 400°F) makes it relatively easy to remove water from the glycol solution.

As well as absorbing water from the wet gas stream, the glycol solution occasionally carries with it small amounts of methane and other compounds found in the wet gas. In order to decrease the amount of methane and other compounds that would otherwise be lost, flash tank separator-condensers are employed to remove these compounds before the glycol solution reaches the boiler. The flash tank separator consists of a device that reduces the pressure of the glycol solution stream, allowing the methane and other hydrocarbon derivatives to vaporize (*flash*). The glycol solution then travels to the boiler, which may also be fitted with air or water cooled condensers, which serve to capture any remaining organic compounds that may remain in the glycol solution. The regeneration (stripping) of the glycol is limited by temperature: diethylene glycol and triethylene glycol decompose at or even before their respective boiling points. Such techniques as stripping of hot triethylene glycol with dry gas (e.g., heavy hydrocarbon vapors, the *Drizo process*) or vacuum distillation are recommended.

Another absorption process, the Rectisol process, is a physical acid gas removal process using an organic solvent (typically methanol) at subzero temperatures, and characteristic of physical acid gas removal processes, it can purify synthesis gas down to 0.1 ppm v/v total sulfur, including hydrogen sulfide (H_2S) and carbonyl sulfide (COS), and carbon dioxide (CO_2) in the ppm v/v range (Mokhatab et al., 2006; Liu et al., 2010; Abdel-Aal et al., 2016). The process uses methanol as a wash solvent and the wash unit operates under favorable at temperatures below 0°C (32°F). To lower the temperature of the feed gas temperatures, it is cooled against the cold-product streams, before entering the absorber tower. At the absorber tower, carbon dioxide and hydrogen sulfide (with carbonyl sulfide) are removed. By use of an intermediate flash, co-absorbed products such as hydrogen and carbon monoxide are recovered, thus increasing the product recovery rate. To reduce the required energy demand for the carbon dioxide compressor, the carbon dioxide product is recovered in two different pressure steps (medium pressure and lower pressure). The carbon dioxide product is essentially sulfur-free (hydrogen sulfide-free, carbonyl sulfide-free) and water free. The carbon dioxide products can be used for enhanced oil recovery (EOR) and/or sequestration or as pure carbon dioxide for other processes.

3.2. Adsorption

Adsorption is a physical-chemical phenomenon in which the gas is concentrated on the surface of a solid or liquid to remove impurities. It must be emphasized that *adsorption* differs from *absorption* in that absorption is not a physical-chemical surface phenomenon but a process in which the absorbed gas is ultimately distributed throughout the absorbent (liquid). Dehydration using a solid adsorbent or solid-desiccant is the primary form of dehydrating gas stream using adsorption, and usually consists of two or more adsorption towers, which are filled with a solid desiccant (Mokhatab et al., 2006; Abdel-Aal et al., 2016; Speight, 2019). Typical desiccants include activated alumina or a granular silica gel material. A wet gas stream is passed through these towers, from top to bottom. As the wet gas passes around the particles of desiccant material, water is retained on the surface of these desiccant particles. Passing through the entire desiccant bed, almost all of the water is adsorbed onto the desiccant material, leaving the dry gas to exit the bottom of the tower. There are several solid desiccants which possess the physical characteristic to adsorb water from the gas stream. These desiccants are generally used in dehydration systems consisting of two or more towers and associated regeneration equipment.

Molecular sieves – a class of aluminosilicates which produce the lowest water dew points and which can be used to simultaneously sweeten, dry gases and liquids (Mokhatab et al., 2006; Maple, and Williams 2008; Abdel-Aal et al., 2016; Speight, 2019) – are commonly used in dehydrators ahead of plants designed to recover ethane and other higher-boiling hydrocarbon derivatives. These plants operate at very cold temperatures and require very dry feed gas to prevent formation of hydrates. Dehydration to -100°C (-148°F) dew point is possible with molecular sieves. Water dew points less than -100°C (-148°F) can be accomplished with special design and definitive operating parameters (Mokhatab et al., 2006).

Molecular sieves are commonly used to selectively adsorb water and sulfur compounds from light hydrocarbon streams such as liquefied petroleum gas (LPG), propane, butane, pentane, light olefin derivatives, and alkylation feed. Sulfur compounds that can be removed are hydrogen sulfide, mercaptan derivatives, sulfide derivatives, and disulfide derivatives. In the process, the sulfur-containing feedstock is passed through a bed of sieves at ambient temperature. The operating pressure must be high enough to keep the feed in the liquid phase. The operation is cyclic in that the adsorption step is stopped at a predetermined time before sulfur breakthrough occurs. Sulfur and water are removed from the sieves by purging with fuel gas at 205 to 315°C (400 to 600°F).

Solid-adsorbent dehydrators are typically more effective than liquid absorption dehydrators (e.g., glycol dehydrators) and are usually installed as a type of straddle system along gas pipelines. These types of dehydration systems are best suited for large volumes of gas under very high pressure, and are thus usually located on a pipeline downstream of a compressor station. Two or more towers are required due to the fact that after a certain period of use, the desiccant in a particular tower becomes saturated with water. To regenerate and recycle the desiccant, a high-temperature heater is used to heat gas to a very high temperature and passage of the heated gas stream through a saturated desiccant bed vaporizes the water in the desiccant tower, leaving it dry and allowing for further gas stream dehydration.

Although two-bed adsorbent treaters have become more common (while one bed is removing water from the gas, the other undergoes alternate heating and cooling), on occasion, a three-bed system is used: one bed adsorbs, one is being heated, and one is being cooled. An additional advantage of the three-bed system is the facile conversion of a two-bed system so that the third bed can be maintained or replaced, thereby ensuring continuity of the operations and reducing the risk of a costly plant shutdown.

Silica gel (SiO_2) and alumina (Al_2O_3) have good capacities for water adsorption (up to 8% w/w). Bauxite (crude alumina, Al_2O_3) adsorbs up to 6% by weight water, and molecular sieves adsorb up to 15% by weight water. Silica is usually selected for dehydration of sour gas because of its high tolerance to hydrogen sulfide and to protect molecular sieve beds from plugging by sulfur. Alumina *guard beds* serve as protectors by the act of attrition and may be referred to as an *attrition reactor* containing an *attrition catalyst* (Speight, 2000, 2014, 2017) may be placed ahead of the molecular sieves to remove the sulfur compounds. Downflow reactors are commonly used for adsorption processes, with an upward flow regeneration of the adsorbent and cooling using gas flow in the same direction as adsorption flow.

Solid desiccant units generally cost more to buy and operate than glycol units. Therefore, their use is typically limited to applications such as gases having a high hydrogen sulfide content, very low water dew point requirements, simultaneous control of water, and hydrocarbon dew points. In processes where cryogenic temperatures are encountered, solid desiccant dehydration is usually preferred over conventional methanol injection to prevent hydrate and ice formation (Kidnay and Parrish, 2006).

3.3. Cryogenics

Another possibility of biogas drying is by cooling the gas in electrically powered gas coolers, at temperatures below 10°C (50°F), which allows a lot of humidity to be removed. In order to minimize the relative humidity, but not the absolute humidity, the gas can be warmed up again after cooling, in order to prevent condensation along the gas pipelines.

Thus, cryogenic upgrading makes use of the distinct boiling/sublimation points of the different gases particularly for the separation of carbon dioxide and methane. In the process, the raw (untreated) biogas is cooled down to the temperatures where the carbon dioxide in the gas condenses or sublimates and can be separated as a liquid or a solid fraction, while methane accumulates in the gas phase. Water and siloxane derivatives are also removed during cooling of the gas. However, the content of methane in the biogas affects the characteristics of the gas, i.e., higher pressures and/or lower temperatures are needed to condense or sublime carbon dioxide when it is in a mixture with methane. Cooling usually takes place in several steps in order to remove the different gases in the biogas individually and to optimize the energy recovery.

As an example, a hot biogas stream containing water and carbon dioxide can be upgraded by cooling the gas to 40°C (104°F) most of the water is condensed. The remaining water is removed in the carbon dioxide removal step – the carbon dioxide to be removed from the gas stream to meet the specifications. The final concentration of carbon dioxide in the gas stream is determined by the specification of the Wobbe index. For carbon dioxide removal, considering the high partial pressure of carbon dioxide both membranes and physical solvents can be chosen, where membranes are at their maximum scale and physical solvents are at their minimum scale.

The main advantage of the cryogenic process is that it is possible to obtain biogas with high methane content of up to 99% v/v. The main disadvantage is that for the upgrading process is necessary to use many of technological equipment, especially compressors, turbines and heat exchangers. This significant demand for additional equipment can make the cryogenic separation process extremely expensive.

4. Acid Gas Removal

In addition to water removal, one of the most important parts of gas processing involves the removal of hydrogen sulfide and carbon dioxide, which are generally referred to as contaminants. Biogas from some sources contains significant amounts of hydrogen sulfide and carbon dioxide and, analogous to natural gas that contains these same impurities, is usually referred to as *sour gas*. Sour gas is undesirable because the sulfur compounds it contains can be extremely harmful, even lethal, to breathe and the gas can also be extremely corrosive. The process for removing hydrogen sulfide from sour gas is commonly referred to as *sweetening* the gas. Thus, sulfur compounds, mainly hydrogen sulfide, must be removed from the syngas to the best possible extent since they can poison catalysts. The methanation catalyst is particularly prone to sulfur poisoning. Regenerative sorbents (so-called sulfur guards) can be used to reduce sulfur concentrations to the necessary limits, well below 1 ppm v/v. Washing techniques (based on physical or chemical adsorption) can also be implemented, making sulfur recovery via a Claus process possible if economically viable.

There are four general processes used for emission control (often referred to in another, more specific context as flue gas desulfurization: (i) physical adsorption in which a solid adsorbent is used, (ii) physical absorption in which a selective absorption solvent is used, (iii) chemical absorption is which a selective absorption solvent is used, (iv) and catalytic oxidation thermal oxidation (Soud and Takeshita, 1994; Mokhatab et al., 2006; Speight, 2014, 2019).

4.1. Adsorption

Adsorption is a physical-chemical phenomenon in which the gas is concentrated on the surface of a solid or liquid to remove impurities. Activated carbon and molecular sieves are also capable of adsorbing water in addition to the acid gases.

In order to avoid any confusion, it must be emphasized here that *absorption* differs from *adsorption* in that absorption is not a physical-chemical surface phenomenon but a process in which the absorbed gas is ultimately distributed throughout the absorbent (liquid). The process depends only on physical solubility and may include chemical reactions in the liquid phase (*chemisorption*). Common absorbing media used are water, aqueous amine solutions, caustic, sodium carbonate, and nonvolatile hydrocarbon oils, depending on the type of gas to be absorbed. In these processes, a solid with a high surface area is used. Molecular sieves (zeolites) are widely used and are capable of adsorbing large amounts of gas. In practice, more than one adsorption bed is used for continuous operation – one bed is in use while the other is being regenerated.

On the other hand, adsorption is usually a gas-solid interaction in which an adsorbent such as activated carbon (the *adsorbent* or *adsorbing medium*) which can be regenerated upon *desorption* (Mokhatab et al., 2006; Boulinguiez and Le Cloirec, 2009; Speight, 2014, 2019). The quantity of material adsorbed is proportional to the surface area of the solid and, consequently, adsorbents are usually granular solids with a large surface area per unit mass. Activated carbon impregnated with potassium iodide can catalytically react with oxygen and hydrogen sulfide to form water and sulfur. The reaction is best achieved at 50 to 70°C (122 to 158°F) and 100 to 120 psi. The activated carbon adsorption beds must be regenerated or replaced before they become ineffective due to over-saturation.

Regeneration is accomplished by passing hot dry fuel gas through the bed. Molecular sieves are competitive only when the quantities of hydro- gen sulfide and carbon disulfide are low. The captured (adsorbed) gas can be desorbed with hot air or steam either for recovery or for thermal destruction. Adsorber units are widely used to increase a low gas concentration prior to incineration unless the gas concentration is very high in the inlet air stream and the process is also used to reduce problem odors (or obnoxious odors) from gases. There are several limitations to the use of adsorption systems, but it is generally the case that the major limitation is the requirement for minimization of particulate matter and/or condensation of liquids (e.g., water vapor) that could mask the adsorption surface and drastically reduce its efficiency.

4.2. Absorption

Absorption is achieved by dissolution (a physical phenomenon) or by reaction (a chemical phenomenon) (Barbouteau and Galaud, 1972; Mokhatab et al., 2006; Speight, 2014, 2019). In addition to economic issues or constraints, the solvents used for gas processing should have: (i) a high capacity for acid gas, (ii) a low tendency to dissolve hydrogen, (iii) a low tendency to dissolve low-molecular weight hydrocarbon derivatives, (iv) low vapor pressure at operating temperatures to minimize solvent losses, (v) low viscosity, (vi) low thermal stability, (vii) absence of reactivity toward gas components, (viii) low tendency for fouling, and (ix) a low tendency for corrosion, and (x) economically acceptable (Mokhatab et al., 2006; Speight, 2014, 2019).

Noteworthy commercial processes used are the Selexol process, the Sulfinol process, and the Rectisol process (Mokhatab et al., 2006; Speight, 2014, 2019). In these processes, no chemical reaction occurs between the acid gas and the solvent. The solvent, or absorbent, is a liquid that selectively absorbs the acid gases and leaves out the hydro- carbons.

The Selexol process uses a mixture of the dimethyl ether of propylene glycol as a solvent. It is nontoxic and its boiling point is not high enough for amine formulation. The selectivity of the solvent for hydrogen sulfide (H_2S) is much higher than that for carbon dioxide (CO_2), so it can be used to selectively remove these different acid gases, minimizing carbon dioxide content in the hydrogen sulfide stream sent to the sulfur recovery unit (SRU) and enabling regeneration of solvent for carbon dioxide recovery by economical flashing. In the process, a biogas stream is injected in the bottom of the absorption tower operated at 1000 psi. The rich solvent is flashed in a flash drum (flash reactor) at 200 psi where methane is flashed and recycled back to the absorber and joins the sweet (low-sulfur or no-sulfur) gas stream. The solvent is then flashed at atmospheric pressure and acid gases are flashed off. The solvent is then stripped by steam to completely regenerate the solvent, which is recycled back to the absorber. Any hydrocarbon derivatives are condensed and any remaining acid gases are flashed from the condenser drum. This process is used when there is a high acid gas partial pressure and no heavy hydrocarbon derivatives. Diisopropanolamine (DIPA) can be added to this solvent to remove carbon dioxide to a level suitable for pipeline transportation.

Noteworthy commercial processes commercially used are the Selexol, the Sulfinol, and the Rectisol processes. In these processes, no chemical reaction occurs between the acid gas and the solvent. The solvent, or absorbent, is a liquid that selectively absorbs the acid gases and leaves out the hydrocarbon derivatives. In the Selexol process for example, the solvent is dimethyl ether or polyethylene glycol. Raw biogas passes countercurrently to the descending solvent. When the solvent becomes saturated with the acid gases, the pressure is reduced, and hydrogen sulfide and carbon dioxide are desorbed. The solvent is then recycled to the absorption tower.

The Sulfinol process uses a solvent that is a composite solvent, consisting of a mixture of diisopropanolamine (30 to 45% v/v) or methyl diethanolamine (MDEA), sulfolane (tetrahydrothiophene dioxide) (40 to 60% v/v), and water (5 to 15% v/v). The acid gas loading of the Sulfinol solvent is higher and the energy required for its regeneration is lower than those of purely chemical solvents. At the same time, it has the advantage over purely physical solvents that severe product specifications can be met more easily and co-absorption of hydrocarbon derivatives is relatively low. Aromatic compounds, higher molecular weight hydrocarbon derivatives, and carbon dioxide are soluble to a lesser extent. The process is typically used when the hydrogen sulfide-carbon dioxide ratio is greater than 1:1 or where carbon dioxide removal is not required to the same extent as hydrogen sulfide removal. The process uses a conventional solvent absorption and regeneration cycle in which the sour gas components are removed from the feed gas by countercurrent contact with a lean solvent stream under pressure. The absorbed impurities are then removed from the rich solvent by stripping with steam in a heated regenerator column. The hot lean solvent is then cooled for reuse in the absorber. Part of the cooling may be by heat exchange with the rich solvent for partial recovery of heat energy. The solvent reclaimer is used in a small ancillary facility for recovering solvent components from higher boiling products of alkanolamine degradation or from other high-boiling or solid impurities.

The big difference between water scrubbing and the Selexol process is that carbon dioxide and hydrogen sulfide are more soluble in Selexol which results in a lower solvent demand and reduced pumping. In addition, water and halogenated hydrocarbon derivatives (contaminants in biogas from landfills) are removed when scrubbing biogas with Selexol. Selexol scrubbing is always designed with recirculation. Due to formation of elementary sulfur stripping the Selexol solvent with air is not recommended but with steam or inert gas (upgraded biogas or natural gas). Removing hydrogen sulfide on beforehand is an alternative.

4.3. Chemisorption

Chemisorption (chemical absorption) processes are characterized by a high capability of absorbing large amounts of acid gases. They use a solution of a relatively weak base, such as monoethanolamine. The acid gas forms a weak bond with the base which can be regenerated easily. Mono- and diethanolamine derivatives are frequently used for this purpose. The amine concentration normally ranges between 15 and 30%. The gas stream is passed through the amine solution where sulfides, carbonates, and bicarbonates are formed. Diethanolamine is a favored absorbent due to its lower corrosion rate, smaller amine loss potential, fewer utility requirements, and minimal reclaiming needs. Diethanolamine also reacts reversibly with 75% of carbonyl sulfides (COS), while the mono- reacts irreversibly with 95% of the carbonyl sulfide and forms a degradation product that must be disposed in an environmentally acceptable manner.

The ethanolamine process, known as the *Girbotol* process, removes acid gases (hydrogen sulfide and carbon dioxide) from gases. The Girbotol process uses an aqueous solution of ethanolamine ($H_2NCH_2CH_2OH$) that reacts with hydrogen sulfide at low temperatures and releases hydrogen sulfide at high temperatures. The ethanolamine solution fills a tower (the absorber) through which the sour gas is bubbled. Purified gas leaves the top of the tower, and the ethanolamine solution leaves the bottom of the tower with the absorbed acid gases. The ethanolamine solution enters a reactivator towcr where heat drives the acid gases from the solution. Ethanolamine solution, restored to its original condition, leaves the bottom of the reactivator tower to go to the top of the absorber tower, and acid gases are released from the top of the reactivator.

Alkanolamine scrubbers can be used to remove hydrogen sulfide or hydrogen sulfide and carbon dioxide simultaneously (Katz, 1959; Kohl and Riesenfeld, 1985; Maddox et al., 1985; Polasek and Bullin, 1985; Jou et al., 1985; Pitsinigos and Lygeros, 1989; Kohl and Nielsen, 1997; Mokhatab et al., 2006; Speight, 2014; Abdel-Aal et al., 2016; Speight, 2019). For instance, monoethanolamine (MEA), diethanolamine (DEA), methyldiethanolamine (MDEA), diisopropanolamine (DIPA) and triethanolamine (TEA) can be used. The alkanolamine scrubbing process is also carried out in two steps: in the first step involves the amine contactor and the second step involves the regenerator. In a process with diethanolamine and methyldiethanolamine the main reactions in equilibrium in the system are:

$$2H_2O \rightarrow H_3O^+ + OH^-$$
$$CO_2 + 2H_2O \rightarrow HCO_3^- + H_3O^+$$
$$HCO_3^- + H_2O \rightarrow H_3O^+ + CO_3^{2-}$$
$$H_2S + H_2O \rightarrow H_3O^+ + HS^-$$
$$HS^- + H_2O \rightarrow H_3O^+ + S^{2-}$$
$$DEA^+ + H_2O \rightarrow DEA + H_3O^+$$
$$MDEA^+ + H_2O \rightarrow MDEA + H_3O^+$$
$$DEA^+COO^- + H_2O \rightarrow DEA + HCO_3^-$$

The pH is an important parameter because it affects the concentration of the ionic species and therefore the reactions with the amines. Moreover, the temperature and pressure also have a significant effect on the process.

The reactions that involve carbon dioxide and hydrogen sulfide with amines are as follows:

$$DEA + CO_2 + H_2O \rightarrow DEA^+COO^- + H_3O^+$$
$$MDEA + CO_2 + H_2O \rightarrow MDEA^+ + HCO_3^- \text{ (Eq. 14)}$$

$MDEA + H_2S \rightarrow MDEA^+ + HS^-$
$DEA + H_2S \rightarrow DEA^+ + HS^-$

This system is very complex due to the number of reactions involved and other reactions may also occur in the system. Tertiary amines such as DMEA do not react directly with carbon dioxide and hydrogen sulfide with amines is almost instantaneous by proton transfer. Thus, tertiary amines are used for selective hydrogen sulfide removal and for selective hydrogen sulfide and carbon dioxide removal a mixed amine system (tertiary and primary or secondary) is employed. Moreover, other compounds can be mixed with amine solutions.

Thus, treatment of a gas stream to remove the acid gas constituents (hydrogen sulfide and carbon dioxide) is most often accomplished by contact of the gas stream with an alkaline solution. The most commonly used treating solutions are aqueous solutions of the ethanolamine or alkali carbonates, although a considerable number of other treating agents have been developed in recent years (Mokhatab et al., 2006; Speight, 2014, 2019). Most of these newer treating agents rely upon physical absorption and chemical reaction. When only carbon dioxide is to be removed in large quantities or when only partial removal is necessary, a hot carbonate solution or one of the physical solvents is the most economical selection.

The primary process for sweetening sour gas uses an amine (*olamine*) solution to remove the hydrogen sulfide (the *amine process*). The sour gas is run through a tower, which contains the olamine solution. There are two principle amine solutions used, monoethanolamine (MEA) and diethanolamine (DEA). Either of these compounds, in liquid form, will absorb sulfur compounds from the gas stream as it passes through. The effluent gas is virtually free of sulfur compounds, and thus loses its sour gas status. The amine solution used can be regenerated for reuse and, although most sour gas sweetening involves the amine absorption process, it is also possible to use solid desiccants like iron sponge to remove hydrogen sulfide and carbon dioxide (Mokhatab et al., 2006; Abdel-Aal et al., 2016; Speight, 2019), including bio-based iron sponge (Cherosky and Li, 2013).

Diglycolamine (DGA), is another amine solvent used in the Econamine process in which absorption of acid gases occurs in an absorber containing an aqueous solution of diglycolamine, and the heated rich solution (saturated with acid gases) is pumped to the regenerator (Reddy and Gilmartin, 2008). Diglycolamine solutions are characterized by low freezing points, which make them suitable for use in cold climates.

The most well-known hydrogen sulfide removal process is based on the reaction of hydrogen sulfide with iron oxide (often also called the iron sponge process or the dry box method) in which the gas is passed through a bed of wood chips impregnated with iron oxide. The iron oxide is converted to the corresponding sulfur which is reaerated by oxidation (Mokhatab et al., 2006; Speight, 2014, 2019):

$$Fe_2O_3 + 3H_2S \rightarrow Fe_2S_3 + 3H_2O$$
$$2Fe_2S_3 + 3O_2 \rightarrow 2Fe_2O_3 + 6S$$

The iron oxide process (which was implemented during the 19th Century and also referred to as the iron sponge process) is the oldest and still the most widely used batch process for sweetening the biogas (Zapffe, 1963; Anerousis and Whitman, 1984; Mokhatab et al., 2006; Speight, 2006, 2014). The reaction can be exploited by adding the iron chloride to the digester feed material or passing the biogas through a bed of iron oxide-containing material. The iron oxide comes in different forms such as rusty steel wool, iron oxide pellets or wood pellets coated with iron oxide. The iron oxide media needs to be replaced periodically. The regeneration process is highly exothermic and must be controlled to avoid problems.

In the process, the sour gas is passed down through the bed. In the case where continuous regeneration is to be utilized a small concentration of air is added to the sour gas before it is processed. This air serves to continuously regenerate the iron oxide, which has reacted with hydrogen sulfide, which serves to extend the on-stream life of a given tower but probably serves to decrease the total amount of sulfur that a given weight of bed will remove. Ferric hydroxide [$Fe(OH)_3$] can also be used in this process.

The process is usually best applied to gases containing low to medium concentrations (300 ppm v/v) of hydrogen sulfide or mercaptan derivatives. This process tends to be highly selective and does not normally remove significant quantities of carbon dioxide. As a result, the hydrogen sulfide stream from the process is usually high purity. The use of iron oxide process for sweetening sour gas is based on adsorption of the acid gases on the surface of the solid sweetening agent followed by chemical reaction of ferric oxide (Fe_2O_3) with hydrogen sulfide:

$$2Fe_2O_3 + 6H_2S \rightarrow 2Fe_2S_3 + 6H_2O$$

The reaction requires the presence of slightly alkaline water and a temperature below 43°C (110°F) and bed alkalinity (pH: 8 to 10) should be checked regularly, usually on a daily basis. The pH level is be maintained through the injection of caustic soda with the water. If the gas does not contain sufficient water vapor, water may need to be injected into the inlet gas stream.

The ferric sulfide produced by the reaction of hydrogen sulfide with ferric oxide can be oxidized with air to produce sulfur and regenerate the ferric oxide:

$$2Fe_2S_3 + 3O_2 \rightarrow 2Fe_2O_3 + 6S$$
$$2S + 2O_2 \rightarrow 2SO_2$$

The regeneration step is exothermic and air must be introduced slowly so the heat of reaction can be dissipated. If air is introduced quickly the heat of reaction may ignite the bed. Some of the elemental sulfur produced in the regeneration step remains in the bed. After several cycles this sulfur will for a cake over the ferric oxide, decreasing the reactivity of the bed. Typically, after 10 cycles the bed must be removed and a new bed introduced into the vessel.

The iron oxide process is one of several metal oxide-based processes that scavenge hydrogen sulfide and organic sulfur compounds (mercaptan derivatives) from gas streams through reactions with the solid based chemical adsorbent (Kohl and Riesenfeld, 1985). They are typically non-regenerable, although some are partially regenerable, losing activity upon each regeneration cycle. Most of the processes are governed by the reaction of a metal oxide with hydrogen sulfide to form the metal sulfide. For regeneration, the metal oxide is reacted with oxygen to produce elemental sulfur and the regenerated metal oxide. In addition, to iron oxide, the primary metal oxide used for dry sorption processes is zinc oxide.

In the zinc oxide process, the zinc oxide media particles are extruded cylinders 3-4 mm in diameter and 4 to 8 mm in length (Kohl and Nielsen, 1997; Mokhatab et al., 2006; Abdel-Aal et al., 2016; Speight, 2019) and react readily with the hydrogen sulfide:

$$ZnO + H_2S \rightarrow ZnS + H_2O$$

At increased temperatures (205 to 370°C, 400 to 700°F), zinc oxide has a rapid reaction rate, therefore providing a short mass transfer zone, resulting in a short length of unused bed and improved efficiency.

Removal of larger amounts of hydrogen sulfide from gas streams requires a continuous process, such as the *Ferrox* process or the *Stretford* process. The *Ferrox process* is based on the same chemistry as the iron oxide process except that it is fluid and continuous. The *Stretford* process employs a solution containing vanadium salts and anthraquinone disulfonic acid (Maddox, 1974; Mokhatab et al., 2006; Abdel-Aal et al., 2016).

Most hydrogen sulfide removal processes return the hydrogen sulfide unchanged, but if the quantity involved does not justify installation of a sulfur recovery plant (usually a Claus plant) it is necessary to select a process that directly produces elemental sulfur. In the *Beavon-Stretford* process, a hydrotreating reactor converts sulfur dioxide in the off-gas to hydrogen sulfide which is contacted with Stretford solution (a mixture of vanadium salt, anthraquinone disulfonic acid, sodium carbonate, and sodium hydroxide) in a liquid-gas absorber. The hydrogen sulfide reacts stepwise with sodium carbonate and anthraquinone disulfonic acid to produce elemental sulfur, with vanadium serving as a catalyst. The solution proceeds to a tank where oxygen is added to regenerate the reactants. One or more froth or slurry tanks are used to skim the product sulfur from the solution, which is recirculated to the absorber.

Even though the removal of hydrogen sulfide can be achieved by chemical or physical absorption, chemical scrubbers are the most commonly used systems for hydrogen sulfide removal (Mokhatab et al., 2006; Speight, 2014, 2019). The most important processes are iron-based chelation methods and absorption by alkanolamine derivatives. In iron-based chelation processes there are two main reactions:

$$H_2S + 2Fe^{3+}\ Chelant^{n-} \rightarrow S + 2H^+ + 2\ Fe^{2+}\ Chelant^{n-}$$
$$O_2 + 4Fe^{2+}\ Chelant^- + 2H_2O \rightarrow 4Fe^{3+}\ Chelant^- + 4\ OH^-$$

The first step involves the chemical reaction to produce elemental sulfur and in the second step iron(III) is regenerated. Several commercial process are based on iron chelation, such as the Lo-cat and Sulferox processes. Ethylenediamine tetra-acetic acid (EDTA), hydroxyethylenediaminetriacetic acid (HEDTA), diethylenetriamine penta-acetic acid (DTPA), and nitrilotriacetic acid (NTA) are the most conventional ligands used in this process. The operating conditions are typically: 4 to 8°C (39.5 to 46.5°F), pH 4 to pH 8, iron concentration on the order of 1,000 to 10,000 ppm v/v) with a chelant/iron ratio in the range 1.1 to 2.0 (Deshmukh and Shete, 2013; Ramírez et al., 2015).

The process using potassium phosphate is known as phosphate desulfurization, and it is used in the same way as the Girbotol process to remove acid gases from liquid hydrocarbon derivatives as well as from gas streams. The treatment solution is a water solution of potassium phosphate (K_3PO_4), which is circulated through an absorber tower and a reactivator tower in much the same way as the ethanolamine is circulated in the Girbotol process; the solution is regenerated thermally. Processes using ethanolamine and potassium phosphate are now widely used.

4.4. Other Processes

There is a series of alternate processes that involve (i) the use of chemical reactions to remove contaminants from gas streams or (ii) the use of specialized equipment to physically remove contaminants from gas streams.

As example of the first category, i.e., the use of chemical reactions to remove contaminants from gas streams, strong basic solutions are effective solvents for acid gases. However, these solutions are not normally used for treating large volumes of biogas because the acid gases form stable salts, which are not easily regenerated. For example, carbon dioxide and hydrogen sulfide react with aqueous sodium hydroxide to yield sodium carbonate and sodium sulfide, respectively

$$CO_2 + 2NaOH \rightarrow Na_2CO_3 + H_2O$$
$$H_2S + 2NaOH \rightarrow Na_2S + 2H_2O$$

However, a strong caustic solution is used to remove mercaptans from gas and liquid streams. In the *Merox* process, for example, a caustic sol- vent containing a catalyst such as cobalt, which is capable of converting mercaptans (RSH) to caustic insoluble disulfides (RSSR), is used for streams rich in mercaptans after removal of hydrogen sulfide. Air is used to oxidize the mercaptan derivatives to disulfide derivatives. The caustic solution is then recycled for regeneration. The Merox process is mainly used for treatment of refinery gas streams.

As one of the major contaminants in biogas streams, carbon dioxide must optimally be removed as it reduces the energy content of the gas and affect the selling price of the gas. Moreover, it becomes acidic and corrosive in the presence of water that has a potential to damage the pipeline and the equipment system. Hence, the presence of carbon dioxide in biogas remains one of the challenging gas separation problems in process engineering for carbon dioxide/methane systems. Therefore, the removal of carbon dioxide from the biogas through the purification processes is vital for an improvement in the quality of the product (Mokhatab et al., 2006; Speight, 2014, 2019).

Carbonate washing is a mild alkali process (typically the alkali is potassium carbonate, K_2CO_3) for gas processing for the removal of acid gases (such as carbon dioxide and hydrogen sulfide) from gas streams and uses the principle that the rate of absorption of carbon dioxide by potassium carbonate increases with temperature (Mokhatab et al., 2006; Speight, 2014, 2019). It has been demonstrated that the process works best near the temperature of reversibility of the reactions:

$$K_2CO_3 + CO_2 + H_2O \rightarrow 2KHCO_3$$
$$K_2CO_3 + H_2S \rightarrow KHS + KHCO_3$$

The Fluor process uses propylene carbonate to remove carbon dioxide, hydrogen sulfide, carbonyl sulfide, water and higher boiling hydrocarbon derivatives (C_2^+) from gas stream (Abdel-Aal et al., 2016).

Water washing, in terms of the outcome, is almost analogous to (but often less effective than) washing with potassium carbonate (Kohl and Riesenfeld, 1985; Kohl and Nielsen, 1997), and it is also possible to carry out the desorption step by pressure reduction. The absorption is purely physical and there is also a relatively high absorption of hydrocarbon derivatives, which are liberated at the same time as the acid gases. The water scrubbing processes uses the higher solubility of carbon dioxide in water to separate the carbon dioxide from biogas. This process is done under high pressure and removes hydrogen sulfide as well as carbon dioxide. The main disadvantage of this process is that it requires a large volume of water that must be purified and recycled. An analogous process, polyethylene glycol scrubbing, is similar to water scrubbing but it is more efficient. However, the process, does require the regeneration of a large volume of polyethylene glycol.

In *chemical conversion processes*, contaminants in gas emissions are converted to compounds that are not objectionable or that can be removed from the stream with greater ease than the original constituents. For example, a number of processes have been developed that remove hydrogen sulfide and sulfur dioxide from gas streams by absorption in an alkaline solution.

Catalytic oxidation is a chemical conversion process that is used predominantly for destruction of volatile organic compounds and carbon monoxide. These systems operate in a temperature regime on the order of 205 to 595°C (400 to 1100°F) in the presence of a catalyst – in the absence of the catalyst, the system would require a higher operating temperature. The catalysts used are typically a combination of noble metals deposited on a ceramic base in a variety of configurations (e.g., honeycomb-shaped) to enhance good surface contact. Catalytic systems are usually classified on the basis of bed types such as *fixed bed* (or *packed bed*) and *fluid bed* (*fluidized bed*). These systems generally have very high destruction efficiencies for most volatile organic compounds, resulting in the formation of carbon dioxide, water, and varying amounts of hydrogen chloride (from halogenated hydrocarbon derivatives). The presence in emissions of chemicals such as heavy metals, phosphorus, sulfur, chlorine, and most halogens in the incoming air stream act as poison to the system and can foul up the catalyst. Thermal oxidation systems, without the use of catalysts, also involve chemical conversion (more correctly, chemical destruction) and operate at temperatures in excess of 815°C (1500°F), or 220 to 610°C (395 to 1100°F) higher than catalytic systems.

Other processes include the *Alkazid process* for removal of hydrogen sulfide and carbon dioxide using concentrated aqueous solutions of amino acids. The hot potassium carbonate process decreases the acid content of natural and refinery gas from as much as 50% to as low as 0.5% and operates in a unit similar to that used for amine treating. The *Giammarco-Vetrocoke* process is used for hydrogen sulfide and/or carbon dioxide removal. In the hydrogen sulfide removal section, the reagent consists of sodium carbonate (Na_2CO_3) or potassium carbonate (K_2CO_3) or a mixture of the carbonates which contains a mixture of arsenite derivatives and arsenate derivatives; the carbon dioxide removal section utilizes hot aqueous alkali carbonate solution activated by arsenic trioxide (As_2O_3) or selenous acid (H_2SeO_3) or tellurous acid (H_2TeO_3). A word of caution might be added about the last three chemicals which are toxic and can involve stringent environmental-related disposal protocols.

Molecular sieves are highly selective for the removal of hydrogen sulfide (as well as other sulfur compounds) from gas streams and over continuously high absorption efficiency. They are also an effective means of water removal and thus offer a process for the simultaneous dehydration and desulfurization of gas. Gas that has excessively high water content may require upstream dehydration, however (Mokhatab et al., 2006; Speight, 2014; Abdel-Aal et al., 2016; Speight, 2019). The carbon molecular sieve method uses differential adsorption characteristics to separate methane and the carbon molecular sieve method uses differential adsorption characteristics to separate methane and carbon dioxide.

In addition, the *molecular sieve process* is similar to the iron oxide process. Regeneration of the bed is achieved by passing heated clean gas over the bed. As the temperature of the bed increases, it releases the adsorbed hydrogen sulfide into the regeneration gas stream. The sour effluent regeneration gas is sent to a flare stack, and up to 2% v/v of the gas seated can be lost in the regeneration process. A portion of the gas stream may also be lost by the adsorption of hydrocarbon components by the sieve (Mokhatab et al., 2006; Speight, 2014, 2019).

In this process, unsaturated hydrocarbon components, such as olefin derivatives and aromatic derivatives, tend to be strongly adsorbed by the molecular sieve. Molecular sieves are susceptible to poisoning by such chemicals as glycols and require thorough gas cleaning methods before the adsorption step. Alternatively, the sieve can be offered some degree of protection by the use of *guard beds* in which a less expensive catalyst is placed in the gas stream before contact of the gas with the sieve, thereby protecting the catalyst from poisoning. This concept is analogous to the use of guard beds or attrition catalysts in the petroleum industry (Speight, 2000, 2014, 2017).

Carbon molecular sieves are excellent products to separate specifically a number of different gaseous compounds in biogas. Thereby the molecules are usually loosely adsorbed in the cavities of the carbon sieve but not irreversibly bound. The selectivity of adsorption is achieved by different mesh sizes and/or application of different gas pressures. When the pressure is released the compounds extracted from the biogas are desorbed. The process – pressure swing adsorption (PSA) technology – can be used to enrich methane from biogas the molecular sieve is applied which is produced from coke rich in pores in the micrometer range. The pores are then further reduced by cracking of the hydrocarbon derivatives.

In order to reduce the energy consumption for gas compression, a series of vessels are linked together. The gas pressure released from one vessel is subsequently used by the others. Usually four vessels in series are used filled with the molecular sieve which removes at the same time carbon dioxide and any water vapor. After removal of hydrogen sulfide, i.e., using activated carbon and water condensation in a cooler at 4°C, the biogas flows at a pressure of 90 psi into the adsorption unit. The first column cleans the raw gas at 90 psi to an upgraded biogas with a vapor pressure of less than 10 ppm water and a methane content of 96% v/v, or even higher.

In the second column the pressure of 90 psi is first released to approximately 45 psi by pressure communication with column 4, which was previously degassed by a slight vacuum. In a second step the pressure is then reduced to atmospheric pressure. The released gas flows back to the digester in order to recover the methane. The third column is evacuated from 15 psi to 0.15 psi. The desorbed gas consists predominantly of carbon dioxide but also some methane and is therefore normally released to the environment. In order to reduce methane losses the system can be designed with recirculation of the desorbed gases.

A number of other possible impurities, such as organic acids (e.g., formic and acetic acid) and unsaturated and higher boiling hydrocarbon derivatives, are normally not included in purity specifications for synthesis processes because they are already undesirable in upstream process stages of biosyngas production (e.g., compression steps). However, organic compounds present must be below their dew point at pressure of the gas application to prevent condensation and fouling in the system. For organic compounds with sulfur or nitrogen heteroatoms (such as thiophene derivatives and pyridine derivatives) the additional specification applies that they need to be removed below ppm v/v level, as they are intrinsically poisonous for the catalyst.

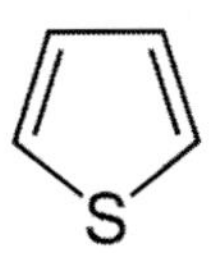

Thiophene

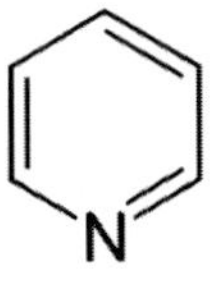

Pyridine

5. Condensable Hydrocarbons Removal

Hydrocarbon derivatives that are higher molecular weight than methane that are present in gas streams are valuable raw materials and important fuels. They can be recovered by lean oil extraction. The first step in this process is to cool the treated gas by exchange with liquid propane after which the cooled gas is then washed with a cold hydrocarbon liquid, which dissolves most of the condensable hydrocarbon derivatives. The uncondensed gas is dry biogas and is composed mainly of methane. Dry biogas may then be used either as a fuel or as a chemical feedstock. Another way to recover any higher molecular weight hydrocarbon constituents from biogas is by using cryogenic cooling (cooling to very low temperatures on the order of -100 to -115°C (-150 to -175°F), which are achieved primarily through lowering the temperatures to below the dew point.

To prevent hydrate formation, biogas streams may be treated with glycols, which dissolve water efficiently. Ethylene glycol (EG), diethylene glycol (DEG), and triethylene glycol (TEG) are typical solvents for water removal. Triethylene glycol is preferable in vapor phase processes because of its low vapor pressure, which results in less glycol loss. The triethylene glycol absorber unit typically contains 6 to 12 bubble-cap trays to accomplish the water absorption. However, more contact stages may be required to reach dew points below -40°C (-40°F). Calculations to determine the number of trays or feet of packing, the required glycol concentration, or the glycol circulation rate require vapor-liquid equilibrium data. In addition, predicting the interaction between triethylene glycol and water vapor in biogas over a broad range allows the designs for ultra-low dew point applications to be made.

One alternative to using bubble-cap trays is the use of the adiabatic expansion of the inlet gas. In the process, the inlet gas is first treated to remove water and acid gases, then cooled via heat exchange and refrigeration. Further cooling of the gas is accomplished through turbo expanders, and the gas is sent to a demethanizer to separate methane from the higher-boiling hydrocarbon derivatives. Improved recovery of the higher-boiling hydrocarbon derivatives could be achieved through better control strategies and by use of on-line gas chromatographic analysis.

Membrane separation process are very versatile and are designed to process a wide range of feedstocks and offer a simple solution for removal and recovery of higher boiling hydrocarbon derivatives from gas streams (Foglietta, 2004; May-Britt, 2008; Basu et al., 2010; Rongwong et al., 2012; Ozturk and Demirciyeva, 2013; Abdel-Aal et al., 2016). There are two membrane separation techniques: (i) high pressure gas separation and (ii) gas-liquid adsorption. The high pressure separation process selectively separates hydrogen sulfide and carbon dioxide from methane. Usually, this separation is performed in three stages and produces methane with a high degree of purity (on the order of 96% v/v methane).

The separation process is based on high-flux membranes that selectively permeates higher boiling hydrocarbon derivatives (compared to methane) and are recovered as a liquid after recompression and condensation. The residue stream from the membrane is partially depleted of higher boiling hydrocarbon derivatives, and is then sent to sales gas stream. Gas permeation membranes are usually made with vitreous polymers that exhibit good selectivity but, to be effective, the membrane must be very permeable with respect to the separation process (Rojey et al., 1997).

Membrane separation is based on the selectivity properties of the membrane material. The driving force is the difference in chemical potential, which includes the effect of temperature and partial pressure régimes. Membranes can be operated at high pressure (>300 psi) or low pressure (120 to 150 psi). Several membranes can be used for biogas upgrading and these can be classified by the type of material: non-polymeric membranes (such as alumina, zeolites, and carbon), ceramic membranes, palladium membranes, and carbon molecular sieve membranes. Nevertheless, polymeric membranes are the most widely used because they are cheaper than inorganic membranes. Moreover, polymeric materials can be easily fabricated into flat sheets or asymmetric hollow fibres, both shapes that are usually used in industrial applications.

The material properties that define the performance are molecular structure, glass transition temperature, crystallinity and degree of crosslinking. However, commercially available polymeric membranes are usually degraded by hydrogen sulfide, siloxanes, and/or volatile organic compounds (Accettola et al., 2008; Ajhar et al., 2010; Gislon et al., 2013). For instance, hydrogen sulfide can cause plasticization of glassy polymeric membranes and this alters the polymeric structure and weakens the mechanical strength of the membrane.

In the high-pressure separation option, pressurized gas (36 bar) is first cleaned over for example an activated carbon bed to remove (halogenated) hydrocarbon derivatives and hydrogen sulfide from the raw gas as well as oil vapor from the compressors. The carbon bed is followed by a particle filter and a heater. The membranes made of acetate-cellulose separate small polar molecules such as carbon dioxide, moisture and the remaining hydrogen sulfide. These membranes are not effective in separating nitrogen from methane.

The raw gas is upgraded in 3 stages to a clean gas with at least 96% v/v methane. The waste gas from the first two stages is recycled and the methane can be recovered. The waste gas from stage 3 (and in part of stage 2) is flared or used in a steam boiler as it still contains 10 to 20% v/v methane. First experiences have shown that the membranes can last up to 3 years which is comparable to the lifetime of membranes for natural gas purification -a primary market for membrane technology - which last typically two to five years.

The clean gas is further compressed up to 3600 psi and stored in steel cylinders in capacities of 276 m3 divided in high, medium and low pressure banks. The membranes are very specific for given molecules, i.e., hydrogen sulfide and carbon dioxide are upgraded in different modules. The utilization of hollow-fiber membranes allows the construction of very compact modules working in cross flow.

The gas-liquid absorption membrane method is a separation technique which was developed for biogas upgrading only recently. The essential element is a microporous hydrophobic membrane separating the gaseous from the liquid phase. The molecules from the gas stream, flowing in one direction, which are able to diffuse through the membrane will be absorbed on the other side by the liquid flowing in counter current.

The absorption membranes work at approx. atmospheric pressure (1 bar) which allows low-cost construction. The removal of gaseous components is very efficient. At a temperature of 25 to 35°C (77 to 95°F) the hydrogen sulfide concentration in the raw gas of 2% v/v is reduced to less than 0.025% v/v (250 ppm v/v) – the absorbent is sodium hydroxide (NaOH) or another alkaline material. Sodium hydroxide saturated with hydrogen sulfide can be used in water treatment to remove heavy metals. The concentrated hydrogen sulfide solution is fed into a Claus reaction or oxidized to elementary sulfur. The biogas is upgraded very efficiently from 55% v/v methane (43% v/v carbon dioxide) to more than 96% v/v methane. The amine solution is regenerated by heating and the carbon dioxide released is pure and can be sold for industrial applications.

The main advantages of membranes are (i) safety and simplicity of operation, (ii) relative ease of maintenance and scale-up, (iii) excellent reliability, (iv) small footprint, and (v) operation without hazardous chemicals. Several operation modes can be employed – as an example, biogas upgrading can be achieved by a two-stage cascade process with recycling and a single stage provided good flexibility for integration into biogas plants.

5.1. Extraction

There are two principle techniques for removing higher-boiling hydrocarbon constituents from gas stream liquids: (i) the absorption method and (ii) the cryogenic expander process. In the latter process (the cryogenic expander process), a turboexpander is used to produce the necessary refrigeration and very low temperatures and high recovery of light components, such as ethane and propane, can be attained. The gas stream is first dehydrated using a molecular sieve followed by cooling of the dry stream. The separated liquid containing most of the heavy fractions is then demethanized, and the cold gases are expanded through a turbine that produces the desired cooling for the process. The expander outlet is a two-phase stream that is fed to the top of the demethanizer column. This serves as a separator in which: (i) the liquid is used as the column reflux and the separator vapors combined with vapors stripped in the demethanizer are exchanged with the feed gas, and (ii) the heated gas, which is partially recompressed by the expander compressor, is further recompressed to the desired distribution pressure in a separate compressor.

The extraction of higher-boiling hydrocarbon constituents from the gas stream produces both cleaner, purer biogas, as well as the valuable hydrocarbon derivatives for use in, say, a petrochemical plant) (Speight, 2019). This process allows for the recovery of approximately 90 to 95% v/v of the ethane originally in the gas stream. In addition, the expansion turbine is able to convert some of the energy released when the gas stream is expanded into recompressing the gaseous methane effluent, thus saving energy costs associated with extracting ethane.

5.2. Absorption

The absorption method of high molecular weight recovery of hydrocarbon derivatives is very similar to using absorption for dehydration (Mokhatab et al., 2006; Speight, 2014, 2019). The main difference is that, in the absorption of higher-boiling hydrocarbon constituents from the gas stream, absorbing oil is used as opposed to glycol. This absorbing oil has an affinity for the higher-boiling hydrocarbon constituents in much the same manner as glycol has an affinity for water. Before the oil has picked up any higher-boiling hydrocarbon constituents, it is termed *lean* absorption oil.

The *oil absorption process* involves the countercurrent contact of the lean (or stripped) oil with the incoming wet gas with the temperature and pressure conditions programmed to maximize the dissolution of the liquefiable components in the oil. The *rich* absorption oil (sometimes referred to as *fat* oil), containing higher-boiling hydrocarbon constituents, exits the absorption tower through the bottom. It is now a mixture of absorption oil, propane, butanes, pentanes, and other higher boiling hydrocarbon derivatives. The rich oil is fed into lean oil stills, where the mixture is heated to a temperature above the boiling point of the higher-boiling hydrocarbon constituents but below that of the oil. This process allows for the recovery of higher boiling constituents from the biogas stream.

The basic absorption process is subject to modifications that improve process effectiveness and even to target the extraction of specific higher-boiling hydrocarbon constituents. In the refrigerated oil absorption method, where the lean oil is cooled through refrigeration, propane recovery can be on the order of 90%+ v/v and approximately 40% v/v of the ethane can be extracted from the gas stream. Extraction of the other, higher-boiling hydrocarbon constituents is typically near-quantitative using this process.

5.3. Fractionation

Fractionation processes are very similar to those processes classed as *liquids removal* processes but often appear to be more specific in terms of the objectives: hence the need to place the fractionation processes into a separate category. The fractionation processes are those processes that are used (i) to remove the more significant product stream first, or (ii) to remove any unwanted light ends from the higher-boiling liquid products.

In the general practice of gas processing, the first unit is a de-ethanizer followed by a depropanizer then by a debutanizer and, finally, a butane fractionator. Thus each column can operate at a successively lower pressure, thereby allowing the different gas streams to flow from column to column by virtue of the pressure gradient, without necessarily the use of pumps. The purification of hydrocarbon gases by any of these processes is an important part of gas processing operations.

Thus, after any higher-boiling hydrocarbon constituents have been removed from the gas stream, they must be separated (fractionated) into the individual constituents prior to sales. The process of fractionation occurs in stages with each stage involving separation of the hydrocarbon derivatives as individual products. The process commences with the removal of the lower-boiling hydrocarbon derivatives from the feedstock. The particular fractionators are used in the following order: (i) the de-ethanizer, which is used to separate the ethane from the stream from the gas stream, (ii) the depropanizer, which is used to separate the propane from the de-ethanized gas stream, (iii) the debutanizer, which is used to separate the butane isomers, leaving the pentane isomers and higher boiling hydrocarbon derivatives in the gas stream, and (iv) the butane splitter or de-isobutanizer, which is used to separate n-butane and iso-butane.

After the recovery of the higher-boiling hydrocarbon constituents, sulfur-free dry gas (methane) may be liquefied for transportation through cryogenic tankers. Further treatment may be required to reduce the water vapor below 10 ppm and carbon dioxide and hydrogen sulfide to less than 100 and 50 ppm v/v, respectively. Two methods are generally used to liquefy the gas: (i) the expander cycle and (ii) mechanical refrigeration. In the expander cycle, part of the gas is expanded from a high transmission pressure to a lower pressure. This lowers the temperature of the gas. Through heat exchange, the cold gas cools the incoming gas, which in a similar way cools more incoming gas until the liquefaction temperature of methane is reached.

In mechanical refrigeration, a multicomponent refrigerant consisting of nitrogen, methane, ethane, and propane is used through a cascade cycle. When these liquids evaporate, the heat required is obtained from the gas, which loses energy/temperature till it is liquefied. The refrigerant gases are recompressed and recycled.

5.4. Enrichment

The gas product must meet specific quality measures in order for the pipeline grid to operate properly. Consequently, biogas which, in most cases contains contaminants, must be processed, i.e., cleaned, before it can be safely delivered to the high-pressure, long-distance pipelines that transport the product to the consuming public. A gas stream that is not within certain specific gravities, pressures, Btu content range, or water content levels will cause operational problems, pipeline deterioration, or can even cause pipeline rupture. Thus, the purpose of *enrichment* is to produce a gas stream for sale. Therefore, the process concept is essentially the separation of higher-boiling hydrocarbon constituents from the methane to produce a lean, dry gas.

As an example, carbon dioxide is to some extent soluble in water and therefore some carbon dioxide will be dissolved in the liquid phase of the digester tank. In upgrading with the in situ methane enrichment process, sludge from the digester is circulated to a desorption column and then back to the digester. In the desorption process, carbon dioxide is desorbed by pumping air through the sludge. The constant removal of carbon dioxide from the sludge leads to an increased concentration of methane in the biogas phase leaving the digester.

The gas stream received and transported must (especially in the United States and many other countries) meet the quality standards specified by pipeline. These quality standards vary from pipeline to pipeline and are usually a function of (i) the design of the pipeline system, (ii) the design of any downstream interconnecting pipelines, and (iii) the requirements of the customer. In general, these standards specify that the gas stream should

(i) be within a specific Btu content range, typically 1,035 Btu ft^3 ± 50 Btu ft^3, (ii) be delivered at a specified hydrocarbon dew point temperature level to prevents any vaporized gas liquid in the mix from condensing at pipeline pressure, (iii)) contain no more than trace amounts of elements such as hydrogen sulfide, carbon dioxide, nitrogen, water vapor, and oxygen, (iv) be free of particulate solids and liquid water that could be detrimental to the pipeline or its ancillary operating equipment. Gas processing equipment, whether in the field or at processing/treatment plants, assures that these specifications can be met.

In most cases processing facilities extract contaminants and higher-boiling hydrocarbon derivatives from the gas stream but, in some cases, the gas processors blend some higher-boiling hydrocarbon derivatives into the gas stream in order to bring it within acceptable Btu levels. For instance, in some areas if the produced gas (including coalbed methane) does not meet (is below) the Btu requirements of the pipeline operator, in which case a blend of higher Btu-content gas stream or a propane-air mixture is injected to enrich the heat content (Btu value) prior for delivery to the pipeline.

The number of steps and the type of techniques used in the process of creating a pipeline-quality gas stream most often depends upon the source and makeup of the wellhead production stream. Among the several stages of gas processing are: (i) gas-oil separation, (ii) water removal, (iii) liquids removal, (iv) nitrogen removal, (v) acid gas removal, and (vi) fractionation.

In many instances pressure relief at the wellhead will cause a natural separation of gas from oil (using a conventional closed tank, where gravity separates the gas hydrocarbon derivatives from the heavier oil). In some cases, however, a multi-stage gas-oil separation process is needed to separate the gas stream from the crude oil. These gas-oil separators are commonly closed cylindrical shells, horizontally mounted with inlets at one end, an outlet at the top for removal of gas, and an outlet at the bottom for removal of oil. Separation is accomplished by alternately heating and cooling (by compression) the flow stream through multiple steps. However, the number of steps and the type of techniques used in the process of creating a pipeline-quality gas stream most often depends upon the source and makeup of the gas stream. In some cases, several of the steps may be integrated into one unit or operation, performed in a different order or at alternative locations (lease/plant), or not required at all.

6. Tar Removal

Tar is a highly viscous liquid product (typically produced in thermal processes) that condenses in the low temperature zones of the equipment (the pyrolysis unit or the gasifier) thereby interfering with the gas flow and leading to system disruption. It is, perhaps, the most undesirable product of the process and a high residual tar concentrations in the can gas prevent utilization (Devi et al., 2003; Torres et al., 2007; Anis and Zainal, 2011; Huang et al., 2011). Methods for reduction or elimination of tar can be divided into two broad groups: (i) in situ tar reduction also called primary tar reduction, which focus on reducing or avoiding tar formation in the reactor, and (ii) post gasification tar reduction also called or secondary tar reduction, which strips the tar from the product gas. A combination of in situ and post-gasification tar reductions can prove to be more effective than either of the single stages alone. The two basic post-gasification methods are physical removal and cracking (catalytic or thermal).

6.1. Physical Methods

In a process for the physical removal of tar, the tar is treated tar as dust particles or mist and the tar is condensed before separation. Physical tar removal can be accomplished by cyclones, barrier filters, wet electrostatic precipitators (ESPs) or wet scrubbers. The selection of the equipment application depends on the load concentrations of particulate matter (PM) and tar, particle size distribution and particulate tolerance of downstream users.

Cyclone collectors are the most common of the inertial collector class and are effective in removing coarser fractions of particulate matter and operate by contacting the particles in the gas stream with a liquid. In principle the particles are incorporated in a liquid bath or in liquid particles which are much larger and therefore more easily collected. In the process, the particle-laden gas stream enters an upper cylindrical section tangentially and proceeds downward through a conical section. Particles migrate by centrifugal force generated by providing a path for the carrier gas to be subjected to a vortex-like spin. The particles are forced to the wall and are removed through a seal at the apex of the inverted cone. A reverse-direction vortex moves upward through the cyclone and discharges through a top center opening. Cyclones are often used as primary collectors because of their relatively low efficiency (50 to 90% is usual).

Cyclones may not be efficient when removing small tar droplets and barrier filters are porous material which can capture certain amount of tar when the product gas passes through the filters. As an aid, catalyst grains can be integrated as a fixed bed inside the filter to promote the simultaneous removal of particulate matter and tar (Rapagna et al., 2010).

Fabric filters are typically designed with non-disposable filter bags. As the gaseous (dust-containing) emissions flow through the filter media (typically cotton, polypropylene, fiberglass, or Teflon), particulate matter is collected on the bag surface as a dust cake. Fabric filters operate with collection efficiencies up to 99.9% although other advantages are evident but there are several issues that arise during use of such equipment.

Wet scrubbers are devices in which a counter-current spray liquid is used to remove particles from an air stream. Device configurations include plate scrubbers, packed bed scrubbers, orifice scrubbers, venturi scrubbers, and spray towers, individually or in various combinations. Wet scrubbers can achieve high collection efficiencies at the expense of prohibitive pressure drops. The *foam scrubber* is a modification of a wet scrubber in which the particle-laden gas is passed through a foam generator, where the gas and particles are enclosed by small bubbles of foam.

Other methods include use of high-energy input *venturi scrubbers* or electrostatic scrubbers where particles or water droplets are charged, and flux force/condensation scrubbers where a hot humid gas is contacted with cooled liquid or where steam is injected into saturated gas. In the latter scrubber the movement of water vapor toward the cold water surface carries the particles with it (*diffusiophoresis*), while the condensation of water vapor on the particles causes the particle size to increase, thus facilitating collection of fine particles.

Electrostatic precipitators operate on the principle of imparting an electric charge to particles in the incoming air stream, which are then collected on an oppositely charged plate across a high-voltage field. Particles of high resistivity create the most difficulty in collection. Conditioning agents such as sulfur trioxide (SO_3) have been used to lower resistivity. Important parameters include design of electrodes, spacing of collection plates, minimization of air channeling, and collection-electrode rapping techniques (used to dislodge particles). Techniques under study include the use of high-voltage pulse energy to enhance particle charging, electron-beam ionization, and wide plate spacing. Electrical precipitators are capable of efficiencies >99% under optimum conditions, but performance is still difficult to predict in new situations.

Wet electrostatic precipitator units have a high collection efficiency (on the order of 90%) over the entire range of particle size down to 0.5 mm with a very low-pressure drop. Wet scrubbers can achieve a high collection efficiency (also on the order of 90%). as well.

6.2. Thermal Methods

Cracking methods are more advantageous in terms of recovering the energy content in the tar by converting the high molecular weight constituents of the tar into lower molecular weight (gaseous) products such as hydrocarbon derivatives and hydrogen at high temperature (up to 1200°C, 2190°F) or catalytic reactions (up to 800°C, 1470°F). Catalytic cracking is commercially used in many refineries plants and has been demonstrated to be one of the most effective processes for the conversion of high molecular weight products (such as crude oil residua, heavy oi, extra heavy oil, and tar sand bitumen) into gases and lower molecular weight distillable and liquids (Parkash, 2003; Gary et al., 2007; Speight, 2014; Hsu and Robinson, 2017; Speight, 2017).

Non-metallic catalysts such as dolomite ($CaCO_3.MgCO_3$), calcite ($CaCO_3$), zeolite and metallic catalysts such as nickel-based catalysts, nickel/molybdenum-based catalysts, nickel/cobalt-based catalysts, molybdenum-based catalysts, platinum-based catalysts, and ruthenium-based catalysts Mo, NiO, Pt and Ru have been applied to various tar conversion processes.

Tar removal can also be achieved by catalytic reforming which has the advantage of keeping the carbon contained in the tars available for further conversion to fuel. Another option to reduce the tar content of the syngas is the use of catalytic bed material in the gasification reactor. Olivine sand has been shown to effectively reduce the tar content in synthesis gas from steam gasification.

7. Other Contaminant Removal

The types and the amounts of the contaminants that occur in biogas are, like those contaminants presented above, depend on the composition of the biomass feedstock and the type of process used to produce the gas as well as the individual process parameters. In addition to the gas from anaerobic digesters, the found in landfill gas vary widely from landfill site to landfill site as the items disposed of also vary widely (i.e., there is site specificity). Moreover, there is variation in how frequently gas is extracted and in what volumes, and the stage of decomposition of the waste. This section presents some of the contaminants that are not always presented in other works related to contaminant presence and removal. For example, siloxanes derivatives generally occur to a greater extent in biogas from landfill sites compared to the presence of these chemicals in biogas from other sources.

7.1. Nitrogen Removal

Nitrogen may often occur in sufficient quantities in gas streams and, consequently, lower the heating value of the gas. Thus several plants for *nitrogen removal* from gas streams have been built, but it must be recognized that nitrogen removal requires liquefaction and fractionation of the entire gas stream, which may affect process economics. In some cases, the nitrogen-containing gas stream is blended with a gas having a higher heating value and sold at a reduced price depending upon the thermal value (Btu/ft^3).

For high flow-rate gas streams, a cryogenic process is typical and involves the use of the different volatility of methane (b.p.-161.6°C/-258.9°F) and nitrogen (b.p. -195.7°C/-320.3°F) to achieve separation. In the process, a system of compression and distillation columns drastically reduces the temperature of the gas mixture to a point where methane is liquefied and the nitrogen is not. On the other hand, for smaller volumes of gas, a system utilizing pressure swing adsorption (PSA) is a more typical method of separation.

The pressure swing adsorption process is a fixed-bed adsorption process that is carried out at constant temperature and variable pressure. In the process, carbon dioxide is separated from the biogas by adsorption on a surface under elevated pressure. The adsorbing material, usually activated carbon or zeolites, is regenerated by a sequential decrease in pressure before the column is reloaded again. . An upgrading plant, using this technique, typically has four, six or nine vessels working in parallel. When the adsorbing material in one vessel becomes saturated the raw gas flow is switched to another vessel in which the adsorbing material has been regenerated. During regeneration the pressure is decreased in several steps. The gas that is desorbed during the first and eventually the second pressure drop may be returned to the inlet of the raw gas, since it will contain some methane that was adsorbed together with carbon dioxide. The gas desorbed in the following pressure reduction step is either led to the next column or if it is almost entirely methane free it is released to the atmosphere. If hydrogen sulfide is present in the raw gas, it will be irreversibly adsorbed on the adsorbing material. In addition, water present in the raw gas can destroy the structure of the material. Therefore hydrogen sulfide and water needs to be removed before the PSA-column.

Thus, the *pressure swing adsorption process* enables the separation of carbon dioxide from biogas. Nevertheless, pressure swing adsorption can be used for the separation of other compounds such as nitrogen, oxygen, and carbon monoxide. Several packing materials can be used for the removal of carbon dioxide and these include zeolite, silica gel and activate carbon. In order to be effective continuously the pressure swing adsorption process is composed of various fixed-bed columns running in alternative cycles of adsorption, regeneration, and pressure build-up (Nikolić et al., 2009; Alonso-Vicario et al., 2010; Santos et al., 2013). In some cases, the regeneration step is carried out under vacuum and this technology is known as vacuum swing adsorption (VSA), which requires a vacuum pump at the outlet.

Moreover, the regeneration step can be carried out by increasing the temperature to enhance the desorption of compounds such as hydrogen sulfide (which is strongly adsorbed on the packing material). On increasing the temperature at constant pressure the process is also known as temperature swing adsorption (TSA). The temperature swing adsorption process is used for gas sweetening. Biogas must be fed dry into pressure swing adsorption, vacuum swing adsorption, and temperature swing adsorption processes and the water content must therefore be reduced prior to these treatments.

The pressure swing adsorption operation can be based on equilibrium or kinetic separation. The more strongly adsorbed and faster diffusing compounds are retained on the packing material in equilibrium and kinetic separation, respectively. Natural zeolites have been used for hydrogen sulfide and carbon dioxide removal by pressure swing adsorption with thermal desorption working at 25°C (77°F) in the pressure range 45 to 100 psi. However, variations in the concentration of carbon dioxide can affect the performance of the pressure swing adsorption process. For example, an increase of 5% in the flow rate of the carbon dioxide can lead to an increase in the temperature at the top of the packed bed – thus, the use of a thermocouple could enhance the control strategy.

Also, in pressure swing adsorption method, methane and nitrogen can be separated by using an adsorbent with an aperture size very close to the molecular diameter of the larger species (the methane) which allows nitrogen to diffuse through the adsorbent. This results in a purified gas stream that is suitable for pipeline specifications. The adsorbent can then be regenerated, leaving a highly pure nitrogen stream. The pressure swing adsorption method is a flexible method for nitrogen rejection, being applied to both small and large flow rates.

7.2. Ammonia Removal

Biomass with a relatively high nitrogen content will generate a product gas that contains ammonia (NH_3) and hydrogen cyanide (HCN) (Yuan et al., 2010) which, in turn, will generate nitrogen oxides (NOx) during combustion. In the case of the gasification of biomass (to produce biogas), the concentration of ammonia in the product gas depends not only on the nature of the biomass feedstock used but also on the gasifier design parameters and operating conditions.

Ammonia can be removed by wet scrubbing technology which has been widely adopted in the existing biomass gasification processes (Dou et al., 2002; Proell et al., 2005) and the ammonia cab ve removed when the gas is dried. As a result, a separate cleaning step is therefore usually not necessary.

Compared with wet scrubbing technology, hot-gas cleanup technology, preferably employing catalysts, is more advantageous with respect to energy efficiencies as it eliminates the needs of cooling the product gas and re-heating again for the syngas applications (Torres et al., 2007). Catalytic processes effectively remove ammonia by converting it to nitrogen, hydrogen, and water. Nickel-Ni-based catalysts have higher activity, while other catalysts do not have good potentials for ammonia.

7.3. Particulate Matter Removal

Particulate matter that is present in the product gas can also be a serious problem for some end-users and catalysts used for cleaning product gases have been demonstrated to be negatively affected by particulate matter (Gustafsson et al., 2011).

Particulate matter can be removed using standard technologies such as cyclones, filters and separators, which also reduces the tar content of the gas stream, the extent of tar removal depending on the particle separation technology used. The presence of alkali is of particular importance, because alkali can form silicates with low melting temperatures that may negatively affect the filter operation.

7.4. Siloxane Removal

Siloxane compounds are a subgroup of silicone derivatives containing silicon-oxygen (Si-O) bonds with organic radicals which are widely used for a variety of industrial processes. Siloxancs are used in products such as deodorants and shampoos, and can therefore be found in biogas from sewage sludge treatment plants and in landfill gas. When siloxanes are burned, silicon oxide, a white powder, is formed which can create a problem in gas engines. Although most siloxanes disperse into the atmosphere where they are decomposed, some end up in wastewater, but more generally, siloxane derivatives generally end up as a significant component in the sludge.

As sludge undergoes anaerobic digestion, it may be subjected to temperatures up to 60°C (140°F). At this point the siloxanes contained in the sludge will volatize and become an unwanted constituent of the resulting biogas. This problem can be exacerbated by the fact that silicone-based anti-foaming agents are frequently added to the anaerobic digesters and these silicones sometimes biodegrade into siloxanes. Unfortunately, when siloxanes gasses are burned, they are usually converted into silicon dioxide particles, which are chemically and physically similar to sand.

Currently, there are six primary technologies for removing siloxanes from biogas and include the following process options: (i) activated carbon is widely used to remove organic substances from gases and liquids due to its superior adsorbent properties, (ii) activated alumina (Al_2O_3) absorbs siloxanes from biogas; when the alumina becomes saturated, the absorption capability can be recovered by passing a regeneration gas through it, (iii) refrigeration with condensation in combination can be used to selectively remove specific compounds by lowering the temperature or pressure of the gas, and then allowing the compound to precipitate out to a liquid, and then settle out, (iv) synthetic resins remove volatile materials through adsorption. They can be specially formulated to remove specific classes of compounds, (v) liquid absorbents are used by a small number of landfill operators to treat biogas prior to use in combustion devices such as gas turbines, and (vi) membrane technology is a relatively recent development for siloxane removal; however, membranes are subject to acid deterioration from the acidic content usually found in raw biogas. Of these technologies, activated carbon appears to be among the most dominant in the industry.

7.5. Alkali Metal Salt Removal

Compared with fossil fuels, biomass is rich in alkali salts that typically vaporize at high gasifier temperatures but condense downstream below 600°C (1110°F). Methods are available to strip the alkali contents since condensation of alkali salts can cause serious corrosion problems. The alkali will condense onto fine solid particles and can be subsequently captured in a cyclone, electrostatic precipitators, or filters when the gas temperature is below 600°C (1110°F). The hot gas can also be passed through a bed of active bauxite to remove alkali when cooling of gas is not permitted.

7.6. Biological Methods

Biological processes have mainly been used to remove hydrogen sulfide (Syed et al., 2006). Hydrogen sulfide can be oxidized by microorganisms of the species Thiobacillus and Sulfolobus. The degradation requires oxygen and therefore a small amount of air (or pure oxygen if levels of nitrogen should be minimized) is added for biological desulfurization to take place. The degradation can occur inside the digester and can be facilitated by immobilizing the microorganisms occurring naturally in the digestate. An alternative is to use a trickling filter which the biogas passes through when leaving the digester. In the trickling filter the microorganisms grow on a packing material. Biogas with added air meets a counter flow of water containing nutrients. The sulfur containing solution is removed and replaced when the pH drops below a certain level. Both methods are widely applied, however they are not suitable when the biogas is used as vehicle fuel or for grid injection due to the remaining traces of oxygen. An alternative system is available in which where the absorption of the hydrogen sulfide is separated from the biological oxidation to sulfur – hence, the biogas flow remains free of oxygen. There has also been focus on the removal of siloxane derivatives and carbon dioxide (Jensen and Webb, 1995; Soreanu, et al., 2011).

As another example, natural bacteria can convert hydrogen sulfide into elemental sulfur in the presence of oxygen and iron. This can be done by introducing a small amount (two to five per cent) of air into the head space of the digester. As a result, deposits of elemental sulfur will be formed in the digester. Even though this situation will reduce the hydrogen level, it will not lower it below that recommended for pipeline-quality gas. This process may be optimized by a more sophisticated design where air is bubbled through the digester feed material. It is critical that the introduction of the air be carefully controlled to avoid reducing the amount of biogas that is produced.

In this section, the main biological processes for the removal of hydrogen sulfide are presented.

7.6.1. Biofiltration

Traditional gas cleaning and air pollution control technologies for pollutant gases, such as adsorption, absorption and combustion, were developed to treat high concentration waste gas streams associated with process emissions from stationary point sources (Mokhatab et al., 2006). Although these technologies rely on established physico-chemical principles to achieve effective control of gaseous pollutants, in many cases the control technique yields products which require further treatment before disposal or recycling of treatment materials. In the case of treatment of dilute waste gas streams, however, these traditional methods are relatively less effective, more expensive and wasteful in terms of energy consumption and identification of alternative control measures is warranted. A suitable alternate air pollution control technology is biofiltration, which utilizes naturally occurring microorganisms supported on a stationary bed (filter) to continuously treat contaminants in a flowing waste gas stream (Allen and Uang, 1992; Accettola et al., 2008).

Biofiltration by definition is the aerobic degradation of pollutants from (in the current context) biogas in the presence of a carrier media. The early development work on biofiltration technology concentrated on organic media, such as, peat, compost, and wood bark. In general terms, organic compounds are degraded to carbon dioxide and water, while inorganic compounds, such as, sulfur compounds are oxidized to form oxygenated derivatives. The formation of these acidic compounds can lead to a lowering of pH of the filtration media; which in turn impacts on the performance of the system. Removal of the oxidized compound from the media is an important consideration in the design of biofiltration systems. Biofiltration, like many processes based on bio-treatment in the crude oil refining and gas processing industries(El-Gendy and Speight, 2016; Speight and El-Gendy, 2018), biofiltration is successfully emerging as a reliable, low cost option for a broad range of air treatment applications. It is now becoming apparent that biological treatment will play a far more significant role in achieving environmental control on air emissions.

In the biofiltration process, three types of bioreactor designs are usually considered: the biofilter, (ii) the biotrickling filter, and (iii) the bioscrubber. The main differences between these systems concern their design and mode of operation: microorganism conditioning, the nature of the fluid phase (gas or liquid), and the presence or absence of stationary solid phases. Nevertheless, biogas desulfurization has been carried out by the biotrickling filter and the bioscrubber.

The biofilter is a pollution control technique which involves using a bioreactor that contains living material to capture and biologically degrade pollutants such as hydrogen sulfide. Common uses include processing wastewater, capturing harmful chemicals or silt from surface runoff, and the macrobiotic oxidation of contaminants in gas streams. The technology finds greatest application in treating malodorous compounds and water-soluble volatile organic compounds (VOCs). Compounds treated are typically mixed VOCs and various sulfur compounds, including hydrogen sulfide. Very large airflows may be treated and although a large area (footprint) has typically been required.

In the process, the polluted gas flow is purified with biofiltration by conducting the gas flow upward through a filter bed that consists from biological material, e.g., compost, tree bark or peat. The filter material is a carrier of a thin water film in which micro-organism live. The pollution in the gas flow is held back by ad -and absorption on the filter material, and then broken by present micro-organism. The filter material serves as a supplier of necessary nutrients. The products of the conversion are carbon-dioxide, sulfate, and nitrate. The dry weight of the filter varies typically from 40 to 60 %. To reduce desiccation of the bed the gas flow must has to be saturated with water. For this reason polluted gas flow is moistened before it goes through the biofilter, which is achieved by using a pre-scrubber. The relative humidity of the gas must be 95%. In practice it is always better to apply a moistener to protect the biofilter against dehydration.

A biotrickling filter is a packed bed bioreactor with immobilized biomass Montebello et al., 2012; Fernández et al., 2013a, 2013b; Montebello et al., 2013). The gas flows through a fixed bed co- or counter-currently to a mobile liquid phase. Synthetic carriers are usually used and these include plastic, ceramic, lava rocks, polyurethane foam, etc. The synthetic carrier does not provide any nutrients so the liquid mobile phase must contain nutrients for the growth and maintenance of the biomass. Programmed or continuous discharge of recirculation medium help to remove the oxidation products. The hydrogen sulfide must be transferred from the gas to liquid phase and the degradation is finally carried out in the biofilm.

According to the microbial cycling of sulfur, the biological oxidation of hydrogen sulfide to sulfate is one of the major reactions involved. There are numerous sulfur-oxidizing microorganisms, but hydrogen sulfide is exclusively oxidized by prokaryotes. In any case, the biotrickling filter is typically inoculated with aerobic active sludge and during the process a specific biomass population is developed.

A biotrickling filter can be operated under aerobic or anoxic conditions. The removal of hydrogen sulfide from biogas has been mainly studied under aerobic conditions and the overall reaction is:

$$HS^- + 2O_2 \rightarrow SO_4^{2-} + 2H^+$$

In summary, two compounds can be produced; sulfate and elemental sulfur, and the production ratio of sulfate (SO_4^{2-}) will be dependent on the oxygen-hydrogen sulfide ratio and the trickling liquid velocity. Air is used to supply oxygen and must, therefore, be supplied in excess. However, control of the air supply is an important issue for safety and operational reasons – the lower and upper explosive limits for methane in air are 5% v/v and 15% v/v, respectively.

On the other hand, biogas can be diluted (mainly by the nitrogen content in air) and undesired residual oxygen can be found in the outlet stream. Therefore, a high oxygen supply is a drawback because the calorific power is also decreased and, moreover, a lack of oxygen increases clogging problems caused by the formation of elemental sulfur. The biogas dilution depends on the efficiency in the mass transfer and the concentration of hydrogen sulfide. An alternative to the direct supply of air mixed with the biogas stream involves the use of venture-based devices to increase the oxygen mass transfer. However, the installation of an aerated liquid recirculation system could produce hydrogen sulfide stripping of the dissolved sulfide.

The anoxic biotrickling filter is based on dissimilatory nitrate reduction. Dissimilatory nitrate reduction is carried out by certain bacteria that can use nitrate and/or nitrite as electron acceptors instead of oxygen:

- Complete denitrification *vs.* partial hydrogen sulfide oxidation
$5H_2S + NO_3^- \rightarrow 5S + N_2 + 4H_2O + 2OH^-$

- Complete denitrification *vs.* complete hydrogen sulfide oxidation
$5H_2S + 8NO_3^- \rightarrow 5SO_4^{2-} + 4N_2 + 4H_2O + 2H_+$

- Partial denitrification *vs.* partial hydrogen sulfide oxidation
$H_2S + 2NO_3^- \rightarrow S + 4NO_2^- + H_2O$

- Partial denitrification *vs.* complete hydrogen sulfide oxidation
$H_2S + 4NO_3^- \rightarrow SO_4^{2-} + 4NO_2^- + 2H_+$

7.6.2. Bioscrubbing

Bioscrubber systems have been used for hydrogen sulfide removal from biogas (Nishimura and Yoda, 1997). The bioscrubber involves a two-stage process with an absorption tower and a bioreactor, in which the sulfide is oxidized to sulfur and/or sulfate. For example, the Shell-Paques THIOPAQ® process employs alkaline conditions to produce elemental sulfur. In the first step of this process, the hydrogen sulfide is absorbed into an alkaline solution by reaction with hydroxyl and bicarbonate ions. In the second step the hydrosulfide is oxidized to elemental sulfur under oxygen-limiting conditions. Thus:

$H_2S + OH^- \rightarrow HS^- + H_2O$
$H_2S + HCO_3^- \rightarrow HS^- + CO_2 + H_2O$
$H_2S + CO_3^{2-} \rightarrow HS^- + HCO_3^-$

7.6.3. Bio-Oxidation

Bio-oxidation (biological oxidation) is one of the most used methods of desulfurisation, based on injection of a small amount of air (2 to 8 % v/v) into the raw biogas. This way, the hydrogen sulfide is biologically oxidized either to solid free sulfur (Figure 7.29) solid) or to liquid sulfurous acid (H_2SO_3):

$$2H_2S + O_2 \rightarrow 2H_2O + 2S$$
$$2H_2S + 3O_2 \rightarrow 2H_2SO_3$$

In practice, the precipitated sulfur is collected and added to the storage tanks where it is mixed with digestate, in order to improve the fertilizer properties of digestate. Biological desulfurization is frequently carried out inside the digester and, for this kind of desulfurization, oxygen and *Sulfobacter oxydans* bacteria must be present, to convert hydrogen sulfide into elementary sulfur, in the presence of oxygen. Typically, *Sulfobacter oxydans* is present inside the digester (does not have to be added) as the anaerobic digester substrate contains the necessary nutrients for their metabolism. In the process, the air is injected directly in the headspace of the digester and the reactions occur in the reactor headspace, on the floating layer (if existing) and on reactor walls. Due to the acidic nature of the products there is the risk of corrosion. The process is dependent of the existence of a stable floating layer inside the digester and the process often takes place in a separate reactor.

Chemical biogas desulfurisation can take place outside of digester, using e.g., a base (usually sodium hydroxide). Another chemical method to reduce the content of hydrogen sulfide is to add commercial ferrous solution (Fe^{2+}) to the feedstock. Ferrous compounds bind sulfur in an insoluble compound in the liquid phase, thereby preventing the production of gaseous hydrogen sulfide.

8. TAIL GAS CLEANING

Tail gas cleaning (also called tail gas treating) involves the removal of the remaining sulfur compounds from gases remaining after sulfur recovery. Tail gas from a typical Claus process, whether a conventional Claus or one of the extended versions of the process (Parkash, 2003; Mokhatab et al., 2006; Gary et al., 2007; Speight, 2014; Hsu and Robinson, 2017; Speight, 2017, 2019) usually contains small but varying quantities of carbonyl sulfide, carbon disulfide, hydrogen sulfide, and sulfur dioxide as well as sulfur vapor. In addition, there may be hydrogen, carbon monoxide, and carbon dioxide in the tail gas. In order to remove the rest of the sulfur compounds from the tail gas, all of the sulfur bearing species must first be converted to hydrogen sulfide which is then absorbed into a solvent and the clean gas vented or recycled for further processing.

It is necessary to develop and implement reliable and cost-effective technologies to cope with the various sales specifications and environmental requirements. In response to this trend, several new technologies are now emerging to comply with the most stringent regulations. The typical sulfur recovery efficiencies for a Claus plant is on the order of 90 to 96% w/w for a two-stage plant and 95 to 98% w/w for a three-stage plant. Most environmental agencies require the sulfur recovery efficiency to be in the range of 98.5 to 99.9+% w/w and, as a result, there is the need to reduce the sulfur content of the Claus plant tail gas need to be reduced further.

Moreover, a sulfur removal process must be very precise, since some biogas streams may contain only a small quantity of sulfur-containing compounds that must be reduced several orders of magnitude. Most consumers of gas require less than 4 ppm v/v in the gas – a characteristic feature of a gas stream that contains hydrogen sulfide is the presence of carbon dioxide (generally in the range of 1 to 4% v/v). In cases where the gas stream does not contain hydrogen sulfide, there may also be a relative lack of carbon dioxide.

8.1. Claus Process

The Claus process is not so much a gas cleaning process but a process for the disposal of hydrogen sulfide, a toxic gas that originates in most gas streams. Burning hydrogen sulfide as a fuel gas component or as a flare gas component is precluded by safety and environmental considerations since one of the combustion products is the highly toxic sulfur dioxide (SO_2), which is also toxic. As described above, hydrogen sulfide is typically removed from the refinery light ends gas streams through an olamine process after which application of heat regenerates the olamine and forms an acid gas stream. Following from this, the acid gas stream is treated to convert the hydrogen sulfide elemental sulfur and water. The conversion process utilized in most modern refineries is the Claus process, or a variant thereof.

The Claus process involves combustion of approximately one-third of the hydrogen sulfide to sulfur dioxide and then reaction of the sulfur dioxide with the remaining hydrogen sulfide in the presence of a fixed bed of activated alumina, cobalt molybdenum catalyst resulting in the formation of elemental sulfur:

$$2H_2S + 3O_2 \rightarrow 2SO_2 + 2H_2O$$
$$2H_2S + SO_2 \rightarrow 3S + 2H_2O$$

Different process flow configurations are in use to achieve the correct hydrogen sulfide/sulfur dioxide ratio in the conversion reactors.

In a split-flow configuration, one-third split of the acid gas stream is completely combusted and the combustion products are then combined with the non-combusted acid gas upstream of the conversion reactors. In a once-through configuration, the acid gas stream is partially combusted by only providing sufficient oxygen in the combustion chamber to combust one-third of the acid gas.

Two or three conversion reactors may be required depending on the level of hydrogen sulfide conversion required. Each additional stage provides incrementally less conversion than the previous stage. Overall, conversion of 96 to 97% v/v of the hydrogen sulfide to elemental sulfur is achievable in a Claus process. If this is insufficient to meet air quality regulations, a Claus process tail gas treater is utilized to remove essentially the entire remaining hydrogen sulfide in the tail gas from the Claus unit. The tail gas treater may employ employs a proprietary solution to absorb the hydrogen sulfide followed by conversion to elemental sulfur.

8.2. SCOT Process

The SCOT (Shell Claus Off-gas Treating) unit is a most common type of tail gas unit and uses a hydrotreating reactor followed by amine scrubbing to recover and recycle sulfur, in the form of hydrogen, to the Claus unit) (Nederland, 2004).

Early SCOT units used diisopropanolamine in the Sulfinol (Sulfinol-D) solution. Methyl diethanolamine-based Sulfinol (Sulfinol-M) was used later to enhance hydrogen sulfide removal and to allow for selective rejection of carbon dioxide in the absorber. To achieve the lowest possible hydrogen sulfide content in the treated gas, the Super-SCOT configuration was introduced. In this version, the loaded Sulfinol-M solution is regenerated in two stages. The partially stripped solvent goes to the middle of the absorber, while the fully stripped solvent goes to the top of the absorber. The solvent going to the top of the absorber is cooled below that used in the conventional SCOT process.

In the process, tail gas (containing hydrogen sulfide and sulfur dioxide) is contacted with hydrogen and reduced in a hydrotreating reactor to form hydrogen sulfide and water. The catalyst is typically cobalt/molybdenum on alumina. The gas is then cooled in a water contractor. The hydrogen sulfide-containing gas enters an amine absorber which is typically in a system segregated from the other refinery amine systems. The purpose of segregation is two-fold: (i) the tail gas treater frequently uses a different amine than the rest of the plant, and (ii) the tail gas is frequently cleaner than the refinery fuel gas (in regard to contaminants) and segregation of the systems reduces maintenance requirements for the SCOT unit. Amines chosen for use in the tail gas system tend to be more selective for hydrogen sulfide and are not affected by the high levels of carbon dioxide in the off-gas.

However, the SCOT process can be configured in various ways. For example, it can be integrated with the upstream acid gas removal unit if the same solvent is used in both units. Another configuration has been used to cascade the upstream gas cleanup with the

SCOT unit. A hydrogen sulfide-lean acid gas from the upstream gas treating unit is sent to a SCOT process with two absorbers. In the first absorber, the hydrogen sulfide-lean acid gas is enriched, while the second absorber treats the Claus tail gas. A common stripper is used for both SCOT absorbers and, in this latter configuration, different solvents could be used in both the upstream and the SCOT units.

A hydrotreating reactor converts sulfur dioxide in the off-gas to hydrogen sulfide that is then contacted with a Stretford solution (a mixture of a vanadium salt, anthraquinone disulfonic acid, sodium carbonate, and sodium hydroxide) in a liquid-gas absorber (Abdel-Aal et al., 2016). The hydrogen sulfide reacts stepwise with sodium carbonate and the anthraquinone sulfonic acid to produce elemental sulfur, with vanadium serving as a catalyst. The solution proceeds to a tank where oxygen is added to regenerate the reactants. One or more froth or slurry tanks are used to skim the product sulfur from the solution, which is recirculated to the absorber. Other tail gas treating processes include: (i) caustic scrubbing, (ii) polyethylene glycol treatment, (iii) Selectox process, and (iv) a sulfite/bisulfite tail gas treating (Mokhatab et al., 2006; Speight, 2014, 2019).

REFERENCES

Abatzoglou, N. and Boivin, S. 2009. A Review of Biogas Purification Processes. Biofuels *Bioprod. Biorefin*, 3(1): 42-71.

Abdel-Aal, H. K., Aggour, M. A., and Fahim, M. A. 2016. *Petroleum and Gas Field Processing*. CRC Press, Taylor & Francis Publishers, Boca Raton Florida.

Abraham, H. 1945. *Asphalts and Allied Substances*. Van Nostrand Scientific Publishers, New York.

Accettola, F., Guebitz, G. M. and Schoeftner, R. 2008. Siloxane Removal from Biogas by Biofiltration: Biodegradation Studies. *Clean Technol. Environ. Policy*, 10(2): 211-218.

Ajhar, M., Travesset, M., Yüce, S. and Melin, T. 2010. Siloxane Removal from Landfill and Digester Gas – A Technology Overview. *Bioresour. Technol.*, 101(9): 2913-2923.

Allen, R., and Yang, Y. 1992. Biofiltration: An Air Pollution Control Technology for Hydrogen Sulfide Emissions. In: *Industrial Environmental Chemistry: Waste minimization in Industrial Processes and Remediation of Hazardous Waste.* D. T. Sawyer and A. E. Martell Editors). Springer, Boston, Massachusetts. Page 273-287.

Alonso-Vicario, A., Ochoa-Gómez, J. R., Gil-Río, S., Gómez-Jiménez-Aberasturi, O., Ramírez-López, C. A. and Torrecilla-Soria, J. 2010. Purification and Upgrading of Biogas by Pressure Swing Adsorption on Synthetic and Natural Zeolites. *Microporous Mesoporous Mater.,* 134(1–3): 100-107.

Andriani, D., Wresta, A., Atmaja, T. D. and Saepudin, A. 2014. A Review on Optimization Production and Upgrading Biogas through CO Removal Using Various Techniques. *Appl. Biochem. Biotechnol,* 172(4): 1909-1928.

Anis, S., and Zainal, Z. A. 2011. Tar reduction in biomass producer gas via mechanical, catalytic and thermal methods: a review. *Renew Sustain Energy Rev.*, 15: 2355-2377.

ASTM. 2018. *Annual Book of Standards*, ASTM International, West Conshohocken, Pennsylvania.

ASTM D3246. 2018. Standard Test Method for Sulfur in Petroleum Gas by Oxidative Microcoulometry. *Annual Book of Standards*, ASTM International, West Conshohocken, Pennsylvania.

Barbouteau, L., and Dalaud, R. 1972. In *Gas Purification Processes for Air Pollution Control.* G. Nonhebel (Editor). Butterworth and Co., London, United Kingdom. Chapter 7.

Basu, S., Khan, A. L., Cano-Odena, A., Liu, C. and Vankelecom, I. F. 2010. Membrane-based Technologies for Biogas Separations. *Chem. Soc. Rev,* 39(2): 750-768.

Boulinguiez, B. and Le Cloirec, P. 2009. Biogas Pre-upgrading by Adsorption of Trace Compounds onto Granular Activated Carbons and an Activated Carbon Fiber Cloth. *Water Sci. Technol,* 59(5): 935-944.

Cherosky, P. and Li, Y. 2013. Hydrogen Sulfide Removal from Biogas by Bio-based Iron Sponge. *Biosyst. Eng.,* 114(1): 55-59.

Curry, R. N. 1981. *Fundamentals of Natural Gas Conditioning.* PennWell Publishing Co., Tulsa, Oklahoma.

Deshmukh, M. G. and Shete, A. 2013. Oxidative Absorption of Hydrogen Sulfide Using Iron-Chelate Based Process: Chelate Degradation. *J. Anal. Bioanal. Tech*, 88(3): 432-36.

Devi, L., Ptasinski, K. J., and Jenssen, F. J. J. G. 2003. A Review of the Primary Measures for Tar Elimination in Biomass Gasification Processes. *Biomass Bioenergy*, 24: 125-140.

Dou, B., Zhang, M., and Gao, J. 2002. High-temperature removal of NH3, Organic Sulfur, HCl, and Tar Component from Coal-Derived Gas. *Ind. Eng. Chem. Res*, 41: 4195-4200.

El-Gendy, N. Sh., and Speight, J. G. 2016. *Handbook of Refinery Desulfurization.* CRC Press, Taylor & Francis Group, Boca Raton, Florida.

Fernández, M., Ramírez, M., Gómez, J. M. and Cantero, D. 2013a. Biogas Biodesulfurization in an anoxic Biotrickling Filter Packed with Open-Pore Polyurethane Foam. *J. Hazard. Mater*, 264: 529-535.

Fernández, M., Ramírez, M., Pérez, R. M., Gómez, J. M. and Cantero, D. 2013b. Hydrogen Sulfide Removal From Biogas by an Anoxic Biotrickling Filter Packed with Pall Rings. *Chem. Eng. J*, 225: 456-463.

Forbes, R. J. 1958a. *A History of Technology*, Oxford University Press, Oxford, United Kingdom.

Forbes, R. J. 1958b. *Studies in Early Petroleum Chemistry.* E. J. Brill, Leiden, Netherlands.

Forbes, R. J. 1959. *More Studies in Early Petroleum Chemistry*. E. J. Brill, Leiden, Netherlands:

Forbes, R. J. 1964. *Studies in Ancient Technology*. E. J. Brill, Leiden, Netherlands.

Gary, J. G., Handwerk, G. E., and Kaiser, M. J. 2007. *Petroleum Refining: Technology and Economics*. 5th Edition. CRC Press, Taylor & Francis Group, Boca Raton, Florida.

Gislon, P., Galli, S. and Monteleone, G. 2013. Siloxanes Removal from Biogas by High Surface Area Adsorbents. *Waste Manag*, 33(12): 2687-2693.

Gustafsson, E., Lin, L., and Seemann, M. 2011 Characterization of Particulate Matter in the Hot Product Gas from Indirect Steam Bubbling Fluidized Bed Gasification of Wood Pellets. *Energy Fuels*, 25: 1781–1789.

Hoiberg A. J. 1960. *Bituminous Materials: Asphalts, Tars and Pitches*, I & II. Interscience, New York.

Hsu, C. S., and Robinson, P. R. (Editors). 2017. *Handbook of Petroleum Technology*. Springer International Publishing AG, Cham, Switzerland.

Huang, J., Schmidt, K. G., Bian, Z. 2011. Removal and Conversion of Tar in Syngas from Woody Biomass Gasification for Power Utilization Using Catalytic Hydrocracking. *Energies* 4: 1163-1177.

Jensen, A. B. and Webb, C. 1995. Treatment of H_2S-containing Gases – A Review of Microbiological Alternatives. *Enzyme Microb. Tech*, 17(1): 2-10.

Jou, F. Y., Otto, F. D., and Mather, A. E. 1985. In *Acid and Sour Gas Treating Processes*. S. A. Newman (Editor). Gulf Publishing Company, Houston, Texas. Chapter 10.

Katz, D. K. 1959. *Handbook of Natural Gas Engineering*. McGraw-Hill Book Company, New York.

Kidnay, A. J., and Parrish, W. R. *2006 Fundamentals of Natural Gas Processing*. CRC Press, Taylor & Francis Group, Boca Raton, Florida.

Kohl, A. L., and Nielsen, R. B., 1997. *Gas Purification*. Gulf Publishing Company, Houston, Texas.

Kohl, A. L., and Riesenfeld, F. C. 1985. *Gas Purification*. 4th Edition, Gulf Publishing Company, Houston, Texas.

Maddox, R. N. 1982. *Gas Conditioning and Processing*. Volume 4. Gas and Liquid Sweetening. Campbell Publishing Co., Norman, Oklahoma.

Maddox, R. N., Bhairi, A., Mains, G. J., and Shariat, A. 1985. In *Acid and Sour Gas Treating Processes*. S. A. Newman (Editor). Gulf Publishing Company, Houston, Texas. Chapter 8.

May-Britt, H. 2008. Membranes in Gas Separation. In: *Handbook of Membrane Separations*. CRC Press, Taylor & Francis Group, Boca Raton, Florida. Page 65-106.

Mody, V., and Jakhete, R. 1988. *Dust Control Handbook*. Noyes Data Corp., Park Ridge, New Jersey.

Mokhatab, S., Poe, W. A., and Speight, J. G. 2006. *Handbook of Natural Gas Transmission and Processing*. Elsevier, Amsterdam, Netherlands.

Montebello, A. M., Fernández, M., Almenglo, F., Ramírez, M., Cantero, D. and Baeza, M. 2012. Simultaneous Methyl Mercaptan and Hydrogen Sulfide Removal in the Desulfurization of Biogas in Aerobic and Anoxic Biotrickling Filters. *Chem. Eng. J,* 200/202(0): 237-246.

Montebello, A. M., Bezerra, T., Rovira, R., Rago, L., Lafuente, J. and Gamisans, X. 2013. Operational Aspects, pH Transition and Microbial Shifts of a H2S Desulfurizing Biotrickling Filter with Random Packing Material. *Chemosphere*, 93(11): 2675-2682.

Newman, S. A. 1985. *Acid and Sour Gas Treating Processes*. Gulf Publishing, Houston, Texas.

Nikolić, D., Kikkinides, E. S. and Georgiadis, M.C. 2009. Optimization of Multibed Pressure Swing Adsorption Processes. *Ind. Eng. Chem. Res,* 48(11): 5388-5398.

Niesner, J., Jecha, D. and Stehlík, P. 2013. Biogas Upgrading Technologies: State of the Art Review in the European Region. *Chem. Eng. Trans*, 35: 517-522.

Nishimura, S. and Yoda, M. 1997. Removal of Hydrogen Sulfide from an Anaerobic Biogas Using a Bio-Scrubber. *Water Sci. Technol*, 36(6-7): 349-356.

Ozturk, B. and Demirciyeva, F. 2013. Comparison of Biogas Upgrading Performances of Different Mixed Matrix Membranes. *Chem. Eng. J,* 222: 209-217.

Parkash, S. 2003. *Refining Processes Handbook.* Gulf Professional Publishing, Elsevier, Amsterdam, Netherlands.

Pitsinigos, V. D., and Lygeros, A. I. 1989. Predicting H_2S-MEA Equilibria. *Hydrocarbon Processing*, 58(4): 43-44.

Polasek, J., and Bullin, J. 1985. In *Acid and Sour Gas Treating Processes*. S. A. Newman (Editor). Gulf Publishing Company, Houston, Texas. Chapter 7.

Proell, T., Siefert, I. G., and Friedl, A. 2005. Removal of NH3 from Biomass Gasification Producer Gas by Water Condensing in an Organic Solvent Scrubber. *Ind. Eng. Chem. Res,* 44: 1576-1584.

Ramírez M., Gómez J. M. and Cantero D. 2015. Biogas: Sources, Purification and Uses. In: *Hydrogen and Other Technologies*. U. C. Sharma, S. Kumar, R. Prasad (Editors). Studium Press LLC, New Delhi, India. Chapter: 13, page 296-323.

Rapagna, S., Gallucci, K., and Marcello, M. D. 2010. Gas Cleaning, Gas Conditioning and Tar Abatement by Means of a Catalytic Filter Candle in a Biomass Fluidized-Bed Gasifier. *Bioresour. Technol.,* 101: 7123-7130.

Reddy, S., and Gilmartin, J. 2008. Fluor's Econamine Plus Technology for Post-Combustion CO2 Capture. *Proceedings. GPA Gas Treatment Conference*, Amsterdam, Netherlands. February 20-22. https://www.fluor.com/SiteCollection Documents/FluorEFG-forPost-CombustionCO2CaptureGPAConf-Feb2008.pdf.

Rongwong, W., Boributh, S., Assabumrungrat, S., Laosiripojana, N. and Jiraratananon, R. 2012. Simultaneous Absorption of CO_2 and H_2S from Biogas by Capillary Membrane Contactor. *J. Membrane Sci.*, 392/393: 38-47.

Ryckebosch, E., Drouillon, M. and Vervaeren, H. 2011. Techniques for Transformation of Biogas to Biomethane. *Biomass Bioenergy*, 35(5): 1633-1645.

Santos, M. P. S., Grande, C. A. and Rodrigues, A. E. 2013. Dynamic Study of the Pressure Swing Adsorption Process for Biogas Upgrading and Its Responses to Feed Disturbances. *Ind. Eng. Chem. Res.*, 52(15): 5445-5454.

Soreanu, G., Béland, M., Falletta, P., Edmonson, K., Svoboda, L. and Al-Jamal, M. 2011. Approaches Concerning Siloxane Removal from Biogas – A Review. *Canadian Biosys. Eng.*, 53: 8.1-8.18.

Soud, H., and Takeshita, M. 1994. *FGD Handbook*. No. IEACR/65. International Energy Agency Coal Research, London, United Kingdom.

Speight, J. G. 1978. *Personal Observations at Archeological Digs at the Cities of Babylon*, Calah, Nineveh, and Ur. College of Science, University of Mosul, Iraq.

Speight, J. G. 2008. *Synthetic Fuels Handbook: Properties, Processes, and Performance.* McGraw-Hill, New York.

Speight, J. G. (Editor), 2011. *The Biofuels Handbook. Royal Society of Chemistry,* London, United Kingdom.

Speight, J. G. 2012. *Shale Oil Production Processes*. Gulf Professional Publishing, Elsevier, Oxford, United Kingdom.

Speight, J. G. 2014. *The Chemistry and Technology of Petroleum* 4th Edition. CRC-Taylor and Francis Group, Boca Raton, Florida.

Speight, J. G. 2015. *Handbook of Petroleum Product Analysis* 2nd Edition. John Wiley & Sons Inc., Hoboken, New Jersey.

Speight, J. G. 2017. *Handbook of Petroleum Refining*. CRC Press, Taylor & Francis Group, Boca Raton, Florida.

Speight, J. G. 2018. *Handbook of Natural Gas Analysis.* John Wiley & Sons Inc., Hoboken, New Jersey.

Speight, J. G., and El-Gendy, N. Sh. 2018. *Introduction to Petroleum Biotechnology*. Gulf Professional Publishing Company, Elsevier, Cambridge, Massachusetts.

Speight, J. G. 2019. *Natural Gas: A Basic Handbook* 2nd Edition. Gulf Publishing Company, Elsevier, Cambridge, Massachusetts.

Syed, M., Soreanu, G., Falletta, P. and Béland, M. 2006. Removal of Hydrogen Sulfide from Gas Streams Using Biological Processes – A Review. *Canadian Biosys. Eng.*, 48: 2.1-2.14.

Torres, W., Pansare, S. S., and Goodwin, J. G. 2007. Hot Gas Removal of Tars, Ammonia, and Hydrogen Sulfide from Biomass Gasification Gas. *Catal. Rev.,* 49: 407-56.

Yuan, S., Zhou, Z., and Li, J. 2010. HCN and NH3 Released from Biomass and Soybean Cake Under Rapid Pyrolysis. *Energy Fuels*, 24: 6166-6171.

Chapter 8

BIOGAS APPLICATIONS

1. INTRODUCTION

Biogas is comprised primarily of methane (50 to 70% v/v) and carbon dioxide (30 to 50% v/v), with trace amounts of other particulates and contaminants depending upon the composition of the biomass feedstock from which the gas is produced and the process parameters (Table 8.1) (Chapter 1). Consequently, cleaned biogas that is free of contaminants (Chapter 7) is lighter than air and has an ignition temperature in the range of 650 to 750°C. (1200 to 1380°F.). It is an odorless and colorless gas that burns with a clear blue flame similar to that of natural gas. Raw biogas (that has not been cleaned and all contaminants removed) has a calorific value on the order of 537 to 700 Btu/ft^3 compared to commercial quality natural gas with a calorific value on the order of 1,011 Btu/ft^3.

During the last three decades plants for production and utilization of biogas have been developed and, furthermore, biogas gas acts as a substitute for fossil fuels such as oil, coal and gas that all contribute to the greenhouse effect (Figure 8.1). Biogas, consisting primarily of methane and carbon-dioxide is a by-product of the anaerobic decomposition and the conversion organic wastes. However, there is always danger associated with many types of gases. In the current context, methane is a colorless and odorless gas which is lighter than air and hence can sometimes go undetected and accumulate to an explosive concentration from 5 to 15% v/v in air (Table 8.2). Thus, utilization of any gas containing methane can increase the risk of explosion.

In fact, in terms of biogas production and composition, lignocellulosic biomass is a promising candidate for anaerobic digestion, to the biogas potential from solid biomass via thermochemical conversion pathways (also called synthesis gas or syngas). Thus, the methane generation potential is expected to be much higher if lignocellulosic biomass resources are used in anaerobic digesters (Chapter 3).

Table 8.1. Typical Ranges of the Constituents in Biogas

Compound	Formula	Composition, % v/v
Methane	CH_4	50-75
Carbon dioxide	CO_2	25-50
Nitrogen	N_2	0-10
Hydrogen	H_2	0-1
Hydrogen sulfide	H_2S	0-3
Oxygen	O_2	0-5
Other contaminants	variable	0-3

Table 8.2. General Properties of Methane

Chemical formula	CH_4
Molar mass	16.04
Appearance	Colorless gas
Odor	Odorless
Density	0.657 g/liter (gas, 25°C, 77°F, 14.7 psi) 0.717 g/liter (gas, 0°C, 32°F, 14.7 psi) 422.62 g/liter (liquid, -162°C, -260°F)
Melting point	−182.5 °C; −296.4 °F; 90.7 K
Boiling point	−161.50 °C; −258.70 °F; 111.65 K[3]
Solubility in water	22.7 mg/liter
Solubility	Soluble in ethanol, diethyl ether, benzene, toluene, methanol, acetone.
Lower explosive limit in air	5% v/v
Upper explosive limit in air	15% v/v

Biogas can be burned to produce both heat and electricity, usually with a reciprocating engine or a microturbine often in a cogeneration arrangement where the electricity and waste heat generated are used to warm the digesters or to heat buildings. Electricity produced by the gas from anaerobic digesters is considered to be renewable energy and may attract subsidies. Also, it is believed that biogas does not contribute to increasing atmospheric carbon dioxide concentrations because the gas is not released directly into the atmosphere and the carbon dioxide comes from an organic source with a short carbon cycle (Posch et al., 2010).

Typically, biogas requires treatment to refine it for use as a fuel (Chapter 7) (Niemczewska, 2012). Hydrogen sulfide, a toxic product formed from sulfates in the feedstock, is released as a trace component of the biogas (Chapter 1, Chapter 7). If the levels of hydrogen sulfide in the gas are high, gas scrubbing and cleaning equipment (such as gas treating using the olamine process) will be needed to process the biogas to within regionally accepted levels. Alternatively, the addition of ferrous chloride ($FeCl_2$) to the digestion tanks inhibits hydrogen sulfide production (Chapter 7).

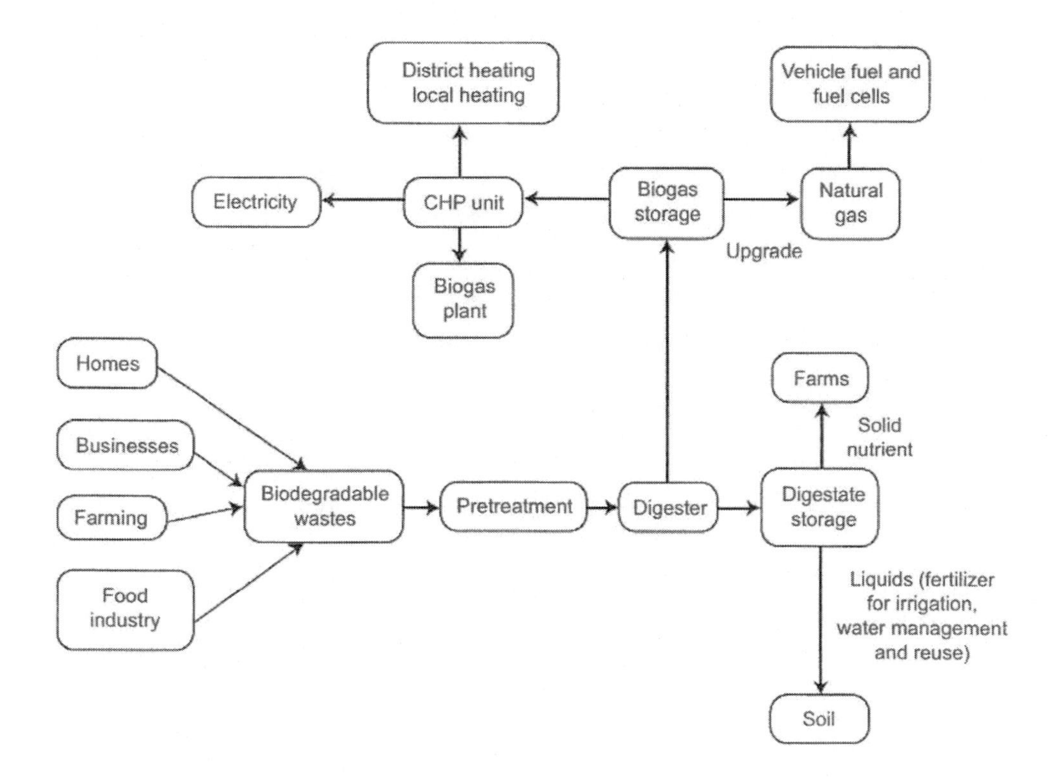

Figure 8.1. Products and Use of Biogas.

Volatile siloxane derivatives can also occur as contaminate the biogas, especially biogas from landfills, especially since the use of siloxane derivatives is increasing, for example, in household industrial cleaning products because volatile methyl siloxane derivatives (VMSs) solvents are aroma free and widely available. In digestion facilities accepting these materials as a component of the feedstock, low-molecular-weight siloxanes volatilize into biogas. When this gas is combusted in a gas engine, turbine, or boiler, siloxanes are converted into silicon dioxide (SiO_2), which deposits internally in the machine, increasing wear and tear. Practical and cost-effective technologies to remove siloxanes and other biogas contaminants are available at the present time (Chapter 7). In certain applications, *in situ* treatment can be used to increase the methane purity by reducing the offgas carbon dioxide content, purging the majority of it in a secondary reactor.

As a recall, biogas, as produced by any of the methods described in earlier chapters (i.e., by anaerobic digestion, Chapter 3, from landfills, Chapter 4, by pyrolysis, Chapter 5, or by gasification, Chapter 6) is typically saturated with water vapor and contains, in addition to methane (CH_4) and carbon dioxide (CO_2), various amounts of hydrogen sulfide (H_2S) as well as other contaminants that vary in amounts depending upon the composition of the biomass feedstocks and the process parameters employed for production of the gas (Chapter 1) (Weiland, 2010). The properties of the contaminants are such that failure to remove them will make the gas unusable and even poisonous to the user. For example, hydrogen sulfide is a toxic gas, with a specific, unpleasant odor, similar to rotten eggs, forming acidic products in combination with the water vapors in the biogas which can also result in corrosion of equipment. To prevent this, and similar example of the adverse effects of other contaminants, the biogas must be dried and any contaminants removed before the biogas is sent to one or more conversion units or for sales.

Typically, in the gas processing industry (also called the *gas cleaning industry* or the *gas refining industry*), the feedstock is not used directly as fuel because of the complex nature of the feedstock and the presence of one of more of the aforementioned impurities that are corrosive or poisonous. It is, therefore, essential that any gas should be contaminant free when it enters any one of the various reactors (Parkash, 2003; Gary et al., 2007; Speight, 2014; Hsu and Robinson, 2017; Speight, 2017). Typically, the primary raw biogas has been subjected to chemical and/or physical changes (refining) after being produced. Thus for many applications the quality of biogas has to be improved (Chapter 7). In addition, landfill gas (Chapter 4) often contains significant amounts of halogenated compounds which need to be removed prior to use. Occasionally the oxygen content is high when air is allowed to invade the gas producer during collection of the landfill gas or when the biomass from which the gas is produced contains an abundance of oxygen-containing organic constituents.

As more efficient bioenergy technologies are developed, fossil fuel inputs will be reduced at which time biomass and its by-products will be used as sources for fueling a variety of energy needs. The energy value of biomass from plant matter originally comes from solar energy through the process known as photosynthesis. In nature, all biomass ultimately decomposes to its elementary molecules with the release of heat. During conversion processes such as combustion, biomass releases its energy, often in the form of heat, and the carbon is re-oxidized to carbon dioxide to replace that which was absorbed while the plant was growing. Essentially the use of biomass for energy is the reversal of photosynthesis.

The choice in the final means for the utilization of biogas has considerable impact on the design of the equipment and the equipment requirements for biogas processing and storage as well as the economics of the biogas conversion system. The biogas may be applied in direct combustion systems (boilers, turbines, or fuel cells) for producing space heating, water heating, drying, absorption cooling, and steam production. The gas used directly in gas turbines and fuel cells may produce electricity. An alternate choice in biogas conversion is the use in stationary or mobile internal combustion engines which may results in shaft horsepower, cogeneration of electricity, and/or vehicular transportation. A final opportunity exists for sale of the biogas through injection into a natural gas pipeline.

Biogas can be used readily in all applications designed for natural gas such as direct combustion including absorption heating and cooling, cooking, space and water heating, drying, and gas turbines. It may also be used in fueling internal combustion engines and fuel cells for production of mechanical work and/or electricity. If the contaminants are removed up to adequate standards, the cleaned gas may be injected into gas pipelines and provide illumination and steam production. Finally, through a catalytic chemical oxidation methane can be used in the production of methanol (CH_3OH) production.

The next section presents a selection of methods by which bio-gas can be used to provide energy in place of fossil fuel sources such as natural gas and liquid fuels.

2. TREATMENT AND STORAGE

Although biogas treatment (cleaning) has been presented in a previous chapter (Chapter 7), a refresher on the various cleaning methods is worth mention here so that this chapter (like the other chapters in this book) can serve as a stand-alone chapter relating to biogas applications.

The hydrogen sulfide contained in biogas caused odors, corrosiveness, and sulfur emissions when the gas is burned. High levels of sulfide in biogas may require removal to protect equipment if the gas is to be used in internal combustion engines, turbines, or fuel cells. The concentration of hydrogen sulfide in the gas is a function of the process feedstock ad the process parameters as well as the content of inorganic sulfate minerals. Also, wastes which are high in proteins containing sulfur-based amino acids (methionine and cysteine) can significantly influence the concentration of hydrogen sulfide in the biogas.

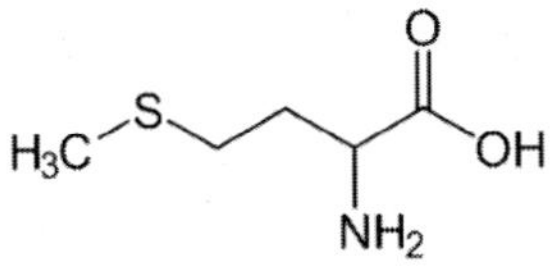

Methionine

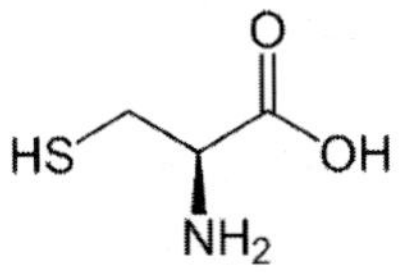

Cysteine

For instance, layer poultry waste containing feathers made of keratin may produce biogas sulfide levels up to 20,000 ppm v/v. Also, sulfate present in the waste, either from an industrial source (such as wood pulping) or from seawater (marine aquiculture) will be reduced by sulfate reducing bacteria in the digester and end up contributing to sulfide levels in the gas.

The treatment of biogas may include removal of components including hydrogen sulfide, water, mercaptan derivatives, carbon dioxide, trace organic compounds, and particulate matter. Due to the corrosive nature of hydrogen sulfide (especially when water is also present), removal processes for this component are well developed and include both dry and wet removal processes (Chapter 7). In a wet process the biogas is passed up-flow through a stripping tower where the aqueous solutions are sprayed in counter-current flow. The tower is generally separated by distribution trays which maximize contact between the biogas and the solution. For small-scale biogas producers, an alternative to the wet absorption systems described above is dry adsorption or chemisorption. Several dry processes are available, using particles of either activated carbon, molecular sieve, iron sponge or other iron-based, granular compounds to remove sulfide from the gas phase to the solid phase. These are sometimes referred to as dry oxidation processes because elemental sulfur or oxides of sulfur are produced (and can be recovered) during oxidative regeneration of the catalyst.

In addition to those aqueous absorbents described for removal of hydrogen sulfide, there are many chemical solutions commercially available which can be used concurrently to remove carbon dioxide and hydrogen sulfide concurrently. In general, these processes employ either solvation solutions where the objective is to dissolve carbon dioxide and hydrogen sulfide in the liquid, or solutions which react chemically to alter the ionic character of these gases and therefore drive them into solution. Solutions of the former category include the solvents and the latter include the alkanolamine derivatives and alkaline salts. Also, there are membrane materials which are specially formulated to selectively separate carbon dioxide from methane. The permeability of the membrane is a direct function of the chemical solubility of the target compound in the membrane. To separate two compounds such as carbon dioxide and methane, one gas must have a high solubility in the membrane while the other is insoluble. Accordingly, rejection (separation) efficiencies are typically quite high when the systems are operated as designed.

Biogas is not typically produced at the time or in the quantity needed to satisfy the conversion system load that it serves. When this occurs, storage systems are employed to smooth out variations in gas production, gas quality and gas consumption. The storage component also acts as a reservoir, allowing downstream equipment to operate at a constant pressure.

A wide variety of materials have been used in making biogas storage vessels. Medium- and high-pressure storage vessels are usually constructed of mild steel while low-pressure storage vessels can be made of steel, concrete and plastics. Each material possesses advantages and disadvantages that the system designer must consider. The newest reinforced plastics feature polyester fabric which appears to be suitable for flexible digester covers. The delivery pressure required for the final biogas conversion system affects the choice for biogas storage.

3. USES

Biogas can be combusted to provide heat, electricity, or both. And, in addition, can be upgraded to pure methane – also called biomethane or renewable natural gas – by removing water, carbon dioxide, hydrogen sulfide, and other trace elements. This upgraded biogas is comparable to conventional natural gas, and thus can be injected into the pipeline grid or used as a transportation fuel in a compressed or liquefied form. Renewable natural gas is considered a drop-in fuel for the natural gas vehicles currently on the road and can qualify as an advanced biofuel (Pruthvira, 2016). It can also be a source for renewable hydrogen, which can be used in stationary fuel cells and fuel cell electric vehicles. The methane content of biogas is the usable portion of the gas and determines its calorific value.

The gas must be desulfurized and dried before utilization to prevent damage of the gas utilization units. Biogas produced by co-fermentation of manure with energy crops or harvesting residues can contain levels of hydrogen sulfide between 100 and 3,000 ppm v/v. A maximum value of less than 1 ppm v/v is defined for both the sum of the nitrogen-containing and sulfur-containing compounds. Although for low-pressure methanol process a level as low as 0.1 ppm v/v of total sulfur is required. For the halides and alkaline metals a lower level of less than 10 ppb v/v is necessary. Solid contaminants must be removed essentially completely because these types of contaminants foul the system and may obstruct fixed-bed reactors.

3.1. Power Generation

Biosynthesis gas is a combustible gas and can be used for the production of electricity in all prime movers from steam cycles, to gas engines, turbines (combined cycle), as well as fuel cells.

Theoretically, biogas can be converted directly into electricity by using a fuel cell. However, this process requires very clean gas and expensive fuel cells. Therefore, this option is still a matter for research and is not currently a practical option. The conversion of biogas to electric power by a generator set is much more practical. In contrast to natural gas, biogas is characterized by a high knock resistance and hence can be used in combustion motors with high compression rates.

Biogas utilization has been established for at least one hundred years and the generation of electricity from biogas can be achieved with the use of a variety of different technologies, including internal combustion engines, turbines, micro turbines, Stirling engines (external combustion engine), organic Rankine cycle engines, and fuel cells. Internal combustion (reciprocating) engines and gas turbines are the most widely used, with micro turbine technology being used at smaller landfills and in niche applications.

The organic Rankine cycle (ORC) is based on the same thermodynamic cycle as the classical traditional Rankine cycle, with the exception of the working fluid. Organic oils with a lower boiling temperature than water are usually the working fluid used in the organic Rankine cycle, which enables operation at relatively low temperatures. The organic Rankine cycle is presently used for geothermal and biomass combustion application.

A motivation, however, to operate a gas turbine with a combined cycle on bio-synthesis gas would be the possibility to use large-scale technologies that are available from coal gasification. Gas Turbines are usually used where the volume of biogas is sufficient to generate more than 5 MW. Simple-cycle gas turbines applicable to biogas energy projects typically achieve efficiencies of 20 to 28% at full load operations. However, these efficiencies drop substantially when the unit is running at partial load. Combined-cycle configurations which recover the waste heat in the gas turbine exhaust to make additional electricity can boost the efficiency of the system to approximately 40% at full load operation. Gas turbines have lower nitrogen oxides emission rates.

In this way it is possible to reach relatively large-scale implementation of biomass-based electricity production without the requirement to go through the complete development of the existing small-scale biomass gasification technologies. The basic technology is also suitable for dedicated biomass firing and, by converting a coal plant into a plant that can also accommodate biomass feedstocks plant, a large green electricity production capacity can be realized. Considering the ambitious green energy targets of many countries and national governments, the possibility for short term implementation may be more important than smaller production capacities but with higher energetic efficiencies.

Methane and power produced in anaerobic digestion facilities can be used to replace the energy that is currently derived from fossil fuels, and hence reduce emissions of greenhouse gases, because the carbon in biodegradable material is part of a carbon cycle. The carbon released into the atmosphere from the combustion of biogas has been removed by plants for them to grow in the recent past, usually within the last decade, but more typically within the last growing season. If the plants are regrown, taking the carbon out of the atmosphere once more, the system has been calculated to be carbon neutral, provided that all of the necessary assumptions are in order. In contrast, carbon in fossil fuels has been sequestered in the earth for many millions of years, the combustion of which increases the overall levels of carbon dioxide in the atmosphere.

Biogas from sewage sludge treatment is sometimes used to run a gas engine to produce electrical power, some or all of which can be used to run the sewage works. Some waste heat from the engine is then used to heat the digester. The waste heat is, in general, enough to heat the digester to the required temperatures.

Biogas to combined heat and power (CHP) is an energy production method and Biogas may be used for electricity production on sewage works, inside a combined heat and power gas engine, in which the waste warmth in the engine is easily employed for heating the digester cooking space heating water heating and process heating. If compressed, it may replace compressed gas to be used in automobiles, where it may fuel automobile engines or fuel cells and it is an infinitely more effective displacer of carbon dioxide compared to normal use within on-site combined heat and power plants.

The gas is usually used in combined heat and power systems using gas or dual fuel engines. Combined heat and power systems use both the power producing ability of a fuel and the inevitable waste heat. Some combined heat and power systems produce primarily heat, and electrical power is secondary (bottoming cycle). Other combined heat and power systems produce primarily electrical power and the waste heat is used to heat process water (topping cycle). In either case, the overall (combined) efficiency of the power and heat produced and used gives a much higher efficiency than using the fuel (biogas) to produce only power or heat.

The efficiency of installations using biogas in the combined heat and power unit is higher, and their number is growing, whereas the thermal efficiency is always higher than electrical efficiency for all combined heat and power units. The energy use of biogas plants in cogeneration can involve the application of reciprocating gas engines, gas turbines and microturbines. Combined heat and power plants using gas-powered reciprocating engines usually produce hot water or saturated steam. Heat is recovered from the heat exchanger on the engine casing, oil cooler and exhaust heat exchanger. Combined heat and power operation can also be used in gas turbines, involving a simple cycle gas turbine with a heat recovery heat exchanger which recovers the heat in the turbine exhaust and converts it to useful thermal energy, typically in the form of steam or hot water.

In combined heat and power plants the product gas is fired on a gas engine. Modified gas engines can run without problems on most product gases even those from air-blown gasification that have calorific values on the lower end of the energy range. Typically, the energetic output is one-third electricity and two-third heat. The main technical challenge in the implementation of integrated biomass gasification combined heat and power plants has been, and still is, the removal of "tar" from the product gas.

Regarding the presence of tar in product gas, it may be stated that tar is equivalent to a major economic penalty in biomass gasification. Tar aerosols and deposits lead to more frequent maintenance and repair of especially gas cleaning equipment and resultantly lower plant capacity factors. This leads to a decrease of revenues or to higher investments, as some equipment will be installed in duplicate to overcome standstills. Furthermore, removal of tar components from the process wastewater requires considerable investments that can even be dramatic as some tar components show poisoning behavior in biologic wastewater treatment systems (e.g., phenol).

Alternatives to the common motor combined heat and power stations are micro-gas turbines and fuel cells. Micro-gas turbines have a good part loading efficiency and long maintenance intervals – the main advantage over reciprocal engines is the availability of the exhaust heat which still has at least 270°C (520°F) after the recuperator. This opens new ways of using the heat for process steam production. Fuel cells result in a higher electric efficiency but need an efficient gas cleaning, because the catalyst for converting methane into hydrogen and the catalyst inside the fuel cell are very sensitive to impurities. The various fuel cell types are operated at temperatures between 80 and 800°C (176 and 1470°F). The investment costs are much higher than for engine-driven combined heat and power.

Microturbines can also be used in combined heat and power systems. In these applications, the waste heat from microturbine exhaust is used to produce hot water (up to 93°C, 199°F). This option can replace a relatively expensive fuel, such as propane, needed to heat water in colder climates to meet space-heating requirements. The sale or use of microturbine waste heat can significantly enhance project economy. Hot water can be used to heat building space, to drive absorption cooling, and to supply other thermal energy needs in a building or industrial process.

The reciprocating internal combustion engine represents the most widely used technology for electricity generation from biogas. The reason is mainly both the power and the economic feasibility of the system. These engines represent a prevalent and consolidated technology, and the related economic risks are very low compared to other technologies.

Gas-powered reciprocating engines, also called gas-powered internal combustion engines, are the modified versions of medium- and high-speed engines powered by liquid fuels. The modifications applied in gas-fueled engines typically include: changes in the shape of head and the top part of pistons, adding a gas and liquid fuel system, expansion of the engine cooling system and the exhaust heat removal system.

Biogas is mainly utilized in engine-based combined heat and power plants, whereas micro-gas turbines and fuel cells are expensive alternatives which need further development work for reducing the costs and increasing their reliability. Gas upgrading and utilization as renewable vehicle fuel or injection into the natural gas grid is of increasing interest because the gas can be used in a more efficient way. Combined heat and power stations (CHPs) which are mainly used for the utilization of biogas need mostly levels of hydrogen sulfide below 250 ppm v/v, in order to avoid excessive corrosion and expensive deterioration of lubrication oil.

The most commonly used conversion technology in biogas applications is the internal combustion engine which usually produces between 800 kilowatts (kW) to megawatts (MW) of power and requires a sustainable flow rate of the biogas. Greater efficiencies are achieved in combined heat and power (CHP) applications where waste heat is recovered from the engine's exhaust to make low pressure steam. Thousands of engines are operated on sewage works, landfill sites and biogas installations. Gas engines do have comparable requirements for gas quality as boilers except that the hydrogen sulfide should be lower to guarantee a reasonable operation time of the engine. Otto engines designed to run on gasoline are far more susceptible to hydrogen sulfide than the more robust diesel engines. For large scale applications diesel engines are therefore standard. Occasionally, organic silica compounds in the gas can create abrasive problems. If so, they should be removed.

Gas-Otto motors are developed specifically for using biogas according to the Otto principle. The engines (lean burn engines) are operated with air surplus, in order to minimize carbon monoxide emissions. This leads to lower gas consumption and reduced motor performance, compensated by using an exhaust turbo charger. Gas-Otto motors require biogas with minimum 45% v/v methane content. Smaller engines are usually Otto engines. For higher electrical performance, adapted diesel aggregates are used. They are equipped with spark plugs. Both engines are named Gas-Otto Engines since their basic operation is based on the Otto principle. These engines can be operated with biogas or natural gas which is useful during the start-up of the biogas plant when the heat is used for digester heating.

In practice, the scale of integrated biomass gasification combined heat and power plants will be limited by the local heat demand (up to approximately. 20 MW). For electricity production on larger scales, integrated gasification combined cycles (IGCC) are preferred in which the gas is fired on a gas turbine. As a gas turbine requires a pressurized feed gas the biomass gasification should be carried out at the pressure of the turbine (typically 5-20 bar) or the atmospherically generated product gas must be pressurized. The first route is preferred as in that case only dust removal from the gas stream and cooling to the turbine inlet temperature (400-500°C) is required, whereas in the alternative route the product gas must be completely cleaned and cooled down to allow compression.

3.2. Grid Injection

Biogas grid-injection is the injection of biogas into the natural gas grid. The raw biogas has to be previously upgraded to biomethane (Köppel et al., 2009). This upgrading implies the removal of contaminants such as hydrogen sulfide or siloxanes, as well as the carbon dioxide. Several technologies are available for this purpose, the most widely implemented being pressure swing adsorption, water or amine scrubbing (absorption processes) and, in recent years, membrane separation (Chapter 7). As an alternative, the electricity and the heat can be used for on-site generation, resulting in a reduction of losses in the transportation of energy.

There are several advantages of using the gas grid for distribution of biomethane. One important advantage is that the grid connects the production site of biomethane, which is usually in rural areas, with more densely populated areas. This enables the gas to reach new customers. It is also possible to increase the biogas production at a remote site, without concerns about utilization of heat excess. Grid injection means that the biogas plant only needs a small combined heat and power unit for the process energy or a biogas burner.

3.3. Fuel

Anaerobic digestion (Chapter 3) is an alternate method for treating biodegradable waste to produce fuel gas and subsequently a reduced volume of waste is disposed. Biogas production has a considerable role in waste management. It is not 100% v/v greenhouse-gas-free, nevertheless, it does not contribute to global warming.

Biogas can be combusted directly to produce heat. In this case, there is no need to scrub the hydrogen sulfide in the biogas. Usually the process utilize dual-fuel burner and the conversion efficiency is 80 to 90%. The main components of the system are anaerobic digester, biogas holder, pressure switch, booster fan, solenoid valve, dual fuel burner and combustion air blower. The methane can be combusted cleanly and can provide the desired energy with limited levels of emission of carbon dioxide into the atmosphere. The carbon released from biogas can be absorbed by photosynthetic plants adding less total atmospheric carbon than the burning of fossil fuels. The use of biomethane lowers water, soil and air pollution not only because it eliminates fossil fuel related pollution but the risk of potentially devastating accidents is also remarkably reduced. As an alternate source of both heat and electricity, biomethane helps preserve forests and biodiversity by providing reduced levels of harmful greenhouse gas.

Biogas is an alternate gaseous biofuel (Borjesson and Mattiasson, 2008) which can be produced from several organic resources and waste. The most straightforward use of biogas is for thermal (heat) energy. In areas where fuels are scarce, small biogas systems can provide the heat energy for basic cooking and water heating. Gas lighting systems can also use biogas for illumination. Conventional gas burners are easily adjusted for biogas by simply changing the air-to-gas ratio. The demand for biogas quality in gas burners is low, only requiring a low gas pressure and maintaining levels of hydrogen sulfide below 100 ppm v/v to achieve a dew point of 150°C (300°F).

Biogas can be used as a vehicle fuel for cars, buses and trucks, providing it is upgraded by removing carbon dioxide, water and hydrogen sulfide. Water scrubbing, chemical scrubbing and PSA are the most widely used techniques for upgrading biogas to vehicle fuel quality. The gas must also be odorized and pressurized to approximately 3000 psi before it can be used as vehicle fuel. In fact, gasoline vehicles can use biogas as a fuel provided the biogas is upgraded to natural gas quality in vehicles that have been adjusted to using natural gas. Most vehicles in this category have been retro-fitted with a gas tank and a gas supply system in addition to the normal gasoline-fueled system. However, dedicated vehicles (using only biogas) are more efficient than these retro-fits.

Upgrading of biogas with injection into the grid or for the utilization as vehicle fuel has become increasing importance because the gas can be used in a more energy-efficient way throughout the whole year. All gas contaminants as well as carbon dioxide must be removed, and the upgraded gas must have a methane content of more than 95% v/v in order to fulfill the quality requirements of the different gas appliances.

Utilization of methane in the transport sector is widely distributed in Sweden and Switzerland. The upgraded biogas is stored at 200 to 250 bars in gas bottles. Various technologies can be applied for increasing the methane content. The most common methods of removing carbon dioxide from biogas are water scrubbing. Water scrubbing uses the higher solubility of carbon dioxide in water to separate the carbon dioxide from biogas. This process is done under high pressure and removes hydrogen sulfide as well as carbon dioxide. The main disadvantage of this process is that it requires a large volume of water that must be purified and recycled. Scrubbing with polyethylene glycol is similar to water scrubbing but is more efficient. However, the process also requires the regeneration of a large volume of polyethylene glycol.

Less frequently used are chemical washing by alkanol amines such as monoethanolamine or dimethylethanolamine as well as membrane technologies and cryogenic separation at low temperature. When removing carbon dioxide from the gas stream, small amounts of methane are also removed. These methane losses must be low level for both environmental and economic reasons since methane is a greenhouse gas 23 times stronger than carbon dioxide.

Thus, after cleaning and/or upgrading with the above-mentioned technologies (Chapter 7), the biogas (transformed into biomethane) can be used as vehicle fuel in adapted vehicles. The simplest and often most cost-effective use of biogas is its application as a fuel for boilers or industrial process (e.g., drying operations, kiln operations as well as cement and asphalt production). In these projects, the gas is piped directly to a nearby customer where it is used employing new or existing combustion equipment as a replacement or supplementary fuel. Only limited condensate removal and filtration treatment is required but some modifications of existing combustion equipment might be necessary.

The sulfurous acid formed in the condensate leads to heavy corrosion and it is recommended to use a corrosion-resistant type of stainless steel for the chimneys or condensation burners and high temperature resistant plastic chimneys. Most of the modern boilers have tin-laminated brass heat exchangers which corrode even faster than iron chimneys. Where possible, cast iron heat exchangers should be utilized. It is also advised to condense the water vapor in the raw gas. Water vapor can cause problems in the gas nozzles. Removal of water will also remove a large proportion of the hydrogen sulfide thereby reducing the potential for corrosion as well as other issues related to stack gas-dew point issues.

The energy users' energy requirements becomes paramount when the evaluation of the sale of biogas for direct use is considered. Because no economical way to store biogas exists, all the gas that is recovered must be used when available, or it is essentially lost along with associated revenue opportunities. When a plant does not have an adequate gas flow to support the entire needs of a facility, biogas can still be used to supply a portion of the needs. For example, in some facilities, only one piece of equipment (e.g., a main boiler) or set of burners are dedicated to biogas burning. These facilities might also have equipment that can use biogas along with other fuels. As biogas is typically a wet gas often containing trace corrosive compounds, the fuel train and possibly some burner "internals" should be replaced with corrosion-resistant materials. Stainless steel has typically been the material selected for that purpose.

A potential problem for boilers is the accumulation of siloxanes. The presence of siloxanes in biogas causes a white substance (similar to talcum powder) to build up on the boiler tubes. Where the material collects and the amount accumulated is likely to be a function of the velocity patterns in the boiler and the siloxane concentrations in the biogas. Operators' experiences to date indicate that annual cleaning is sufficient to avoid operational problems related to silicon oxide accumulation. Boiler operators may also choose to install a gas treatment system to reduce the amount of siloxanes in the biogas prior to delivery to the boiler.

Gas turbines represent the second most frequently used technology for biogas energy production, even though the number of installations is significantly lower than that of the ICEs. Compared with an internal combustion engine of the same size, a gas turbine features lower generation efficiency and a markedly lower power to heat ratio (cogeneration ratio). On the other hand, a gas turbine is significantly lighter (e.g., a 1 MW turbine weighs approx. 1 tonne, whereas an ICE of the same size – approx. 10 tonnes) and smaller. In a gas turbine, the only source of heat is the exhaust gas which can still be converted to useful energy.

Cogeneration projects using biogas generate both electricity and thermal energy, usually in the form of steam or hot water. Several cogeneration systems utilizing biogas have been installed at industrial operations, using both engines and turbines. In the system, electrical power is generated with the use of a gas turbine system which is comprised of the standard components: an air compressor, a combustion chamber and a gas expander. The exhaust heat from the gas turbine system is directed through a heat recovery steam generator. Cool water is also feed into the heat recovery steam generator and exchanges heat with the hot exhaust gases producing steam for useful applications.

Trigeneration is the conversion of a fuel into three useful energy products: electricity, hot water or steam, and chilled water. The system produces electrical power, steam and cooling with the use of an absorption chiller system.

Gas turbines operate based on a thermodynamic Brayton Cycle. The term "gas" refers to the atmospheric air that is taken into the engine and used as the working medium in the energy conversion process. This atmospheric air is first drawn into the engine where it is compressed, heated, and then expanded for power generation. The power produced by the expansion turbine and consumed by a compressor is proportional to the absolute temperature of the gas passing through the device.

A gas turbine that uses biogas is very similar to a natural gas turbine, except that it requires twice the number of fuel-regulating valves and injectors, which is the effect of low heating value. The biogas-based turbine systems require more inspections, cleaning and general maintenance and requires a higher level of biogas treatment for the removal of siloxanes.

Microturbines are small combustion turbines that can be used in stationary power generation applications. The basic components of a microturbine are the compressor, turbine generator, and recuperator. In a microturbine, the combustion air (inlet air) is compressed using a compressor and then it is preheated in the recuperator using heat from the turbine exhaust in order to increase overall efficiency. These are small gas turbines which can be used for power generation and cogeneration. The efficiency of these quiet operating machines range between 20 to 30% and their exhaust heat can be used to generate low-pressure steam or hot water. These turbines can function from a minimum of 35% v/v methane in the feed has and are therefore ideal for use with biogas. In addition nitrogen oxide emissions from microturbines are even lower than the conventional gas turbine system.

In biogas micro-turbines, air is pressed into a combustion chamber at high pressure and mixed with biogas. The air-biogas mixture is burned causing the temperature increase and the expanding of the gas mixture. The hot gases are released through a turbine, which is connected to the electricity generator. The heated air and biogas are burned in the combustion chamber, and the release of heat causes the expansion of the gas. The expanding gas, sent through a gas turbine, turns the generator, which then produces electricity.

Traditional gas or diesel internal-combustion engines mix fuel and air inside the cylinder and the mixture is ignited causing combustion that pushes against the piston. The Stirling engine works differently. It contains a working gas (which may be air or an inert gas such as helium or hydrogen) that is sealed inside the engine and used over and over. Rather than burning the fuel inside the cylinder, the Stirling engine uses external heat to expand the gas contained inside the cylinder. As it expands, the gas pushes against the piston. The Stirling engine then recycles the captive working gas by cooling and compressing it, then reheating it again to expand and drive the pistons which, in turn, drive a generator and produce electricity.

The Stirling engine is quite an efficient engine producing reduced emissions from combustion products since the combustion does not take place within a cylindrical chamber as in the case with conventional internal combustion engines. Instead the combustion takes place externally. The Stirling cycle engine is based on the principle of the Stirling cycle.

The cycle consists of four internally reversible processes in series; isothermal compression, constant volume heating, isothermal expansion and finally constant volume cooling. The high thermal efficiency of the engine is achieved through a regenerator employed in the process cycle. A regenerator (or heat exchanger) allows for the air exiting the compression process to be preheated before the commencement of the combustion process. Heat from the exhaust of the expansion process is used to preheat the air.

The Stirling motor operates without internal combustion, based on the principle that temperature changes of gases result in volume changes. The pistons of the engine are moved by gas expansion caused by heat injection from an external energy source. The necessary heat can be provided from various sources such as a gas burner, running on biogas. In order to use Stirling engines for biogas some technical adaption is necessary. Due to external combustion, also biogas with lower methane content can be used. Stirling motors have the advantage of being tolerant of fuel composition and quality. They are, however, relatively expensive and characterized by low efficiency and the use of these motors may be limited to a number of very specific applications.

An alternative way of complementing the above list of technologies can be the application of a technology to convert heat energy into electricity using a system built on the basis of the organic Rankine cycle (ORC) process, which is not significantly different from the traditional Rankine cycle (used for steam turbine power plants) with water circulation. The only difference is the working fluid – the ORC process utilizes an organic fluid of high molecular mass, rather than water.

The utilization of biogas as vehicle fuel uses the same engine and vehicle configuration as natural gas. In total there are more than 1 million natural gas vehicles all over the world, this demonstrates that the vehicle configuration is not a problem for use of biogas as vehicle fuel. However, the gas quality demands are strict. With respect to these demands the raw biogas from a digester or a landfill has to be upgraded. Through upgrading a gas can be obtained which: (i) has a higher calorific value in order to reach longer driving distances, (ii) has a regular/constant gas quality to obtain safe driving, (iii) does not enhance corrosion due to high levels of hydrogen sulfide, ammonia and water, (iv) does not contain mechanically damaging particles, (v) does not give ice-clogging due to a high water content, and (vi) has a declared and assured quality.

Finally, the most straightforward application of product gas is co-firing in existing coal power plants by injecting the product gas in the combustion zone of the coal boiler. Co-firing percentages up to 10% (on energy basis) are feasible without the need for substantial modifications of the coal boiler. Critical issue in co-firing is the impact of the biomass ash on the quality of the boiler fly and bottom ash. The application of the fly and bottom ash in construction and cement production often sets the specifications for the amount and type of biomass that can be co-fired.

3.4. Fuel Cells

A fuel cell is an electrochemical device that relies on a continuous feed of a fuel, usually hydrogen, to produce direct current (DC) electricity and heat. The chemical energy of a reaction is directly converted into electrical energy and, as a result, high electrical efficiency can be achieved. At the anode hydrogen, methane, or methanol is oxidized, while at the cathode oxygen in air, oxygen, or carbon dioxide is reduced.

Fuel cells are considered the small-scale power plants of the future for production of power and heat with efficiencies exceeding 60% and low emissions. One of the largest digester/fuel cell units is located in Washington State. The fuel cell, located at the South Treatment Plant in Renton, WA, can consume about 154,000 ft^3 of biogas a day to produce up to 1 megawatt (1,000,000 watts) of electricity. That's enough to power 1,000 households, but it's being used instead for the operation of the plant.

Thus, fuel cells are power generating systems that produce DC electricity by combining fuel and oxygen (from the air) in an electrochemical reaction. There is no intermediate process which first converts fuel into mechanical energy and heat. Therefore fuel cells have extremely low emissions. The reaction is similar to a battery however, fuel cells do not store the energy with chemicals internally.

The application of product gas in fuel cells for the production of electricity is still in its early development. In theory, fuel cells have the potential to achieve higher electrical efficiencies compared to simple combustion systems and gas engines. However, in practice this is only relevant on small-scale where combustion/steam-cycles suffer from relatively high losses. A combination that is often mentioned for its extremely high electric efficiency, is the fuel cell downstream a gasifier functioning as the combustion chamber for a (micro) gas turbine. This implies that the fuel cell must operate under pressure. In a simpler system, the fuel cell can be connected to a gasifier to both have a high electric efficiency and a high overall efficiency in a combined heat and power system.

A fuel cell burns hydrogen and produces electricity directly through one or more electrochemical reactions. Depending on the type of fuel cell, also carbon monoxide, hydrogen, methane, and other fuels can be converted. A solid oxide fuel cell can handle all these molecules and therefore is generally considered a good match with product gases from low temperature biomass gasification. Furthermore, methane (and other hydrocarbons) in product gas can increase the performance of a solid oxide fuel cell because it is reformed into hydrogen in the anode of the fuel cell. Since reforming is an endothermal reaction, it provides effective cooling of the fuel cell, which should otherwise be done in a way that generally increases complexity and costs and decreases overall efficiency.

In a first step the fuel is transformed into hydrogen either by a catalytic steam reforming conversion or by a (platinum) catalyst. The hydrogen is converted to direct electrical current and the by-products of the reaction are water and carbon dioxide. Fuel cells demonstrate relatively constant efficiencies over a wide range of loads. There are five types of fuel cells, classified by the type of electrolyte used in the fuel cell: alkaline fuel cell (AFC), phosphoric acid fuel cell (PAFC), molten carbonate fuel cell (MCFC), solid oxide fuel cell (SOFC), and proton exchange membrane fuel cell (PEM) (Carrette et al., 2000; St-Pierre and Wilkinson, 2001; Van Herle et al., 2004; Barclay, 2006).

The *polymer-electrolyte-membrane* fuel cell can be used for biogas. Due to operating temperature of 80°C, the heat can be fed directly into a heat/warm water network. The type of electrolyte used influence the service life of PEM, which is very sensitive to impurities in the fuel gas, including carbon dioxide. For this reason, gas cleaning is very important.

The *phosphoric acid fuel cell*, frequently used with natural gas worldwide. Compared to other fuel cells, the electrical efficiency is low but the advantage is that PAFC is less sensitive to the presence of carbon dioxide and carbon monoxide in the gas. The *molten carbonate fuel cell* uses a fluid carbon flow as electrolyte.

Molten carbonate fuel cells operate at relative high temperatures (650°C, 1200°F) which enables the molten carbonate fuel cells to operate with higher impurities concentration than low temperature fuel cells. In the molten carbonate fuel cells, no combustion occurs, and thus, the concentrations of nitrogen oxides and carbon monoxide are very low. The following are the reactions which take place within the molten carbonate fuel cell are:

Anode Reaction:
$2CO_3^{2-} + 2H_2 \rightarrow 2H_2O + 2CO_2 + 4e^-$

Cathode Reaction:
$2CO_2^- + O_2 + 4e^- \rightarrow 2CO_3^{2-}$

Overall Reaction:
$2CO_2 + O_2 + 4H_2 \rightarrow 2H_2O + 2CO_2$

Pollutants are produced during the fuel treatment, in particular for the steam reforming of the methane embedded in the biogas. The molten carbonate fuel cell is insensitive to carbon monoxide and tolerates carbon dioxide concentrations up to 40% v/v. Due to its operation temperature of 600 to 700°C (1110 to 1290°F), conversion of methane into hydrogen, also called reforming, can take place inside the cell. Its dissipated heat can for example be used in a downstream turbine.

The *solid oxide fuel cell* is another type of high-temperature fuel cell, operating at 750 to 1000°C (1380 to 1830°F).

The solid oxide fuel cell is a large capacity fuel cell capable of providing as much as 50 MW or more of electrical power. These cells can operate with temperatures in the range of 800 to 1000°C (1470 to 1830°F) with an electric efficiency on the order of 45% to 60%. The chemical reactions which takes place within the fuel cell are:

Anode Reaction:
$2H_2 + 2O^{2-} \rightarrow 2H_2O + 4e^-$
$2CO + 2O^{2-} \rightarrow 2CO_2 + 4e^-$

Cathode Reaction:
$2O_2 + 4e^- \rightarrow 2O^{2-}$

Overall Reaction
$2H_2 + O_2 \rightarrow 2H_2O$
$2CO + O_2 \rightarrow 2CO_2$

The solid oxide fuel cell has a high electrical efficiency and the reforming of methane to hydrogen can take place within the cell. The use of biogas is suited due to its low sensitivity to sulfur. The solid oxide fuel cell has been used with biogas, with carbon dioxide as the reforming agent, and this will allow the partial conversion of methane into syngas at the anodes of the solid oxide fuel cell (Rasmussen and Hagen, A. 2010). Therefore, the feeding of biogas directly into the solid oxide fuel cell avoids the need for an external reform step. The presence of impurities such as hydrogen sulfide in the fuel is a critical issue as is the presence of other impurities, even in trace amounts (Rasi et al., 2007, 2011). The maximum hydrogen sulfide concentration for a 24 hour period in the solid oxide fuel cell is below 7 to 9 ppm v/v. However, under some circumstances that are process specific, inhibition by hydrogen sulfide can occur at concentrations on the order of 1 ppm v/v hydrogen sulfide.

3.5. Synthetic Natural Gas

Whereas high-temperature gasification processes yield bio-synthesis gas with high concentrations of carbon monoxide and little methane, interest in synthetic natural gas (SNG) production is concentrated on gasification processes that yield product gases with high methane contents. Synthetic natural gas is a gas with similar properties as natural gas but produced by methanation of hydrogen and carbon monoxide in the product gas from a gasification unit. Methanation is the catalytic reaction of carbon monoxide and/or carbon dioxide with hydrogen, forming methane and water:

$CO + 3H_2 \rightarrow CH_4 + H_2O$

The key process step for the production of synthetic natural gas from biomass is the gasification unit (Chapter 6). In the early nineteenth century, coal gasification was used for the production of so-called town gas for street lamps. Later, gasification was used during World War II to produce liquid fuels from coal, and the process underwent a revival during the oil crises of the 1970s and 1980s. The choice of gasifier used and the choice of gasification medium depend to a large extent on both the feedstock properties and the potential application of the resulting product gas/syngas. For fluidized bed gasification, for example, the maximum operating temperature is limited by the melting temperature of the mineral ash, since higher temperatures would cause sintering of the bed.

Consecutive and side reactions (such as the shift conversion and the hydrogenation of carbon) make the calculation of equilibrium conditions very complex. The methanation reactions of both carbon monoxide and carbon dioxide are highly exothermic. Such high heat releases strongly affect the process design of the methanation plant since it is necessary to prevent excessively high temperatures in order to avoid catalyst deactivation and carbon deposition. The highly exothermic reaction generally creates a problem for the design of methane synthesis plants: either the temperature increase must be limited by recycling of reacted gas or steam dilution, or special techniques such as isothermal reactors or fluidized beds, each with indirect cooling by evaporating water, must be used.

The methanation process is a very well-known and technically important as a catalytic purification step for the removal of trace carbon monoxide (typically below several % v/v) from process gases, especially for hydrogen production. Methanation of gases with a high content of carbon monoxide, such as the product gas from a gasification process, is not well established and there are no commercial catalysts available (i.e., the Dakota coal-to-synthetic natural gas plant uses a non-commercial catalyst).

For the removal of carbon monoxide from synthesis gases, catalysts with usually <15% w/w nickel are used predominantly. For synthetic natural gas production, catalysts with high nickel content are preferred, similar to those used in reforming naphtha to a methane-rich gas. Catalysts based on ruthenium have been tried repeatedly for methanation but have not found the broad application of nickel-based catalysts. Catalytic activity is affected seriously even by very low concentrations of catalyst poisons in gases to be reformed. Such catalyst poisons are sulfur, arsenic, copper, vanadium, lead, and chlorine or halogens in general. Precaution must be taken with nickel-containing catalysts to prevent formation of highly poisonous nickel carbonyl [$Ni(CO)_4$]. In practical operation of methanation plants, temperatures below 200°C (309°F) at the nickel catalyst must be avoided.

3.6. Fischer-Tropsch Synthesis

In a previous chapter (Chapter 6), there has been mention of the use of the gasification process to convert carbonaceous feedstocks, such as crude oil residua, tar sand bitumen, coal, oil shale, and biomass, into the starting chemicals for the production of petrochemicals. This is one of the potential processes where biogas may be the best fit leading to a variety of products.

Thus, another application of bio-synthesis gas is found in catalytic synthesis processes (such as the Fischer-Tropsch process). However, the catalysts for this process are very sensitive to small amounts of impurities in the gas stream and, in general, the required gas purity specifications depend on economic considerations, i.e., additional investments in gas cleaning versus accepting decreasing production due to poisoning of catalysts or higher maintenance intervals due to fouling of the system.

The chemistry of the gasification process is based on the thermal decomposition of the feedstock and the reaction of the feedstock carbon and other pyrolysis products with oxygen, water, and fuel gases such as methane and is represented by a sequence of simple chemical reactions (Table 10.1). However, the gasification process is often considered to involve two distinct chemical stages: (i) devolatilization of the feedstock to produce volatile matter and char, (ii) followed by char gasification, which is complex and specific to the conditions of the reaction – both processes contribute to the complex kinetics of the gasification process (Sundaresan and Amundson, 1978).

The Fischer-Tropsch process is a catalytic chemical reaction in which carbon monoxide (CO) and hydrogen (H_2) in the synthesis are converted into hydrocarbon derivatives of various molecular weights. The process can be represented by the simple equation:

$$(2n+1)H_2 + nCO \rightarrow C_nH_{(2n+2)} + nH_2O$$

In this equation, n is an integer. Thus, for n = 1, the reaction represents the formation of methane, which in most gas-to-liquids (GTL) applications is considered an undesirable byproduct.

The Fischer-Tropsch process conditions are usually chosen to maximize the formation of higher molecular weight hydrocarbon liquid fuels which are higher value products. There are other side reactions taking place in the process, among which the water-gas-shift reaction is predominant:

$$CO + H_2O \rightarrow H_2 + CO_2$$

Depending on the catalyst, temperature, and type of process employed, hydrocarbon derivatives ranging from methane to higher molecular paraffin derivatives and olefin derivatives can be obtained. Small amounts of low molecular weight oxygenated derivatives (such as alcohol derivatives and organic acid derivatives) are also formed. Typically, Fischer-Tropsch liquids are (unless the process is designed for the production of other products) hydrocarbon products (which vary from naphtha-type liquids to wax) and non-hydrocarbon products. The production of non-hydrocarbon products requires adjustment of the feedstock composition and the process parameters.

In the non-selective catalytic Fischer-Tropsch synthesis one mole of carbon monoxide reacts with two moles of hydrogen to form mainly paraffin straight-chain hydrocarbons (C_nH_{2n}) with minor amounts of branched paraffin derivatives (such as 2-methyl paraffin derivatives), unsaturated hydrocarbon derivatives (such as α-olefins), and primary alcohol derivatives. Undesirable side reactions include methanation, the Boudouard reaction, coke deposition, oxidation of the catalyst, or carbide formation.

Briefly, the Boudouard reaction is the redox reaction of a chemical mixture of carbon monoxide and carbon dioxide at a given temperature and is the disproportionation of carbon monoxide into carbon dioxide and carbon, or the reverse reaction (i.e., the reaction of carbon dioxide and carbon to produce carbon monoxide):

$$2CO \rightleftharpoons CO_2 + C$$

Typical operation conditions for Fischer-Tropsch synthesis are temperatures of 200 to 350°C (390 to 660°F) and pressures between 375 and 900 psi. In the exothermic Fischer-Tropsch reaction, approximately 20% of the chemical energy is released as heat. The process is represented simply as:

$$nCO + (2n+1)H_2 \rightarrow H[\text{-}CH_2\text{-}]_nH + nH_2O$$

Fischer-Tropsch processes can be used to produce either a light synthetic crude oil (syncrude) and light olefins or heavy waxy hydrocarbons. The syncrude can be refined to a high quality sulfur and aromatic liquid product and specialty waxes or, if hydrocracked and/or isomerized, to produce excellent diesel fuel, lube oils, and naphtha, which is an ideal feedstock for cracking to olefins. For direct production of gasoline and light olefins, the Fischer-Tropsch process is operated at high temperature (330–350°C), for production of waxes and/or diesel fuel, at low temperatures (220 to 250°C, 430 to 480°F).

Several types of catalysts can be used for the Fischer-Tropsch synthesis - the most important are based on iron (Fe) or cobalt (Co). Cobalt catalysts have the advantage of a higher conversion rate and a longer life (over five years). The Co catalysts are in general more reactive for hydrogenation and produce therefore less unsaturated hydrocarbons (olefins) and alcohols compared to iron catalysts. Iron catalysts have a higher tolerance for sulfur, are cheaper, and produce more olefin products and alcohols. The lifetime of the Fe catalysts is short and in commercial installations generally limited to eight weeks [23].

Catalysts for Fischer-Tropsch synthesis can be damaged with impurities as ammonia, hydrogen cyanide, hydrogen sulfide, and carbonyl sulfide. These impurities poison the catalysts and the presence of hydrogen chloride causes corrosion of catalysts and equipment. Also alkaline metals are deposited on the catalyst and tar constituents are deposited on the catalyst which cause poisoning of catalyst and contaminate the products. Particles (dust, soot, ash) cause fouling of the reactor. The removal limit is based on an economic optimum determined by catalyst stand-time and investment in gas cleaning but, generally, all of the impurities should be removed to a concentration below 1 ppm v/v.

The other products that are worthy of consideration is bio-oil (pyrolysis oil, bio-crude) and biochar. Bio-oil is the liquid product produced by the thermal decomposition (pyrolysis or destructive distillation) of biomass (Chapter 5) at temperatures on the order of 500°C (930°F). The product can vary from a light tarry material to a free-flowing liquid – both require further refining to produce specification grade fuels (Parkash, 2003; Gary et al., 2007; Speight, 2014; Hsu and Robinson, 2017; Speight, 2017). Also, depending upon the circumstances, the oil is also potentially of interest as a feedstock for a gasification process.

Biochar is the charcoal-like (carbonaceous) residue that is produced das a result of the pyrolysis of biomass (Chapter 5). It is created using a pyrolysis process in which biomass is heated in a low oxygen environment. Once the pyrolysis reaction has begun, it is self-sustaining, requiring no outside energy input. Byproducts of the process include gas (variable yield and composition, depending on the feedstock and the process parameters) and bio-oil (variable yield and composition, depending on the constituents of the feedstock and the process parameters. The biochar is often presented as a beneficial product to enhance the anaerobic digestion process (Chapter 3) and for use as a fertilizer, it is also eminently suitable (when biogas is the desired product) to be used as a feedstock for gasification to produce bio-synthesis gas (Chapter 6).

3.7. Hydrogen Production

Hydrogen is not only a necessity in crude oil refineries (Parkash, 2003; Gary et al., 2007; Speight, 2014, 2016; Hsu and Robinson, 2017; Speight, 2017) but is also an important alternative energy vector and a bridge to a sustainable energy future. Hydrogen is not an energy source. Hydrogen is not a primary source of energy that exists freely in nature but it is a secondary form of energy that has to be manufactured like electricity.

Hydrogen can be produced from a wide variety of primary energy sources and different production technologies. Approximately half of all the hydrogen as currently produced is obtained from thermo-catalytic and gasification processes using natural gas as a starting material, heavy oils and naphtha make up the next largest source, followed by coal. As an alternate source of hydrogen, biomass can be considered as the best option and has the largest potential, which meets energy requirements and could insure fuel supply in the future. Biomass and biomass-derived fuels can be used to produce hydrogen sustainably and the gasification of biomass (Chapter 6) offers the earliest and most economical route for the production of renewable hydrogen (Balat and Kirtay, 2010).

Hydrogen can be produced from a wide variety of primary energy sources and different production technologies. Most hydrogen is currently produced from nonrenewable sources such as oil, natural gas, and coal. About half of all the hydrogen as currently produced is obtained from thermocatalytic and gasification processes using natural gas as a starting material, heavy oils and naphtha make up the next largest source (Parkash, 2003; Gary et al., 2007; Speight, 2014, 2016; Hsu and Robinson, 2017; Speight, 2017), followed by coal (Balat and Kirtay, 2010; Speight, 2013,) and only 4% and 1% is generated from water using electricity and biomass, respectively (Mohan et al., 2007; Das et al., 2008).

Gasification of biomass has been identified as a possible system for producing renewable hydrogen, which is beneficial to exploit biomass resources, to develop a highly efficient clean way for large-scale hydrogen production. Biomass gasification can be considered as a form of pyrolysis, which takes place in higher temperatures and produces a mixture of gases with a hydrogen content ranging 6 to 6.5% v/v. The synthetic gas produced by the gasification of biomass is made up of hydrogen, carbon monoxide, methane, carbon dioxide, oxygen, nitrogen and tar constituents. When gasifying biomass, tar that is formed together with the synthetic gas is difficult to remove using a physical dust removal method (Chapter 7). The product distribution and gas composition depends on many factors including the gasification temperature and the reactor type. The most important gasifier types are fixed bed (updraft or downdraft fixed beds), fluidized bed, and entrained flow gasifiers. All of these gasifiers should also include significant gas conditioning along with the removal of tar constituents and inorganic impurities as well as the subsequent removal of carbon monoxide and the production of hydrogen by water gas shift reaction (Demirbaş, 2008; Yoon et al., 2010):

$$CO + H_2O \rightleftharpoons CO_2 + H_2$$

Hydrogen can by produced by a reforming process, such as methane reforming and, as a result of the composition (i.e., the methane content) of biogas, a dry reforming can be used for hydrogen production. The product of dry reforming is synthesis gas, which is a mixture carbon monoxide and hydrogen with a low carbon monoxide-hydrogen ratio:

$$CH_4 + CO_2 \rightarrow 2CO + 2H_2$$

A wide variety of supported metal catalysts can be used. Metal catalysts (Co, Mo, Cu Ni) provide high carbon dioxide conversion at temperatures over 700°C (1290°F), while supported metals (Pd, Pt, Rh) allow the process to be carried out at lower temperatures (500 to 700°C, 930 to 1290°F).

In the direct reforming process other reactions also occur and these include steam methane reforming and the water-gas shift reaction:

$$CH4 + H_2O \rightarrow CO + 3H_2 \quad \text{(Direct reforming reaction)}$$
$$CO + H_2O \rightarrow CO_2 + H_2 \quad \text{(Water gas shift reaction)}$$

The steam-methane reforming process reduces the production of carbon on the catalysts and the water has shift reaction is a means of produce higher yields of hydrogen. Catalytic preferential oxidation of carbon monoxide (CO-PROX process) is one of the most suitable methods for carbon monoxide removal (Mishra and Prasad, 2011):

$$2CO + O_2 \rightarrow 2CO_2$$
$$2H_2 + O_2 \rightarrow 2H_2O$$

In this process, carbon monoxide is transformed into carbon dioxide in the first reaction up to a carbon monoxide concentration of less than 10 ppm v/v. The second reaction is undesirable due to the consumption of hydrogen. The process is highly exothermic and temperature control is essential to avoid hydrogen consumption. Finally, the carbon dioxide must be removed by conventional techniques (such as by pressure swing adsorption or by a membrane-based process).

3.8. Other Uses

As the interest in biogas increases, there are other uses for which the gas can be introduced into the market place.

3.8.1. Fertilizer and Soil Conditioner

While there are suitable inorganic substitutes for the nutrients nitrogen, potassium and phosphorous from organic fertilizer, there is no artificial substitute for other substances such as protein, cellulose, and lignin which contribute to increasing the permeability and the hygroscopic nature of the soil while preventing erosion and improving agricultural conditions in general. Organic substances also constitute the basis for the development of the microorganisms responsible for converting soil nutrients into a form that can be readily incorporated by plants (see also Section 3.8.3).

The solid, fibrous component of the digested material can be used as a soil conditioner to increase the organic content of soils. Digester liquor can be used as a fertilizer to supply vital nutrients to soils instead of chemical fertilizers that require large amounts of energy to produce and transport. The use of manufactured fertilizers is, therefore, more carbon-intensive than the use of anaerobic digester liquor fertilizer. In countries such as Spain, where many soils are organically depleted, the markets for the digested solids can be equally as important as the biogas.

Also, due to the decomposition and breakdown of parts of its organic content, digested sludge (sometimes referred to as digestate) provides fast-acting nutrients that easily enter into the soil solution, thus becoming immediately available to the plants. They simultaneously serve as primary nutrients for the development of soil organisms, such as the replenishment of microorganisms lost through exposure to air in the course of spreading the sludge over the fields.

The humic matter and humic acids present in the sludge contribute to a more rapid humification, which in turn helps reduce the rate of erosion (due to rain and dry scatter) while increasing the nutrient supply of the soil. The humic content is especially important in low-humus tropical soils. The relatively high proportion of stable organic building blocks such as lignin and certain cellulose compounds contributes to an unusually high formation rate of stable humus (particularly in the presence of argillaceous matter).

The amount of stable humus formed with digested sludge amounts to twice the amount that can be achieved with decayed manure. It has also been shown that earthworm activity is stimulated more by fertilizing with sludge than with barnyard manure. Digested sludge decelerated the irreversible bonding of soil nutrients with the aid of its ion-exchanger contents in combination with the formation of organo-mineral compounds. At the same time, the buffering capacity of the soil increases, and temperature fluctuations are better compensated.

In addition, the elevated ammonium content of digested sludge helps reduce the rate of nitrogen washout as compared to fertilizers containing substantial amounts of more water-soluble nitrate derivatives and nitrite derivatives (manure, compost). Soil nitrogen in nitrate or nitrite form is also subject to higher nitrogen losses than is ammonium, which first requires nitrification in order to assume a nitrogen form. It takes longer for ammonium to seep into deeper soil strata, in part because it is more easily adsorbed by argillaceous bonds.

However, some of the ammonium becomes fixed in a non-interchangeable form in the intermediate layers of clay minerals. The nitrogen-efficiency of digested sludge may be regarded as comparable to that of chemical fertilizers. In addition to supplying nutrients, sludge also improves soil quality by providing organic mass. The porosity, pore-size distribution and stability of soil aggregates are becoming increasingly important as standards of evaluation in soil-quality analyses.

3.8.2. Cooking Gas

The most common method for utilization of biogas in developing countries is for cooking and lighting, especially in rural areas. The availability of animal residues for biogas generation gives a viable alternative for cooking, lighting fuel and a useful fertilizer. Also, biogas technology is gaining additional motivation through various subsidy programs for market incentive and development of renewable energies.

As an example, the burning of manure and plant residue is a considerable waste of plant nutrients since the farmers in these countries need fertilizer in order to maintain cropland productivity. Nonetheless, many small farmers continue to burn potentially valuable fertilizers, even though they cannot afford to buy chemical fertilizers. At the same time, the amount of technically available nitrogen, potassium and phosphorous in the form of organic materials is around eight times as high as the quantity of chemical fertilizers actually consumed in developing countries. Especially for small farmers, biogas technology is a suitable tool for making maximum use of scarce resources since, after extraction of the energy content of manure and other organic waste material, the resulting sludge remains as a fertilizer, which supports general soil quality as well as higher crop yields.

Conventional gas burners and gas lamps can easily be adjusted to biogas by changing the air to gas ratio. Burning biogas in a boiler is an established and reliable technology. Low demands are set on the biogas quality for this application. Pressure usually has to be on the order of 120 to 375 psi and, furthermore it is recommended to reduce the level of hydrogen sulfide to less than 1000 ppm v/v, which allows the dew point to be maintained at approximately 150°C 300°F).

In these rural areas, by using a bio-digester, which produces the bacteria required for decomposing, cooking gas is generated. The organic garbage such as fallen leaves, kitchen waste, and food waste are fed into a crusher unit, where the mixture is conflated with a small amount of water. The mixture is then fed into the bio-digester, where the bacteria decomposes it to produce cooking gas which is then piped to a kitchen stove. A two cubic meter bio-digester can produce 2 cubic meter of cooking gas and the most notable advantage of using a bio-digester is the sludge (often referred to as the digestate), which is a rich organic manure that can be used as fertilizer.

3.8.3. Use of Digestate

Agricultural biogas production is an integrated element of modern agriculture, which takes into consideration not only economic costs and benefits of agricultural activities, but also socio-economic and environmental benefits. Agricultural biogas production provides intertwined agricultural, economic and environmental benefits and for this reason, the promoters of the biogas development in Europe, after the oil crisis, were the organic farmers, interested in anaerobic digestion not only for renewable energy generation, but as a way to improve fertilizer quality of the animal manure (see also Section 3.8.1).

Because of the increased availability of nitrogen, digestate can be integrated in the fertilization plant of the farm, as it is possible to calculate its fertilizer effects in the same way as for mineral fertilizers. Digestate has lower atomic carbon/nitrogen ratio, compared to raw manure. The lower atomic carbon/nitrogen ratio means that the digestate has a better short term nitrogen-fertilization effect. When the value of the atomic carbon/nitrogen ratio is high, micro-organisms take hold in the soil, as they successfully compete with the plant roots for the available nitrogen.

Digestate is more homogenous, compared to raw slurry, with an improved N-P balance. It has a declared content of plant nutrients, allowing accurate dosage and integration in fertilization plans of farms. Digestate contains more inorganic nitrogen, easier accessible to the plants, than untreated slurry. N-efficiency will increase considerably and nutrient losses by leaching and evaporation will be minimized if digestate is used as fertilizer in conformity with good agricultural practice. For optimum utilization of digestate as fertilizer, the same practice criteria are valid, like in the case of utilization of untreated slurry and manure.

Due to its higher homogeneity and flow properties, digestate penetrates in soil faster than raw slurry. Nevertheless, application of digestate as fertilizer involves risks of nitrogen losses through ammonia emissions and nitrate leaking. Depending on the crop, experience shows that, in Europe, the best time for digestate application is during vigorous vegetative growth. Application as top-fertilizer on crops in full vegetation offers little concern about loss of e.g., nitrogen as nitrate into ground water, since the main part is absorbed immediately by the plants. Danish experience shows that by application of digestate as top-fertilizer, a part of nutrients are even absorbed through the leaves.

Digestate has a high water content and consequently high volume. Thus, conditioning of digestate is focused on the reduction of the volume and to concentrate the nutrients. This is particularly important if digestate has to be transported away from the areas where there is an excess of nutrients from animal manure but not sufficient land available for their application. The nutrients in excess must be transported to other areas in an economic and efficient way. Digestate conditioning aims to reduce volume and by this the nutrient transportation costs as well as to reduce emissions of pollutants and objectionable odors.

The use of the digestate as organic fertilizer can affect sustainability in a variety of ways. It can substitute mineral fertilizers and thereby reduce the environmental and carbon footprint and costs of fertilizer application. Generally, the composition of the digestate depends on the feedstock and the relative composition of nitrogen (N), phosphorous (P), potassium (P) and sulfur (S) can vary significantly. Negative environmental effects of digestate application include nitrogen losses through nitrous oxide and ammonia emissions and washout into the groundwater as nitrate (NO_3^-), as well as methane emissions, but these effects can be minimized by appropriate storage and application. Digestate from slurry should have higher fertilizer value (through mineralisation and availability of nutrients) than undigested slurries and reduce the need for fossil fertilizer. Digestate may be seen as a decarbonized fertilizer.

4. THE FUTURE

Biogas is considered to be a renewable resource because its production-and-use cycle is continuous, and it generates no net carbon dioxide. Organic material grows, is converted and used and then regrows continuously in a repeating cycle. As much carbon dioxide is absorbed from the atmosphere in the growth of the primary bio-resource as is released when the material is completely converted into energy. When carbon dioxide and other minor constituents are confiscated, the product remaining is a purified pipeline-quality natural gas bordering on 100% v/v methane. This biomethane is interchangeable with any auxiliary natural gas, but it is a zero-carbon resource. Integrating this zero-carbon profile with the ultra-low traditional emissions of natural gas makes renewable natural gas a nearly perfect fuel. It is storable, it brags high energy density that is almost emission-free, and it is easily transported over existing infrastructure to serve any natural gas application.

Biogas is copious and is available from sources such as landfills, wastewater treatment facilities, and animal and agricultural waste. If fully consumed, the yield from existing organic waste streams could satisfy about 20% v/v of current natural gas use. Biogas feedstock may also soon be farmed economically. Work is already in progress to diversify low-cost energy crops such as algae and other plant species that could be grown on marginal land to serve as a source of biogas production. Future energy crops could make the potential availability practically limitless. With regard to biomass feedstock optimization as well as the entire anaerobic digestion process there have been several remarkable technological advancements.

One of the most promising concepts is the treatment of the liquid fraction on the farm-site in an upflow anaerobic sludge blanket (UASB) reactor (Chapter 3) while the solid fraction is transported to the centralized biogas plant where wet-oxidation can be implemented to increase the biogas yield of the fiber fraction. Integration of the wet oxidation pre-treatment of the solid fraction leads to a high degradation efficiency of the lignocellulosic solid fraction.

Biogas is currently applied as a heating and electricity fuel but is expected to find more advanced applications as a vehicular fuel gradually. The developing technology should not be viewed as a competition to the already existing conventional energy sources but rather a compliment to what is already existing and a sustainable environmental management scheme for the future. Additionally, to improve the awareness of a sustainable fuel, more demonstration plants should be set up.

The current rate of biogas production specifies that these technologies are going to have major impact on the energy consumption in future. The impact includes reduced release of contaminants to the environment ensures that the battle against global warming. Uncomplicated handling of waste that are generated through agricultural activities will make economic feasibility and spent slurry from biogas production can be used as fertilizer for agricultural crops. In this review paper, the modern research associated with biogas production has been presented. Even though many researchers were contributed to biogas production, there are some gaps that need to be investigated.

In future investigations, analyses of synchronized process of co-digestion with substrates are needed to be executed and also researchers need to concentrate on multi stage anaerobic co-digestion with reduced cost through the selection of appropriate expertise. Purification of biogas is the major task behind the fixation of cost level, which should be analyzed in detail with methods like cryogenic upgradation and In situ methane enrichment (Abatzoglou and Boivin, 2009). Further, the developing methods for upgrading and refinement of the produced biogas will receive major attention due to rapid increment in the price of fossil fuels.

Biogas production in the agricultural sector is a very fast growing market in Europe and finds increased interest in many parts of the world. In the next few decades, bioenergy will be the most significant renewable energy source, because it offers an economical attractive alternative to fossil fuels. The success of biogas production will come from the availability at low costs and the broad variety of usable forms of biogas for the production of heat, steam, electricity, and hydrogen and for the utilization as a vehicle fuel. Many sources, such as crops, grasses, leaves, manure, fruit, and vegetable wastes or algae can be use, and the process can be applied in small and large scales. This allows the production of biogas at any place in the world.

For an increased dissemination of biogas plants, further improvements of the process efficiency, and the development of new technologies for mixing, process monitoring, and process control are necessary. Furthermore, the influence of the microbial community structure on process stability and biogas yield requires further efforts and must be analyzed in more detail. Recent research results have demonstrated that strong variations in the community structures occur during the ongoing fermentation process which influences the process efficiency. Molecular analyses have shown the presence of numerous recently unknown bacteria which may have an important influence on the degradation process. A major potential for increasing the biogas yield has also the pretreatment of substrates and the addition of micronutrients.

Important for the future is also a better process control. Today, there are only few sensors available that are sufficiently robust to monitor online. With the increasing number of biogas plants, also an improvement of the effluent quality is necessary, in order to avoid a contamination of ground water with pathogens and nutrients.

The end utilization of the biogas can influence sustainability in a number of ways, depending on whether the gas is used directly in boilers for heat or in an on-site combined heat and power to generate electricity and heat, or if it is upgraded for use as a vehicle fuel or in an off-site combined heat and power system. The optimum use in terms of environmental sustainability and greenhouse gas emissions can be achieved by replacing fossil fuels, such as coal or diesel.

Finally, in order to improve the economic benefit of biogas production, the future trend will go to integrated concepts of different conversion processes, where biogas production will still be a significant part. In a so-called biorefinery concept (Speight, 2011), close to 100% w/w of the biomass is converted into energy or valuable by-products, making the whole concept more economically profitable and increasing the value in terms of sustainability.

REFERENCES

Abatzoglou, N. and Boivin, S. 2009. A Review of Biogas Purification Processes. *Biofuels Bioprod. Biorefin.*, 3(1): 42-71.

Balat, H., and Kirtay, E. 2010. Hydrogen from Biomass – Present Scenario and Future Prospects. *International Journal of Hydrogen Energy,* 35: 7416-7426.

Barclay, F. J. 2006. *Fuel Cells, Engines and Hydrogen: An Exergy Approach.* John Wiley and Sons Inc. Hoboken, New Jersey.

Borjesson, P. and Mattiasson, B. 2008. Biogas as a Resource-efficient Vehicle Fuel. *Trends Biotechnol.*, 26(1): 7-13.

Carrette, L., Friedrich, K. A. and Stimming, U. 2000. Fuel cells: Principles, Types, Fuels and Applications. *Chem. Phys. Chem.*, 1(4): 163-193.

Demirbaş, A. 2008. Hydrogen production from carbonaceous solid wastes by steam reforming. *Energy Sources Part A,* 30: 924-931.

Gary, J. G., Handwerk, G. E., and Kaiser, M. J. 2007. *Crude oil Refining: Technology and Economics,* 5th Edition. CRC Press, Taylor & Francis Group, Boca Raton, Florida.

Hsu, C. S., and Robinson, P. R. (Editors). 2017. *Handbook of Petroleum Technology.* Springer International Publishing AG, Cham, Switzerland.

Köppel, W., Götz, M. and Graf, F. 2009. *Biogas Upgrading for Injection into the Gas Grid.* GWF, Gas – Erdgas, 150(13): 26-35.

Mishra, A. and Prasad, R. 2011. A Review on the Preferential Oxidation of Carbon Monoxide in Hydrogen Rich Gases. *Bull. Chem. React. Eng. Catal.,* 6(1): 1-14.

Niemczewska, J. 2012. *Characteristics of Utilization of Biogas Technology.* Nafta-Gaz, LXVIII(5): 293-297.

Parkash, S. 2003. *Refining Processes Handbook.* Gulf Professional Publishing, Elsevier, Amsterdam, Netherlands.

Poschl M., Ward, S., Owende, P. 2010. Evaluation of Various Bio-gas Production and Utilization Pathways. *Applied Energy*, 87: 3305- 3321.

Pruthvira, N. B. 2016. Introduction to Biogas & Applications. *International Journal of Advanced Research in Mechanical Engineering & Technology (IJARMET),* 2(4): 7-13. http://ijarmet.com/wp-content/themes/felicity/issues/vol2issue4/pruthviraj.pdf.

Rasi, S., Veijanen, A. and Rintala, J. 2007. Trace Compounds of Biogas from Different Biogas Production Plants. *Energy,* 32(8): 1375-1380.

Rasi, S., Läntelä, J. and Rintala, J. 2011. Trace Compounds Affecting Biogas Energy Utilization – A Review. *Energy Convers. Manage.,* 52(12): 3369-3375.

Rasmussen, J. F. B. and Hagen, A. 2010. The Effect of H2S on the Performance of SOFCs using methane containing fuel. *Fuel Cells,* 10(6):1135-1142.

Speight, J. G. 2011. (Editor). *The Biofuels Handbook.* Royal Society of Chemistry, London, United Kingdom.

Speight, J. G. 2013. *The Chemistry and Technology of Coal* 3rd Edition. CRC Press, Taylor & Francis Group, Boca Raton, Florida.

Speight, J. G. 2014. *The Chemistry and Technology of Petroleum* 5th Edition. CRC Press, Taylor & Francis Group, Boca Raton, Florida.

Speight, J. G. 2016. Hydrogen in Refineries. In: *Hydrogen Science and Engineering: Materials, Processes, Systems, and Technology*. D. Stolten and B. Emonts (Editors). Wiley-VCH Verlag GmbH & Co., Weinheim, Germany. Chapter 1. Page 3-18.

Speight, J. G. 2017. *Handbook of Petroleum Refining*. CRC Press, Taylor & Francis Group, Boca Raton, Florida.

St-Pierre, J. and Wilkinson, D. P. 2001. Fuel Cells: A New, Efficient and Cleaner Power Source. *AIChE Journal,* 47(7): 1482-1486.

Van Herle, J., Membrez, Y. and Bucheli, O. 2004. Biogas as a Fuel Source for SOFC Co-generators. *J. Power Sources*, 127(1-2): 300-312.

Weiland, P. 2010. Biogas Production: Current State and Perspectives. *Appl. Microbiol. Biotechnol.,* 85: 849-860.

Yoon, S. J., Choi, Y. C., and Lee, J. G. 2010. Hydrogen production from biomass tar by catalytic steam reforming. *Energy Convers. Manage.,* 51: 42-47.

Conversion Tables

1. Area

1 square centimeter (1 cm^2) = 0.1550 square inches
1 square meter 1 (m^2) = 1.1960 square yards
1 hectare = 2.4711 acres
1 square kilometer (1 km^2) = 0.3861 square miles
1 square inch (1 $inch^2$) = 6.4516 square centimeters
1 square foot (1 ft^2) = 0.0929 square meters
1 square yard (1 yd^2) = 0.8361 square meters
1 acre = 4046.9 square meters
1 square mile (1 mi^2) = 2.59 square kilometers

2. Concentration Conversions

1 part per million (1 ppm) = 1 microgram per liter (1 μg/L)
1 microgram per liter (1 μg/L) = 1 milligram per kilogram (1 mg/kg)
1 microgram per liter (μg/L) x 6.243 x 10^8 = 1 lb per cubic foot (1 lb/ft^3)
1 microgram per liter (1 μg/L) x 10^{-3} = 1 milligram per liter (1 mg/L)
1 milligram per liter (1 mg/L) x 6.243 x 10^5 = 1 pound per cubic foot (1 lb/ft^3)
I gram mole per cubic meter (1 g mol/m^3) x 6.243 x 10^5 = 1 pound per cubic foot (1 b/ft^3)
10,000 ppm = 1% w/w
1 ppm hydrocarbon in soil x 0.002 = 1 lb of hydrocarbons per ton of contaminated soil

3. Nutrient Conversion Factor

1 pound, phosphorus x 2.3 (1 lb P x 2.3) = 1 pound, phosphorous pentoxide (1 lb P_2O_5)
1 pound, potassium x 1.2 (1 lb K x 1.2 = 1 pound, potassium oxide (1 lb K_2O)

4. Temperature Conversions

°F = (°C x 1.8) + 32

°C = (°F - 32)/1.8

(°F - 32) x 0.555 = °C

Absolute zero = -273.15°C

Absolute zero = -459.67°F

5. Sludge Conversions

1,700 lbs wet sludge = 1 yd^3 wet sludge

1 yd^3 sludge = wet tons/0.85

Wet tons sludge x 240 = gallons sludge

1 wet ton sludge x % dry solids/100 = 1 dry ton of sludge

6. Various Constants

Atomic mass	mu = $1.6605402 \times 10^{-27}$
Avogadro's number	N = 6.0221367×10^{23} mol^{-1}
Boltzmann's constant	k = 1.380658×10^{-23} J K^{-1}
Elementary charge	e = $1.60217733 \times 10^{-19}$ C
Faraday's constant	F = 9.6485309 × 104 C · mol-1
Gas (molar) constant	R = k · N ˜ 8.314510 J· mol-1 · K-1 = 0.08205783 L atm mol^{-1} K^{-1}
Gravitational acceleration	g = 9.80665 m s^{-2}
Molar volume of an ideal gas at 1 atm and 25°C	$V_{ideal\ gas}$ = 24.465 L mol^{-1}
Planck's constant	h = $6.6260755 \times 10^{-34}$ J s
Zero, Celsius scale	0°C = 273.15°K

7. Volume Conversion

Barrels (petroleum, U. S.) to Cu feet multiply by 5.6146

Barrels (petroleum, U. S.) to Gallons (U. S.) multiply by 42

Barrels (petroleum, U. S.) to Liters multiply by 158.98

Barrels (US, liq.) to Cu feet multiply by 4.2109

Barrels (US, liq.) to Cu inches multiply by 7.2765 x 103

Barrels (US, liq.) to Cu meters multiply by 0.1192

Barrels (US, liq.) to Gallons multiply by (U. S., liq.) 31.5

Barrels (US, liq.) to Liters multiply by 119.24

Cubic centimeters to Cu feet multiply by 3.5315 x 10^{-5}

Cubic centimeters to Cu inches multiply by 0.06102

Cubic centimeters to Cu meters multiply by 1.0 x 10^{-6}

Cubic centimeters to Cu yards multiply by 1.308 x 10^{-6}

Cubic centimeters to Gallons (US liq.) multiply by 2.642 x 10^{-4}
Cubic centimeters to Quarts (US liq.) multiply by 1.0567 x 10^{-3}
Cubic feet to Cu centimeters multiply by 2.8317 x 10^{4}
Cubic feet to Cu meters multiply by 0.028317
Cubic feet to Gallons (US liq.) multiply by 7.4805
Cubic feet to Liters multiply by 28.317
Cubic inches to Cu cm multiply by 16.387
Cubic inches to Cu feet multiply by 5.787 x 10^{-4}
Cubic inches to Cu meters multiply by 1.6387 x 10^{-5}
Cubic inches to Cu yards multiply by 2.1433 x 10^{-5}
Cubic inches to Gallons (US liq.) multiply by 4.329 x 10^{-3}
Cubic inches to Liters multiply by 0.01639
Cubic inches to Quarts (US liq.) multiply by 0.01732
Cubic meters to Barrels (US liq.) multiply by 8.3864
Cubic meters to Cu cm multiply by 1.0 x 10^{6}
Cubic meters to Cu feet multiply by 35.315
Cubic meters to Cu inches multiply by 6.1024 x 10^{4}
Cubic meters to Cu yards multiply by 1.308
Cubic meters to Gallons (US liq.) multiply by 264.17
Cubic meters to Liters multiply by 1000
Cubic yards to Bushels (Brit.) multiply by 21.022
Cubic yards to Bushels (US) multiply by 21.696
Cubic yards to Cu cm multiply by 7.6455 x 105
Cubic yards to Cu feet multiply by 27
Cubic yards to Cu inches multiply by 4.6656 x 10^{4}
Cubic yards to Cu meters multiply by 0.76455
Cubic yards to Gallons multiply by 168.18
Cubic yards to Gallons multiply by 173.57
Cubic yards to Gallons multiply by 201.97
Cubic yards to Liters multiply by 764.55
Cubic yards to Quarts multiply by 672.71
Cubic yards to Quarts multiply by 694.28
Cubic yards to Quarts multiply by 807.90
Gallons (US liq.) to Barrels (US liq.) multiply by 0.03175
Gallons (US liq.) to Barrels (petroleum, US) multiply by 0.02381
Gallons (US liq.) to Bushels (US) multiply by 0.10742
Gallons (US liq.) to Cu centimeters multiply by 3.7854 x 10^{3}
Gallons (US liq.) to Cu feet multiply by 0.13368
Gallons (US liq.) to Cu inches multiply by 231
Gallons (US liq.) to Cu meters multiply by 3.7854 x 10^{-3}

Gallons (US liq.) to Cu yards multiply by 4.951×10^{-3}
Gallons (US liq.) to Gallons (wine) multiply by 1.0
Gallons (US liq.) to Liters multiply by 3.7854
Gallons (US liq.) to Ounces (US fluid) multiply by 128.0
Gallons (US liq.) to Pints (US liq.) multiply by 8.0
Gallons (US liq.) to Quarts (US liq.) multiply by 4.0
Liters to Cu centimeters multiply by 1000
Liters to Cu feet multiply by 0.035315
Liters to Cu inches multiply by 61.024
Liters to Cu meters multiply by 0.001
Liters to Gallons (US liq.) multiply by 0.2642
Liters to Ounces (US fluid) multiply by 33.814

8. Weight Conversion

1 ounce (1 ounce) = 28.3495 grams (18.2495 g)
1 pound (1 lb) = 0.454 kilogram
1 pound (1 lb) = 454 grams (454 g)
1 kilogram (1 kg) = 2.20462 pounds (2.20462 lb)
1 stone (English) = 14 pounds (14 lb)
1 ton (US; 1 short ton) = 2,000 lbs
1 ton (English; 1 long ton) = 2,240 lbs
1 metric ton = 2204.62262 pounds
1 tonne = 2204.62262 pounds

9. Other Approximations

14.7 pounds per square inch (14.7 psi) – 1 atmosphere (1 atm)
1 kilopascal (kPa) x 9.8692×10^{-3} = 14.7 pounds per square inch (14.7 psi)
1 yd^3 = 27 ft^3
1 US gallon of water = 8.34 lbs
1 imperial gallon of water – 10 lbs
1 ft^3 = 7.5 gallon = 1728 cubic inches = 62.5 lbs.
1 yd^3 = 0.765 m^3
1 acre-inch of liquid = 27,150 gallons = 3.630 ft^3
1-foot depth in 1 acre (in-situ) = 1,613 x (20 to 25 % excavation factor) = ~2,000 yd^3
1 yd^3 (clayey soils-excavated) = 1.1 to 1.2 tons (US)
1 yd^3 (sandy soils-excavated) = 1.2 to 1.3 tons (US)
Pressure of a column of water in psi = height of the column in feet by 0.434.

GLOSSARY

Abiotic: not associated with living organisms; synonymous with *abiological*.

Abiotic transformation: the process in which a substance in the environment is modified by non-biological mechanisms.

Absorption: the penetration of atoms, ions, or molecules into the bulk mass of a substance.

Acetic acid (CH_3CO_2H): Trivial name for ethanoic acid, formed by the oxidation of ethanol with potassium permanganate.

Acetone (CH_3COCH_3): Trivial name for propanone, formed by the oxidation of 2-propanol with potassium permanganate.

Achiral molecule: A molecule that does not contain a stereogenic carbon; an achiral molecule has a plane of symmetry and is superimposable on its mirror image.

Acid: Traditionally considered any chemical compound that, when dissolved in water, gives a solution with a pH less than 7.0.

Acid anhydride: An organic compound that react with water to form an acid.

Acid/base reaction: A reaction in which an acidic hydrogen atom is transferred from one molecule to another.

Acidic: A solution with a high concentration of H^+ ions.

Acidity: the capacity of the water to neutralize OH^-.

Acidophiles: metabolically active in highly acidic environments, and often have a high heavy metal resistance.

Acids, Bases, and Salts: many inorganic compounds are available as acids, bases, or salts.

Addition reaction: A reaction where a reagent is added across a double or triple bond in an organic compound to produce the corresponding saturated compound.

Adsorbent (sorbent): The solid phase or substrate onto which the sorbate adsorbs.

Adsorption: the retention of atoms, ions, or molecules on to the surface of another substance; the two-dimensional accumulation of an adsorbate at a solid surface. In the case of surface precipitation; also used when there is diffusion of the sorbate into the solid phase.

Aerobe: an organism that needs oxygen for respiration and hence for growth.

Aerobic: in the presence of, or requiring, oxygen; an environment or process that sustains biological life and growth, or occurs only when free (molecular) oxygen is present.

Aerobic bacteria: any bacteria requiring free oxygen for growth and cell division.

Aerobic conditions: conditions for growth or metabolism in which the organism is sufficiently supplied with oxygen.

Aerobic respiration: the process whereby microorganisms use oxygen as an electron acceptor.

Alcohol: The family name of a group of organic chemical compounds composed of carbon, hydrogen, and oxygen. The molecules in the series vary in chain length and are composed of a hydrocarbon plus a hydroxyl group. Alcohol includes methanol and ethanol.

Aldehyde: An organic compound with a carbon bound to a -(C=O)-H group; a compound in which a carbonyl group is bonded to one hydrogen atom and to one alkyl group [RC(=O)H].

Algae: microscopic organisms that subsist on inorganic nutrients and produce organic matter from carbon dioxide by photosynthesis.

Alkali metal: A metal in Group IA on the periodic table; an active metal which may be used to react with an alcohol to produce the corresponding metal alkoxide and hydrogen gas.

Alkalinity: the capacity of water to accept H^+ ions (protons).

Alkaliphiles: organisms that have their optimum growth rate at least 2 pH units above neutrality.

Alkalitolerants: organisms that are able to grow or survive at pH values above 9, but their optimum growth rate is around neutrality or less.

Alkylation: a process for manufacturing high octane blending components used in unleaded petrol or gasoline.

Ambient: the surrounding environment and prevailing conditions.

Amide: An organic compound that contains a carbonyl group bound to nitrogen; the simplest amides are formamide ($HCONH_2$) and acetamide (CH_3CONH_2).

Amine: An organic compound that contains a nitrogen atom bound only to carbon and possibly hydrogen atoms; examples are methylamine, CH_3NH_2; dimethylamine, CH_3NHCH_3; and trimethylamine, $(CH_3)_3N$.

Amino acid: A molecule that contains at least one amine group (-NH2) and at least one carboxylic acid group (-COOH); when these groups are both attached to the same carbon, the acid is an α-amino acid – a α-amino acids are the basic building blocks of proteins.

Amorphous solid: A non-crystalline solid having no well-defined ordered structure.

Ammonia: A gaseous compound of hydrogen and nitrogen, NH_3, with a pungent smell and taste.

Anaerobe: an organism that does not need free-form oxygen for growth. Many anaerobes are even sensitive to free oxygen.

Anaerobic: In the absence of oxygen.

Anaerobic bacteria: Micro-organisms that live and reproduce in an environment containing no free or dissolved oxygen; used for anaerobic digestion.

Anaerobic bacteria: any bacteria that can grow and divide in the partial or complete absence of oxygen.

Anaerobic digestion (anaerobic decomposition): Decomposition of biological wastes by micro-organisms, usually under wet conditions, in the absence of air (oxygen), to produce a gas comprising mostly methane and carbon dioxide. See Decomposition.

Anaerobic bacteria: any bacteria that can grow and divide in the partial or complete absence of oxygen.

Anaerobic respiration: The process whereby microorganisms use a chemical other than oxygen as an electron acceptor; common substitutes for oxygen are nitrate, sulfate, and iron.

Annual removals: The net volume of growing stock trees removed from the inventory during a specified year by harvesting, cultural operations such as timber stand improvement, or land clearing.

Anoxic: an environment without oxygen.

API gravity: a measure of the lightness or heaviness of petroleum that is related to density and specific gravity.

$$API = (141.5/sp\ gr\ @\ 60^{o}F) - 131.5$$

Aquasphere: the water areas of the Earth; also called the hydrosphere.

Aquatic chemistry: the branch of environmental chemistry that deals with chemical phenomena in water.

Aquifer: a water-bearing layer of soil, sand, gravel, rock or other geologic formation that will yield usable quantities of water to a well under normal hydraulic gradients or by pumping.

Aromatic ring: An exceptionally stable planar ring of atoms with resonance structures that consist of alternating double and single bonds, such as benzene:

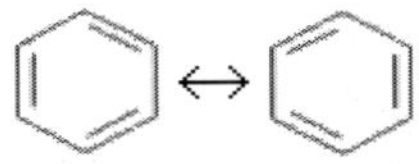

Aromatics: a range of hydrocarbons which have a distinctive sweet smell and include benzene and toluene' occur naturally in petroleum and are also extracted as a petrochemical feedstock, as well as for use as solvents.

Asphaltene (asphaltenes): the brown to black powdery material produced by treatment of petroleum, heavy oil, bitumen, or residuum with a low-boiling liquid hydrocarbon.

Assay: qualitative or (more usually) quantitative determination of the components of a material or system.

Association colloids: colloids which consist of special aggregates of ions and molecules (micelles).

Asymmetric carbon: A carbon atom covalently bonded to four different atoms or groups of atoms.

Atmosphere: the thin layer of gases that cover surface of the Earth; composed of two major components: nitrogen 78.08% and oxygen 20.955 with smaller amounts of argon 0.934%, carbon dioxide 0.035%, neon 1.818×10^{-3}%, krypton 1.14×10^{-4}%, helium 5.24×10^{-4}%, and xenon 8.7×10^{-6}%; may also contain 0.1 to 5% water by volume, with a normal range of 1 to 3%; the reservoir of gases, moderates the temperature of the Earth, absorbs energy and damaging ultraviolet radiation from the sun, transports energy away from equatorial regions and serves as a pathway for vapor-phase movement of water in the hydrologic cycle.

Atomic number: The atomic number is equal to the number of positively charged protons in the nucleus of an atom which determines which identify of the element.

Atomic radius: the relative size of an atom; among the main group of elements, atomic radii mostly decrease from left to right across rows in the Periodic Table; metal ions are smaller than their neutral atoms, and nonmetallic anions are larger than the atoms from which they are formed; atomic radii are expressed in angstrom units of length (Å).

Autotrophs: organisms or chemicals that use carbon dioxide and ionic carbonates for the C that they require.

Avogadro's number: The number of molecules (6.023×10^{23}) in one gram-mole of a substance.

Bacteria: single-celled prokaryotic microorganisms that may be shaped as rods (bacillus), spheres (coccus), or spirals (vibrios, spirilla, spirochetes).

Base: A substance which gives off hydroxide ions (OH^-) in solution.

Basic: Having the characteristics of a base.

Benthic zone: the ecological region at the lowest level of a body of water such as an ocean or a lake, including the sediment surface and some sub-surface layers; organisms living in this zone (benthos or benthic organisms) generally live in close relationship with the substrate bottom; many such organisms are permanently attached to the bottom; because light does not penetrate very deep ocean-water, the energy source for the benthic ecosystem is often organic matter from higher up in the water column which sinks to the depths.

Barrel (bbl): the unit of measure used by the petroleum industry; equivalent to approximately forty-two US gallons or approximately thirty-four (33.6) Imperial gallons or 159 liters; 7.2 barrels are equivalent to one tonne of oil (metric).

Barrel of oil equivalent (boe): The amount of energy contained in a barrel of crude oil, i.e., approximately 5.8 million Btu (6.1 GJ), equivalent to 1,700 kWh; a "petroleum barrel" is a liquid measure equal to 42 U.S. gallons (35 Imperial gallons or 159 litters); approximately 7.2 barrels are equivalent to one tonne of oil (metric).

Base: Traditionally considered any chemical compound that, when dissolved in water, gives a solution with a pH greater than 7.0.

Batch feed: A process by which the reactor is filled with feedstock in discrete amounts, rather than continuously.

Billion: 1×10^9

Bimolecular reaction: The collision and combination of two reactants involved in the rate-limiting step.

Bioaccumulation: The accumulation of substances, such as pesticides, or other chemicals in an organism; occurs when an organism absorbs a chemical – possibly a toxic chemical – at a rate faster than that at which the substance is lost by catabolism and excretion; the longer the biological half-life of a toxic substance the greater the risk of chronic poisoning, even if environmental levels of the toxin are not very high; see Biomagnification.

Bio-augmentation: a process in which acclimated microorganisms are added to soil and groundwater to increase biological activity. Spray irrigation is typically used for shallow contaminated soils, and injection wells are used for deeper contaminated soils.

Biochemical oxygen demand (BOD): an important water-quality parameter; refers to the amount of oxygen utilized when the organic matter in a given volume of water is degraded biologically.

Biochemical conversion: The use of fermentation or anaerobic digestion to produce fuels and chemicals from organic sources.

Biocide: A chemical substance or microorganism intended to destroy, deter, render harmless, or exert a controlling effect on any harmful organism by chemical or biological means.

Biodegradation: the natural process whereby bacteria or other microorganisms chemically alter and break down organic molecules; the breakdown or transformation of a chemical substance or substances by microorganisms using the substance as a carbon and/or energy source.

Biodiesel: A fuel derived from biological sources that can be used in diesel engines instead of petroleum-derived diesel; through the process of transesterification, the triglycerides in the biologically derived oils are separated from the glycerin, creating a clean-burning, renewable fuel.

Biogeochemical cycle: The pathway by which a chemical moves through biotic (biosphere) and abiotic (atmosphere, aquasphere, lithosphere) compartments of the Earth.

Bio-inorganic compounds: natural and synthetic compounds that include metallic elements bonded to proteins and other biological chemistries.

Bioenergy: Useful, renewable energy produced from organic matter – the conversion of the complex carbohydrates in organic matter to energy; organic matter may either be used directly as a fuel, processed into liquids and gasses, or be a residual of processing and conversion.

Bioethanol: Ethanol produced from biomass feedstocks; includes ethanol produced from the fermentation of crops, such as corn, as well as cellulosic ethanol produced from woody plants or grasses.

Biofuels: a generic name for liquid or gaseous fuels that are not derived from petroleum based fossils fuels or contain a proportion of non-fossil fuel; fuels produced from plants, crops such as sugar beet, rape seed oil or re-processed vegetable oils or fuels made from gasified biomass; fuels made from renewable biological sources and include ethanol, methanol, and biodiesel; sources include, but are not limited to: corn, soybeans, flaxseed, rapeseed, sugarcane, palm oil, raw sewage, food scraps, animal parts, and rice.

Biogas: A combustible gas derived from decomposing biological waste under anaerobic conditions. Biogas normally consists of 50 to 60 percent methane. See also landfill gas.

Biomagnification: The increase in the concentration of heavy metals (i.e., mercury) or organic contaminants such as chlorinated hydrocarbons, in organisms as a result of their consumption within a food chain/web; an example is the process by which contaminants such as polychlorobiphenyl derivatives (PCBs) accumulate or magnify as they move up the food chain - PCBs concentrate in tissue and internal organs, and as big fish eat little fish, they accumulate all the PCBs that have been eaten by everyone below them in the food chain; can occur as a result of: (i) persistence, in which the chemical cannot be broken down by environmental processes, (ii) food chain energetics, in which the concentration of the chemical increases progressively as it moves up a food chain, and (iii) a low or non-existent rate of internal degradation or excretion of the substance that is often due to water-insolubility.

Biomass: Any organic matter that is available on a renewable or recurring basis, including agricultural crops and trees, wood and wood residues, plants (including aquatic plants), grasses, animal manure, municipal residues, and other residue materials. Biomass is generally produced in a sustainable manner from water and carbon dioxide by photosynthesis. There are three main categories of biomass - primary, secondary, and tertiary.

Biopower: The use of biomass feedstock to produce electric power or heat through direct combustion of the feedstock, through gasification and then combustion of the resultant gas, or through other thermal conversion processes. Power is generated with engines, turbines, fuel cells, or other equipment.

Bioreactor (digester, biodigester): A device for optimizing the anaerobic digestion of biomass and/or animal manure, and possibly to recover biogas for energy production.

Biorefinery - A facility that processes and converts biomass into value-added products. These products can range from biomaterials to fuels such as ethanol or important feedstocks for the production of chemicals and other materials.

Biomass to liquid (BTL): The process of converting biomass to liquid fuels. Hmm, that seems painfully obvious when you write it out.

Bioremediation: a treatment technology that uses biological activity to reduce the concentration or toxicity of contaminants: materials are added to contaminated environments to accelerate natural biodegradation.

Biosphere: a term representing all of the living entities on the Earth.

Biota: living organisms that constitute the plant and animal life of a region (arctic region, temperate region, sub-tropical region, or tropical region).

Bitumen: also, on occasion, referred to as native asphalt, and extra heavy oil; a naturally occurring material that has little or no mobility under reservoir conditions and which cannot be recovered through a well by conventional oil well production methods including currently used enhanced recovery techniques; current methods involve mining for bitumen recovery.

Bone dry: Having zero percent moisture content. Wood heated in an oven at a constant temperature of 100°C (212°F) or above until its weight stabilizes is considered bone dry or oven dry.

Bottoming cycle: A cogeneration system in which steam is used first for process heat and then for electric power production.

Black liquor: Solution of lignin-residue and the pulping chemicals used to extract lignin during the manufacture of paper.

Breakdown product: a compound derived by chemical, biological, or physical action on a chemical compound; the breakdown is a process which may result in a more toxic or a less toxic compound and a more persistent or less persistent compound than the original compound.

British thermal unit - (Btu) A non-metric unit of heat, still widely used by engineers; One Btu is the heat energy needed to raise the temperature of one pound of water from 15.6 to 16.1°C (60°F to 61°F) at one atmosphere pressure. 1 Btu = 1055 joules (1.055 kJ).

BTEX: The collective name given to benzene, toluene, ethylbenzene and the xylene isomers (*p*-, *m*-, and *o*-xylene); a group of volatile organic compounds (VOCs) found in petroleum hydrocarbons, such as gasoline, and other common environmental contaminants.

BTX: The collective name given to benzene, toluene, and the xylene isomers (*p*-, *m*-, and *o*-xylene); a group of volatile organic compounds (VOCs) found in petroleum hydrocarbons, such as gasoline, and other common environmental contaminants.

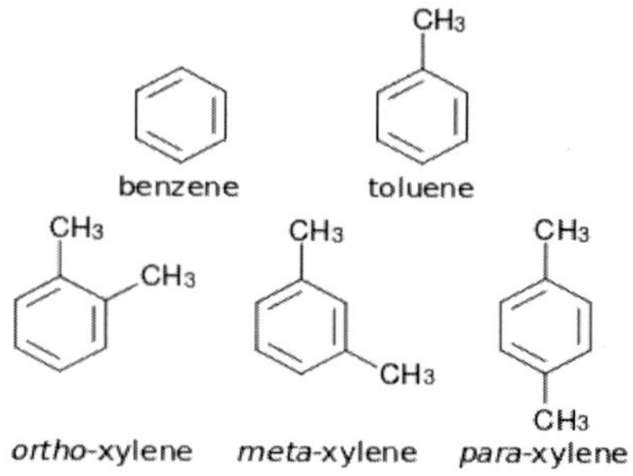

CH3

benzene toluene

CH3 CH3 CH3

CH3

CH3

CH3

ortho-xylene *meta*-xylene *para*-xylene

Buffer solution: A solution that resists change in the pH, even when small amounts of acid or base are added.

Bunker - A storage tank.

Butanol: Though generally produced from fossil fuels, this four-carbon alcohol can also be produced through bacterial fermentation of alcohol.

Calorific value: A measure of the energy released during combustion; the lower calorific value is a measure of the energy released when the water vapor generated during combustion is still in the gas phase while the upper calorific value is a measure of the energy released when water vapor has been condensed. See Wobbe Index.

Carbenium ion: A generic name for carbocation that has at least one important contributing structure containing a tervalent carbon atom with a vacant *p* orbital.

Carbanion: The generic name for anions containing an even number of electrons and having an unshared pair of electrons on a carbon atom (e.g., $Cl3C^{-}$).

Carbon dioxide (CO_2): A product of combustion that acts as a greenhouse gas in the Earth's atmosphere, trapping heat and contributing to climate change.

Carbon monoxide (CO): A lethal gas produced by incomplete combustion of carbon-containing fuels in internal combustion engines. It is colorless, odorless, and tasteless. (As in flavorless, we mean, though it's also been known to tell a bad joke or two.)

Carbon sink: A geographical area whose vegetation and/or soil soaks up significant carbon dioxide from the atmosphere. Such areas, typically in tropical regions, are increasingly being sacrificed for energy crop production.

Carbonyl group: A divalent group consisting of a carbon atom with a double-bond to oxygen; for example, acetone (CH_3-(C=O)-CH_3) is a carbonyl group linking two methyl groups.

Carboxy group (-CO2H or -COOH): A carbonyl group to which a hydroxyl group is attached; carboxylic acids have this functional group.

Carboxylic acid: An organic molecule with a -CO_2H group; hydrogen atom on the -CO_2H group ionizes in water; the simplest carboxylic acids are formic acid (H-COOH) and acetic acid (CH_3-COOH).

Catabolism: The breakdown of complex molecules into simpler ones through the oxidation of organic substrates to *provide* biologically available energy – ATP (adenosine triphosphate) is an example of such a molecule.

Catalysis: the process where a catalyst increases the rate of a chemical reaction without modifying the overall standard Gibbs energy change in the reaction.

Catalyst: A substance that alters the rate of a chemical reaction and may be recovered essentially unaltered in form or amount at the end of the reaction.

Cation exchange: the interchange between a cation in solution and another cation in the boundary layer between the solution and surface of negatively charged material such as clay or organic matter.

Cation exchange capacity (CEC): the sum of the exchangeable bases plus total soil acidity at a specific pH, usually 7.0 or 8.0. When acidity is expressed as salt extractable acidity, the cation exchange capacity is called the effective cation exchange capacity (ECEC), because this is considered to be the CEC of the exchanger at the native pH value; usually expressed in centimoles of charge per kilogram of exchanger (cmol/kg) or millimoles of charge per kilogram of exchanger.

Cellulose: A polysaccharide, polymer of glucose, that is found in the cell walls of plants; a fiber that is used in many commercial products, notably paper.

Cetane number: A measure of the ignition quality of diesel fuel; the higher the number the more easily the fuel is ignited under compression.

Chemical bond: the forces acting among two atoms or groups of atoms that lead to the formation of an aggregate with sufficient stability to be considered as an independent molecular species.

Chemical change: A processes or events that alter the fundamental structure of a chemical.

Chemical dispersion: in relation to oil spills, this term refers to the creation of oil-in-water *emulsions* by the use of chemical dispersants made for this purpose.

Chemical induction (coupling): when one reaction accelerates another in a chemical system there is said to be chemical induction or coupling. Coupling is caused by an intermediate or by-product of the inducing reaction that participates in a second reaction; chemical induction is often *observed* in oxidation-reduction reactions.

Chemical reaction: A process that results in the interconversion of chemical species.

Chemical species: An ensemble of chemically identical molecular entities that can explore the same set of molecular energy levels on the time scale of the experiment; the term is applied equally to a set of chemically identical atomic or molecular structural units in a solid array.

Chemical waste: any solid, liquid, or gaseous waste material that, if improperly managed or disposed of, may pose substantial hazards to human health and the environment.

Chemical weight: The weight of a molar sample as determined by the weight of the molecules (the molecular weight); calculated from the weights of the atoms in the molecule.

Chemotroph: An organism or chemical that uses chemical energy derived from oxidation-reduction reactions for their energy needs.

Chirality: The ability of an object or a compound to exist in right and left-handed forms; a chiral compound will rotate the plane of plane-polarized light.

Chlorinated solvent: a volatile organic compound containing chlorine; common solvents are trichloroethylene, tetrachloroethylene, and carbon tetrachloride.

Chlorofluorocarbon: Gases formed of chlorine, fluorine, and carbon whose molecules normally do not react with other substances; formerly used as spray-can propellants, they are known to destroy the protective ozone layer of the Earth.

Chromatography: A method of chemical analysis where compounds are separated by passing a mixture in a suitable carrier over an absorbent material; compounds with different absorption coefficients move at different rates and are separated.

Cis trans isomers: the difference in the positions of atoms (or groups of atoms) relative to a reference plane in an organic molecule; in a *cis-isomer,* the atoms are on the same side of the molecule, but are on opposite sides in the *trans*-isomer; sometimes called stereoisomers; these arrangements are common in alkenes and cycloalkanes.

Clay: A very fine-grained soil that is plastic when wet but hard when fired; typical clay minerals consist of silicate and aluminosilicate minerals that are the products of weathering reactions of other minerals; the term is also used to refer to any mineral of very small particle size.

Closed-loop biomass: Crops grown, in a sustainable manner, for the purpose of optimizing their value for bioenergy and bioproduct uses. This includes annual crops such as maize and wheat, and perennial crops such as trees, shrubs, and grasses such as switch grass.

Cloud point: the temperature at which paraffin wax or other solid substances begin to crystallise or separate from the solution, imparting a cloudy appearance to the oil when the oil is chilled under prescribed conditions.

CO_2-equivalent (carbon dioxide equivalent): A unit used to standardize measurements of; for example, tonne for tonne, methane is a greenhouse gas that is 21 times more powerful than carbon dioxide in causing the global greenhouse effect and, therefore, one tonne of methane represents 21 tonnes of CO2 equivalent.

Coarse materials: Wood residues suitable for chipping, such as slabs, edgings, and trimmings.

Co-generation: See combined heat and power generation (CHP).

Coke: A hard, dry substance containing carbon that is produced by heating bituminous coal or other carbonaceous materials to a very high temperature in the absence of air; used as a fuel.

Coking: a thermal method used in refineries for the conversion of bitumen and residua to volatile products and coke (see Delayed coking and Fluid coking).

Colligative properties: The properties of a solution that depend only on the number of particles dissolved in it, not the properties of the particles themselves; the main colligative properties addressed at this web site are boiling point elevation and freezing point depression.

Colloidal particles; particles which have some characteristics of both species in solution and larger particles in suspension, which range in diameter form about 0.001 micrometer (μm) to approximately 1 μm, and which scatter white light as a light blue hue observed at right angles to the incident light.

Combination reactions: reactions where two substances combine to form a third substance; an example is two elements reacting to form a compound of the elements and is shown in the general form: A + B → AB; examples include:

$$2Na(s) + Cl_2(g) \rightarrow 2NaCl(s)$$
$$8Fe + S_8 \rightarrow 8FeS$$

Combined heat and power generation (CHP generation, co-generation): The sequential production of electricity and useful thermal energy from a common fuel source. Reject heat from industrial processes can be used to power an electric generator (bottoming cycle). Conversely, surplus heat from an electric generating plant can be used for industrial processes, or space and water heating purposes (topping cycle).

Co-metabolism (cometabolism): the process by which compounds in petroleum may be enzymatically attacked by microorganisms without furnishing carbon for cell growth and division; a variation on biodegradation in which microbes transform a contaminant even though the contaminant cannot serve as the primary energy source for the organisms. To degrade the contaminant, the microbes require the presence of other compounds (primary substrates) that can support their growth.

Complex inorganic chemicals: molecules that consist of different types of atoms (atoms of different chemical elements) which, in chemical reactions, are decomposed with the formation several other chemicals.

Compound: The combination of two or more different elements, held together by chemical bonds; the elements in each compound are always combined in the same proportion by mass (law of definite proportion).

Conjugate acid: A substance which can lose the H^+ ion to form a base.

Conjugate base: A substance which can gain the H^+ ion to form an acid.

Constituent: an essential part or component of a system or group (that is, an ingredient of a chemical mixture); for example, benzene is one constituent of gasoline.

Contaminant: a pollutant unless it has some detrimental effect, can cause deviation from the normal composition of an environment; a pollutant that causes deviations from the normal composition of an environment. Are not classified as pollutants unless they have some detrimental effect.

Conventional crude oil (conventional petroleum): crude oil that is pumped from the ground and recovered using the energy inherent in the reservoir; also recoverable by application of secondary recovery techniques.

Cord: A stack of wood comprising 128 cubic feet (3.62 m^3); standard dimensions are 4 x 4 x 8 feet, including air space and bark. One cord contains approx. 1.2 U.S. tons (oven-dry) = 2400 pounds = 1089 kg.

Cracking: A secondary refining process that uses heat and/or a catalyst to break down high molecular weight chemical components into lower molecular weight products which can be used as blending components for fuels.

Critical pressure: the pressure required to liquefy a gas at its critical temperature; the minimum pressure required to condense gas to liquid at the critical temperature; a substance is still a fluid above the critical point, neither a gas nor a liquid, and is referred to as a supercritical fluid; expressed in atmosphere or psi.

Critical temperature: The temperature above which a gas cannot be liquefied, regardless of the amount of pressure applied; the temperature at the critical point (end of the vapor pressure curve in phase diagram); at temperatures above critical temperature, a substance cannot be liquefied, no matter how great the pressure; expressed in °C.

Cropland: Total cropland includes five components: cropland harvested, crop failure, cultivated summer fallow, cropland used only for pasture, and idle cropland.

Cropland pasture: Land used for long-term crop rotation. However, some cropland pasture is marginal for crop uses and may remain in pasture indefinitely. This category also includes land that was used for pasture before crops reached maturity and some land used for pasture that could have been cropped without additional improvement.

Cull tree - A live tree, 5.0 inches in diameter at breast height (d.b.h.) or larger that is non-merchantable for saw logs now or prospectively because of rot, roughness, or species. (See definitions for rotten and rough trees).

Cultivated summer fallow: cropland cultivated for one or more seasons to control weeds and accumulate moisture before small grains are planted.

Culture: The growth of cells or microorganisms in a controlled artificial environment.

Decomposition: the decay or breaking down of materials into smaller components. See Anaerobic Digestion.

Dedicated energy crops: Crops grown specifically for their fuel value. These include food crops such as corn and sugarcane, and non-food crops such as poplar trees and switchgrass. Currently, two energy crops are under development: short-rotation woody crops, which are fast-growing hardwood trees harvested in 5 to 8 years, and herbaceous energy crops, such as perennial grasses, which are harvested annually after taking 2 to 3 years to reach full productivity.

Degree of completion: The percentage or fraction of the limiting reactant that has been converted to products.

Dehydration reaction (condensation reaction): A chemical reaction in which two organic molecules become linked to each other via covalent bonds with the removal of a molecule of water; common in synthesis reactions of organic chemicals.

Dehydrohalogenation: removal of hydrogen and halide ions from an alkane resulting in the formation of an alkene.

Delayed coking: a coking process in which the thermal reactions are allowed to proceed to completion to produce gaseous, liquid, and solid (coke) products.

Density: the mass (or weight) of a unit volume of any substance at a specified temperature; see also Specific gravity.

Desulfurization: the removal of sulfur or sulfur compounds from a feedstock.

Diesel engine: Named for the German engineer Rudolph Diesel, this internal-combustion, compression-ignition engine works by heating fuels and causing them to ignite; can use either petroleum or bio-derived fuel.

Diesel fuel: A distillate of fuel oil that has been historically derived from petroleum for use in internal combustion engines; also derived from plant and animal sources.

Diesel, Rudolph: German inventor famed for fashioning the diesel engine, which made its debut at the 1900 World's Fair; initially engine to run on vegetable-derived fuels.

Digestate: The treated/ digested effluent from the anaerobic digestion process; also known as anaerobic digester residue, digested biomass, digested slurry.

Digester: An airtight vessel or enclosure in which bacteria decomposes biomass in water to produce biogas.

Denitrification: bacterial reduction of nitrate to nitrite to gaseous nitrogen or nitrous oxides under anaerobic conditions.

Desorption: The release of ions or molecules from solids into solution.

Direct-injection engine: A diesel engine in which fuel is injected directly into the cylinder.

Distillate: Any petroleum product produced by boiling crude oil and collecting the vapors produced as a condensate in a separate vessel, for example gasoline (light distillate), gas oil (middle distillate), or fuel oil (heavy distillate).

Distillation: The primary distillation process which uses high temperature to separate crude oil into vapor and fluids which can then be fed into a distillation or fractionating tower.

Downdraft gasifier: A gasifier in which the product gases pass through a combustion zone at the bottom of the gasifier.

Dry digestion: The customary (but arbitrary) term to refer to a digestion process when using energy crops with a dry matter content of 15 60 16% w/w or higher because the digester contents are generally not pumpable with this water content; see Wet digestion.

Dutch oven furnace: One of the earliest types of furnaces, having a large, rectangular box lined with firebrick (refractory) on the sides and top; commonly used for burning wood.

E85: An alcohol fuel mixture containing 85 percent ethanol and 15 percent gasoline by volume, and the current alternative fuel of choice of the U.S. government.

Ecology: the scientific study of the relationships between organisms and their environments.

Ecological chemistry: the study of the interactions between organisms and their environment that are mediated by naturally occurring chemicals.

Ecology: the study of environmental factors that affect organisms and how organisms interact with these factors and with each other.

Ecosystem: a community of organisms together with their physical environment which can be viewed as a system of interacting and interdependent relationships; this can also include processes such as the flow of energy through trophic levels as well as the cycling of chemical elements and compounds through living and nonliving components of the system; the trophic level of an organism is the position it occupies in a food chain; Ecosystem: a term representing an assembly of mutually interacting organisms and their environment in which materials are interchanged in a largely cyclical manner.

Effluent: The liquid or gas discharged from a process or chemical reactor, usually containing residues from that process.

Electron affinity: the electron affinity of an atom or molecule is the amount of energy released or spent when an electron is added to a neutral atom or molecule in the gaseous state to form a negative ion.

Electron configuration of an atom: The extra-nuclear structure; the arrangement of electrons in shells and subshells; chemical properties of elements (their valence states and reactivity) can be predicted from the electron configuration.

Electron donor: The atom, molecule, or compound that donates electrons (and therefore is oxidized); in bioremediation, the organic contaminant often serves as an electron donor.

Electronegativity: the tendency of an atom to attract electrons in a chemical bond; nonmetals have high electronegativity, fluorine being the most electronegative while alkali metals possess least electronegativity; the electronegativity difference indicates polarity in the molecule.

Elimination: a reaction where two groups such as chlorine and hydrogen are lost from adjacent carbon atoms and a double bond is formed in their place.

Emissions: Substances discharged into the air during combustion, e.g., all that stuff that comes out of your car; waste substances released into the air or water.

Emulsion: a stable mixture of two immiscible liquids, consisting of a continuous phase and a dispersed phase. Oil and water can form both oil-in-water and water-in-oil emulsions. The former is termed a dispersion, while *emulsion* implies the latter. Water-in-oil emulsions formed from petroleum and brine can be grouped into four stability classes: stable, a formal emulsion that will persist indefinitely; meso-stable, which gradually degrade over time due to a lack of one or more stabilizing factors; entrained water, a mechanical mixture characterized by high viscosity of the petroleum component which impedes separation of the two phases; and unstable, which are mixtures that rapidly separate into immiscible layers.

Emulsion stability: generally accompanied by a marked increase in *viscosity* and elasticity, over that of the parent oil which significantly changes behavior. Coupled with the increased volume due to the introduction of brine, emulsion formation has a large effect on the choice of countermeasures employed to combat a spill.

Emulsification: the process of *emulsion* formation, typically by mechanical mixing. In the environment, *emulsions* are most often formed as a result of wave action. Chemical agents can be used to prevent the formation of *emulsions* or to "break" the *emulsions* to their component oil and water phases.

Empirical formula: The simplest whole-number ratio of atoms in a compound.

Emulsan is a polyanionic heteropolysaccharide bioemulsifier produced by *Acinetobacter calcoaceticus* RAG-1; used to stabilize oil-in-water emulsions.

Endergonic reaction: a chemical reaction that requires energy to proceed. A chemical reaction is endergonic when the change in free energy is positive.

Endogenous inhibition: Inhibition that is due to conditions or material created during the process itself that under certain circumstances may inhibit the process, and exogenous inhibition is due to external conditions. See Exogenous inhibition.

Endothermic reaction: A chemical reaction in which heat is absorbed.

Engineered bioremediation: A type of remediation that increases the growth and degradative activity of microorganisms by using engineered systems that supply nutrients, electron acceptors, and/or other growth-stimulating materials.

Enhanced bioremediation: a process which involves the addition of microorganisms (e.g., fungi, bacteria, and other microbes) or nutrients (e.g. oxygen, nitrates) to the subsurface environment to accelerate the natural biodegradation process.

Energy crops: Crops grown specifically for their fuel value; include food crops such as corn and sugarcane, and nonfood crops such as poplar trees and switch grass.

Energy balance: The difference between the energy produced by a fuel and the energy required to obtain it through agricultural processes, drilling, refining, and transportation.

Energy crops: Agricultural crops grown specifically for their energy value.

Energy-efficiency ratio: A number representing the energy stored in a fuel as compared to the energy required to produce, process, transport, and distribute that fuel.

Enhanced recovery: methods that usually involves the application of thermal energy (e.g., steam flooding) to oil recovery from the reservoir.

Enthalpy of formation (ΔH_f): the energy change or the heat of reaction in which a compound is formed from its elements; energy cannot be created or destroyed but is converted from one form to another; the enthalpy change (or heat of reaction) is: $\Delta H = H_2 - H_1$

H_1 is the enthalpy of reactants and H_2 the enthalpy of the products (or heat of reaction); when H_2 is less than H_1 the reaction is exothermic and ΔH is negative, i.e. temperature increases; when H_2 is greater than H_1 the reaction is endothermic and the temperature falls.

Ethanol (ethyl alcohol, alcohol, or grain-spirit): A clear, colorless, flammable oxygenated hydrocarbon; used as a vehicle fuel by itself (E100 is 100% ethanol by volume), blended with gasoline (E85 is 85% ethanol by volume), or as a gasoline octane enhancer and oxygenate (10% by volume).

Entropy: A thermodynamic quantity that is a measure of disorder or randomness in a system; the total entropy of a system and its surroundings always increases for a spontaneous process; the total entropy of a system and its surroundings always increases for a spontaneous process' the standard entropies are entropy values for the standard states of substances.

Environment: the total living and nonliving conditions of an organism's internal and external surroundings that affect an organism's complete life span; the conditions that surround someone or something; the conditions and influences that affect the growth, health, progress, etc., of someone or something; the total living and nonliving conditions (internal and external surroundings) that are an influence on the existence and complete life span of the organism.

Environmental analytical chemistry: the application of analytical chemical techniques to the analysis of environmental samples--in a regulatory setting.

Environmental biochemistry: the discipline that deals specifically with the effects of environmental chemical species on life.

Environmental chemistry: the study of the sources, reactions, transport, effects, and fates of chemical species in water, soil, and air environments, and the effects of technology thereon.

Environmentalist: A person working to solve environmental problems, such as air and water pollution, the exhaustion of natural resources, and uncontrolled population growth.

Environmental pollution: the contamination of the physical and biological components of the Earth system (atmosphere, aquasphere, and geosphere) to such an extent that normal environmental processes are adversely affected.

Environmental science: the study of the environment, its living and nonliving components, and the interactions of these components.

Environmental studies: the discipline dealing with the social, political, philosophical and ethical issues concerning man's interactions with the environment.

Enzyme: a macromolecule, mostly proteins or conjugated proteins produced by living organisms, that facilitate the degradation of a chemical compound (catalyst); in general, an enzyme catalyzes only one reaction type (reaction specificity) and operates on only one type of substrate (substrate specificity); any of a group of catalytic proteins that are produced by cells and that mediate or promote the chemical processes of life without themselves being altered or destroyed.

Enzyme: a macromolecule, mostly proteins or conjugated proteins produced by living organisms, that facilitate the degradation of a chemical compound (catalyst); in general, an enzyme catalyzes only one reaction type (reaction specificity) and operates on only one type of substrate (substrate specificity); any of a group of catalytic proteins that are produced by cells and that mediate or promote the chemical processes of life without themselves being altered or destroyed.

Epoxidation: a reaction wherein an oxygen molecule is inserted in a carbon-carbon double bond and an epoxide is formed.

Epoxides: a subclass of epoxy compounds containing a saturated three-membered cyclic ether. See *Epoxy compounds*.

Epoxy compounds: compounds in which an oxygen atom is directly attached to two adjacent or nonadjacent carbon atoms in a carbon chain or ring system; thus cyclic ethers.

Equilibrium: A state when the reactants and products are in a constant ratio. The forward reaction and the reverse reactions occur at the same rate when a system is in equilibrium.

Equilibrium constant: A value that expresses how far the reaction proceeds before reaching equilibrium. A small number means that the equilibrium is towards the reactants side while a large number means that the equilibrium is towards the products side.

Equilibrium expression: The expression giving the ratio between the products and reactants. The equilibrium expression is equal to the concentration of each product raised to its coefficient in a balanced chemical equation and multiplied together, divided by the concentration of the product of reactants to the power of their coefficients.

Equipment blank: a sample of analyte-free media which has been used to rinse the sampling equipment. It is collected after completion of decontamination and prior to sampling. This blank is useful in documenting and controlling the preparation of the sampling and laboratory equipment.

Ester: A compound formed from an acid and an alcohol; in esters of carboxylic acids, the -COOH group and the -OH group lose a molecular of water and form a -COO- bond (R_1 and R_2 represent organic groups):

$$R_1COOH + R_2OH \rightarrow R_1COOR_2 + H_2O$$

Ether: A compound with an oxygen atom attached to two hydrocarbon groups. Any carbon compound containing the functional group C–O–C, such as diethyl ether (C_2H_5O C_2H_5).

Ethoxy group (CH_3CH_2O-): A two-carbon alkoxy substituent.

Ethylbenzene: A colorless, flammable liquid found in natural products such as coal tar and crude oil; it is also found in manufactured products such as inks, insecticides, and paints; a minor component of JP-8 fuel.

Ethyl group (CH_3CH_2-): a two-carbon alkyl substituent.

Eurkaryotes: microorganisms that have well defined cell nuclei enclosed by a nuclear membrane.

Eutrophication: the growth of algae may become quite high in very productive water, with the result that the concurrent decomposition of dead algae reduces oxygen levels in the water to very low values.

Excess reactant: The excess of a reactant over the stoichiometric amount, with the exception of the limiting reactant; the term may refer to more than one reactant.

Exergy: A combination property of a system and its environment because it depends on the state of both the system and environment; the maximum useful work possible during a process that brings the system into equilibrium with a heat reservoir; when the surroundings are the reservoir, exergy is the potential of a system to cause a change as it achieves equilibrium with its environment and after the system and surroundings reach equilibrium, the exergy is zero; determining exergy is a prime goal of thermodynamics.

Exogenous inhibition: Inhibition that is due to the influence of conditions that are external to the process. See Endogenous inhibition.

Exothermic reaction: A reaction that produces heat and absorbs heat from the surroundings.

Ex-situ bioremediation: a process which involves removing the contaminated soil or water to another location before treatment.

Extent of reaction: The extent to which a reaction proceeds and the material actually reacting can be expressed by the extent of reaction in moles – conventionally relates the feed quantities to the amount of each component present in the product stream, after the reaction has proceeded to equilibrium, through the stoichiometry of the reaction to a term that appears in all reactions.

Facultative anaerobes: microorganisms that use (and prefer) oxygen when it is available, but can also use alternate electron acceptors such as nitrate under anaerobic conditions when necessary.

Fate: the ultimate disposition of the inorganic chemical in the ecosystem, either by chemical or biological transformation to a new form which (hopefully) is non-toxic (degradation) or, in the case of an ultimately persistent inorganic pollutants, by conversion to a less offensive chemicals or even by sequestration in a sediment or other location which is expected to remain undisturbed.

Fatty acids: carboxylic acids with long hydrocarbon side chains; most natural fatty acids have hydrocarbon chains that don't branch; any double bonds occurring in the chain are *cis* isomers – the side chains are attached on the same side of the double bond.

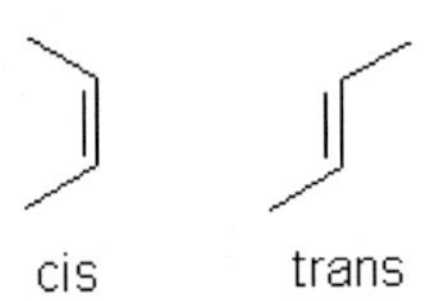

Fauna: all of the animal life of any particular region, ecosystem, or environment; generally, the naturally occurring or indigenous animal life (native animal life).

Feedstock: The biomass used in the creation of a particular biofuel (e.g., corn or sugarcane for ethanol, soybeans or rapeseed for biodiesel).

Fermentation: The process whereby microorganisms use an organic compound as both electron donor and electron acceptor, converting the compound to fermentation products such as organic acids, alcohols, hydrogen, and carbon dioxide; microbial metabolism in which a particular compound is used both as an electron donor and an electron acceptor resulting in the production of oxidized and reduced daughter products; the conversion of carbon-containing compounds by micro-organisms for production of fuels and chemicals such as alcohols, acids or energy-rich gases.

Fiber products; Products derived from fibers of herbaceous and woody plant materials; examples include pulp, composition board products, and wood chips for export.

Fine materials: Wood residues not suitable for chipping, such as planer shavings and sawdust.

Fingerprint: a chromatographic signature of relative intensities used in oil-oil or oil-source rock correlations; mass chromatograms of steranes derivatives or terpane derivatives are examples of fingerprints that can be used for qualitative or quantitative comparison of crude oil.

Flammability limits: a gas mixture will not burn when the composition is lower than the lower flammable limit (LFL); the mixture is also not combustible when the composition is above the upper flammability limit (UFL).

Flammable chemical (flammable substance: a chemical or substance is usually termed flammable if the flash point of the chemical or substance) is below 38°C (100°F.

Flash point: The temperature at which the vapor over a liquid will ignite when exposed to an ignition source. A liquid is considered to be flammable if its *flash point* is less than 60°C (140°F). *Flash point* is an extremely important factor in relation to the safety of spill cleanup operations. Gasoline and other light fuels can ignite under most ambient conditions and therefore are a serious hazard when spilled. Many freshly spilled crude oils also have low *flash points* until the lighter components have evaporated or dispersed.

Flash pyrolysis (fast pyrolysis): A process in which biomass is converted at approximately 500°C (930°F) into a liquid oil/char slurry which can be atomized in an entrained flow gasifier. Another attractive aspect is that the high-energy density slurry can be transported to a central processing facility. See Slow pyrolysis.

Flexible-fuel vehicle (flex-fuel vehicle): A vehicle that can run alternately on two or more sources of fuel; includes cars capable of running on gasoline and gasoline/ethanol mixtures, as well as cars that can run on both gasoline and natural gas.

Flora: the plant life occurring in a particular region or time; generally, the naturally occurring or indigenous plant life (native plant life).

Fluid coking: a continuous fluidised solids process that cracks feed thermally over heated coke particles in a reactor vessel to gas, liquid products, and coke.

Fluidized-bed boiler: A large, refractory-lined vessel with an air distribution member or plate in the bottom, a hot gas outlet in or near the top, and some provisions for introducing fuel; the fluidized bed is formed by blowing air up through a layer of inert particles (such as sand or limestone) at a rate that causes the particles to go into suspension and continuous motion.

Fluids: Liquids; also a generic term applied to all substances that flow freely, such as gases and liquids.

Fly ash: Small ash particles carried in suspension in combustion products.

Forest land: Land at least 10 percent stocked by forest trees of any size, including land that formerly had such tree cover and that will be naturally or artificially regenerated; includes transition zones, such as areas between heavily forested and non-forested lands that are at least 10 percent stocked with forest trees and forest areas adjacent to urban and built-up lands; also included are pinyon-juniper and chaparral areas; minimum area for classification of forest land is 1 acre.

Forest residues: Material not harvested or removed from logging sites in commercial hardwood and softwood stands as well as material resulting from forest management operations such as precommercial thinnings and removal of dead and dying trees.

Forest health: A condition of ecosystem sustainability and attainment of management objectives for a given forest area; usually considered to include green trees, snags, resilient stands growing at a moderate rate, and endemic levels of insects and disease.

Fossil fuel: Solid, liquid, or gaseous fuels formed in the ground after millions of years by chemical and physical changes in plant and animal residues under varying conditions of temperature and pressure. Crude oil, natural gas, coal, and oil shale are fossil fuels.

Fuel cell: A device that converts the energy of a fuel directly to electricity and heat, without combustion.

Fuel cycle: The series of steps required to produce electricity. The fuel cycle includes mining or otherwise acquiring the raw fuel source, processing and cleaning the fuel, transport, electricity generation, waste management and plant decommissioning.

Fuel oil: A heavy residue, black in color, used to generate power or heat by burning in furnaces.

Fuel treatment evaluator (FTE): A strategic assessment tool capable of aiding the identification, evaluation, and prioritization of fuel treatment opportunities.

Fuel wood - Wood used for conversion to some form of energy, primarily for residential use.

Fugitive emissions: Emissions that include losses from equipment leaks, or evaporative losses from impoundments, spills, or leaks.

Functional group: An atom or a group of atoms attached to the base structure of a compound that has similar chemical properties irrespective of the compound to which it is a part; a means of defining the characteristic physical and chemical properties of families of organic compounds.

Functional isomers: compounds which have the same molecular formula that possess different functional groups.

Fungi: non-photosynthetic organisms, larger than bacteria, aerobic and can thrive in more acidic media than bacteria. Important function is the breakdown of cellulose in wood and other plant materials.

Furnace - An enclosed chamber or container used to burn biomass in a controlled manner to produce heat for space or process heating.

Gas: Matter that has no definite volume or definite shape and always fills any space given in which it exists.

Gaseous nutrient injection: a process in which nutrients are fed to contaminated groundwater and soil via wells to encourage and feed naturally occurring microorganisms – the most common added gas is air in the presence of sufficient oxygen, microorganisms convert many organic contaminants to carbon dioxide, water, and microbial cell mass. In the absence of oxygen, organic contaminants are metabolized to methane, limited amounts of carbon dioxide, and trace amounts of hydrogen gas. Another gas that is added is methane. It enhances degradation by co-metabolism in which as bacteria consume the methane, they produce enzymes that react with the organic contaminant and degrade it to harmless minerals.

Gasification: A chemical or heat process used to convert carbonaceous material (such as coal, petroleum, and biomass) into gaseous components such as carbon monoxide and hydrogen.

Gasifier: A device for converting solid fuel into gaseous fuel; in biomass systems, the process is referred to as pyrolitic distillation.

Gasohol: A mixture of 10% anhydrous ethanol and 90% gasoline by volume; 7.5% anhydrous ethanol and 92.5% gasoline by volume; or 5.5% anhydrous ethanol and 94.5% gasoline by volume.

Gas to liquids (GTL): The process of refining natural gas and other hydrocarbons into longer-chain hydrocarbons, which can be used to convert gaseous waste products into fuels.

Gas turbine (combustion turbine): A turbine that converts the energy of hot compressed gases (produced by burning fuel in compressed air) into mechanical power; the used fuel is normally natural gas or fuel oil.

Gel point: The point at which a liquid fuel cools to the consistency of petroleum jelly.

Genetically modified organism (GMO): An organism whose genetic material has been modified through recombinant DNA technology, altering the phenotype of the organism to meet desired specifications.

Geological time: The span of time that has passed since the creation of the Earth and its components; a scale use to measure geological events millions of years ago.

Geometric isomers: Stereoisomers which differ in the geometry around either a carbon-carbon double bond or ring.

Geosphere: a term representing the solid earth, including soil, which supports most plant life.

Global warming: A gradual warming of the Earth due to a series of natural phenomena as well the burning of fossil fuels and industrial pollutants; the exact contribution of each of the phenomena is not known with any degree of certainty.

Glycerol: A small molecule with three alcohol groups ($HOCH_2CH(OH)CH_2OH$); basic building block of fats and oils.

$$\begin{array}{c} HOCH_2 \\ | \\ HOCH_2 \\ | \\ HOCH_2 \end{array}$$

Grassland pasture and range: All open land used primarily for pasture and grazing, including shrub and brush land types of pasture; grazing land with sagebrush and scattered mesquite; and all tame and native grasses, legumes, and other forage used for pasture or grazing; because of the diversity in vegetative composition, grassland pasture and range are not always clearly distinguishable from other types of pasture and range; at one extreme, permanent grassland may merge with cropland pasture, or grassland may often be found in transitional areas with forested grazing land.

Grease car: A diesel-powered automobile rigged post-production to run on used vegetable oil.

Greenhouse effect: The effect of certain gases in the Earth's atmosphere in trapping heat from the sun.

Greenhouse gases: Gases that trap the heat of the sun in the Earth's atmosphere, producing the greenhouse effect. The two major greenhouse gases are water vapor and carbon dioxide. Other greenhouse gases include methane, ozone, chlorofluorocarbons, and nitrous oxide.

Grid: An electric utility company's system for distributing power.

Grid system: An arrangement of power lines connecting power plants and consumers over a large area.

Growing stock: A classification of timber inventory that includes live trees of commercial species meeting specified standards of quality or vigor; cull trees are excluded.

Habitat: The area where a plant or animal lives and grows under natural conditions. Habitat includes living and non-living attributes and provides all requirements for food and shelter.

Hardwoods: Usually broad-leaved and deciduous trees.

Hazardous waste: a potentially dangerous chemical substance that has been discarded, abandoned, neglected, released or designated as a waste material, or one that may interact with other substances to pose a threat.

Haze: a term denoting decreased visibility due to the presence of particles.

Heating value: The maximum amount of energy that is available from burning a substance.

Heavy (crude) oil: oil that is more viscous that conventional crude oil, has a lower mobility in the reservoir but can be recovered through a well from the reservoir by the application of a secondary or enhanced recovery methods.

Hectare: Common metric unit of area, equal to 2.47 acres. 100 hectares = 1 square kilometer.

Herbaceous: Non-woody type of vegetation, usually lacking permanent strong stems, such as grasses, cereals and canola (rape).

Herbicide: A chemical that controls or destroys unwanted plants, weeds, or grasses.

Heteroatom: An element other than carbon and hydrogen that are commonly found in organic molecules, such as nitrogen, oxygen and the halogens.

Heteroatom compounds: chemical compounds that contain nitrogen and/or oxygen and/or sulfur and /or metals bound within their molecular structure(s).

Heterogeneous: varying in structure or composition at different locations in space.'

Heterotroph: an organism that cannot synthesize its own food and is dependent on complex organic substances for nutrition.

Heterotrophic bacteria: bacteria that utilize organic carbon as a source of energy; organisms that derive carbon from organic matter for cell growth.

Heterotrophs: organisms or chemicals that obtain their carbon from other organisms.

Homogeneous: having uniform structure or composition at all locations in space.

Homolog: A compound belonging to a series of compounds that differ by a repeating group; for example, propanol ($CH_3CH_2CH_2OH$), n-butanol ($CH_3CH_2CH_2CH_2OH$), and n-pentanol ($CH_3CH_2CH_2CH_2CH_2OH$) are homologs; they belong to the homologous series of alcohols: $CH_3(CH_2)_nOH$.

Homologous series: compounds which differ only by the number of CH_2 units present.

Humic substances: Dark, complex, heterogeneous mixtures of organic materials that form in the geological systems of the Earth from microbial transformations and chemical reactions that occur during the decay of organic biomolecules, polymers, and resides.

Hydraulic retention time: The time required for the in fluent feed to spent inside the reactor. Thus:

HRT = V/θ

In this equation, θ is the amount of feed inside the reactor and V is the total volume of the reactor. This equation is usually used for the determination of the quantity of the feedstock (sometime referred to as the *influent* used in a particular volume of a reactor.

Hydrocarbonaceous material: a material such as bitumen that is composed of carbon and hydrogen with other elements (heteroelements) such as nitrogen, oxygen, sulfur, and metals chemically combined within the structures of the constituents; even though carbon and hydrogen may be the predominant elements, there may be very few true hydrocarbons (q.v.).

Hydrocarbon compounds: chemical compounds containing only carbon and hydrogen.

Hydrodesulfurization: the removal of sulfur by hydrotreating (q.v.).

Hydrolysis: A chemical transformation process in which a chemical reacts with water; in the process, a new carbon-oxygen bond is formed with oxygen derived from the water molecule, and a bond is cleaved within the chemical between carbon and some functional group.

Hydrophilic: Water loving; the capacity of a molecular entity or of a substituent to interact with polar solvents, in particular with water, or with other polar groups; hydrophilic molecules dissolve easily in water, but not in fats or oils.

Hydrophilic colloids: generally, macromolecules, such as proteins and synthetic polymers, that are characterized by strong interaction with water resulting in spontaneous formation of colloids when they are placed in water.

Hydrophilicity: The tendency of a molecule to be solvated by water.

Hydrophobic: Fear of water; the tendency to repel water.

Hydrophobic colloids: colloids that interact to a lesser extent with water and are stable because of their positive or negative electrical charges.

Hydrophobic effect: The attraction of nonionic, non-polar compounds to surfaces that occurs due to the thermodynamic drive of these molecules to minimize interactions with water molecules.

Hydrophobic interaction: The tendency of hydrocarbons (or of lipophilic hydrocarbon-like groups in solutes) to form intermolecular aggregates in an aqueous medium, and analogous intramolecular interactions.

Hydroprocesses: refinery processes designed to add hydrogen to various products of refining.

Hydrosphere: the water areas of the Earth; also called the aquasphere.

Hydrotreating: the removal of heteroatomic (nitrogen, oxygen, and sulfur) species by treatment of a feedstock or product at relatively low temperatures in the presence of hydrogen.

Hypoxic: A condition of reduced oxygen content of air or a body of water detrimental to aerobic organisms.

Idle cropland: Land in which no crops were planted; acreage diverted from crops to soil-conserving uses (if not eligible for and used as cropland pasture) under federal farm programs is included in this component.

Incinerator: Any device used to burn solid or liquid residues or wastes as a method of disposal.

Inclined grate: A type of furnace in which fuel enters at the top part of a grate in a continuous ribbon, passes over the upper drying section where moisture is removed, and descends into the lower burning section. Ash is removed at the lower part of the grate.

Indirect emissions: emissions that are a consequence of the activities of the reporting entity, but occur at sources owned or controlled by another entity.

Indirect-injection engine: An older model of diesel engine in which fuel is injected into a pre-chamber, partly combusted, and then sent to the fuel-injection chamber.

Indirect liquefaction: Conversion of biomass to a liquid fuel through a synthesis gas intermediate step.

Industrial wood: All commercial round wood products except fuel wood.

Infiltration rate: the time required for water at a given depth to soak into the ground.

Inhibition: The decrease in rate of reaction brought about by the addition of a substance (inhibitor), by virtue of its effect on the concentration of a reactant, catalyst, or reaction intermediate; a component having no effect reduces the effect of another component.

Inoculum: a small amount of material (either liquid or solid) containing bacteria removed from a culture in order to start a new culture.

Inorganic: pertaining to, or composed of, chemical compounds that are not organic, that is, contain no carbon-hydrogen bonds; examples include chemicals with no carbon and those with carbon in non-hydrogen-linked forms.

Inorganic acid: an inorganic compound that elevates the hydrogen concentration in an aqueous solution; alphabetically, examples are:

Carbonic acid (HCO_3): an inorganic acid.

Hydrochloric acid (HCl): a highly corrosive, strong inorganic acid with many uses.

Hydrofluoric acid (HF): an inorganic acid that is highly reactive with silicate, glass, metals, and semi-metals.

Nitric acid (HNO_3): a highly corrosive and toxic strong inorganic acid.

Phosphoric acid: not considered a strong inorganic acid; found in solid form as a mineral and has many industrial uses.

Sulfuric acid: a highly corrosive inorganic acid. It is soluble in water and widely used.

Inorganic base: an inorganic compound that elevates the hydroxide concentration in an aqueous solution; alphabetically, examples are:

Ammonium hydroxide (ammonia water): a solution of ammonia in water.

Calcium hydroxide (lime water): a weak base with many industrial uses.

Magnesium hydroxide: referred to as brucite when found in its solid mineral form.

Sodium bicarbonate (baking soda): a mild alkali.

Sodium hydroxide (caustic soda): a strong inorganic base; used widely used in industrial and laboratory environments.

Inorganic chemistry: The study of inorganic compounds, specifically the structure, reactions, catalysis, and mechanism of action.

Inorganic compound: a compound that consists of an ionic component (an element from the Periodic Table) and an anionic component; a compound that does not contain carbon chemically bound to hydrogen; carbonates, bicarbonates, carbides, and carbon oxides are considered inorganic compounds, even though they contain carbon; a large number of compounds occur naturally while others may be synthesized; in all cases, charge neutrality of the compound is key to the structure and properties of the compound.

Inorganic reaction chemistry: inorganic chemical reactions fall into four broad categories: combination reactions, decomposition reactions, single displacement reactions, and double displacement reactions.

Inorganic salts: inorganic salts are neutral, ionically-bound molecules and do not affect the concentration of hydrogen in an aqueous solution.

Inorganic synthesis, the process of synthesizing inorganic chemical compounds, is used to produce many basic inorganic chemical compounds.

In situ: in its original place; unmoved; un excavated; remaining in the subsurface.

In-situ bioremediation: a process which treats the contaminated water or soil where it was found.

Interfacial Tension: the net energy per unit area at the interface of two substances, such as oil and water or oil and air. The air/liquid interfacial tension is often referred to as surface tension; the SI units for *interfacial tension* are milli-Newtons per meter (mN/m). The higher the *interfacial tension*, the less attractive the two surfaces are to each other and the more size of the interface will be minimized. Low surface tensions can drive the spreading of one fluid on another. The surface tension of an oil, together its viscosity, affects the rate at which spilled oil will spread over a water surface or into the ground.

Intermolecular forces: Forces of attraction that exist between particles (atoms, molecules, ions) in a compound.

Internal Standard (IS): a pure analyte added to a sample extract in a known amount, which is used to measure the relative responses of other analytes and surrogates that are components of the same solution. The *internal standard* must be an analyte that is not a sample component.

Intramolecular: (i) Descriptive of any process that involves a transfer (of atoms, groups, electrons, etc.) or interactions (such as forces) between different parts of the same molecular entity; (ii) relating to a comparison between atoms or groups within the same molecular entity.

Intrinsic bioremediation: A type of bioremediation that manages the innate capabilities of naturally occurring microbes to degrade contaminants without taking any engineering steps to enhance the process.

Inversions: conditions characterized by high atmospheric stability which limit the vertical circulation of air, resulting in air stagnation and the trapping of air pollutants in localized areas.

Ionic bond: A chemical bond or link between two atoms due to an attraction between oppositely charged (positive-negative) ions.

Ionic bonding: Chemical bonding that results when one or more electrons from one atom or a group of atoms is transferred to another. Ionic bonding occurs between charged particles.

Ionic compounds: Compounds where two or more ions are held next to each other by electrical attraction.

Ionic liquids: An ionic liquid is a salt in the liquid state or a salt with a melting point lower than 100°C (212°F); variously called liquid electrolytes, ionic melts, ionic fluids, fused salts, liquid salts, or ionic glasses; powerful solvents and electrically conducting fluids (electrolytes).

Ionic radius: a measure of ion size in a crystal lattice for a given coordination number (CN); metal ions are smaller than their neutral atoms, and nonmetallic anions are larger than the atoms from which they are formed; ionic radii depend on the element, its charge, and its coordination number in the crystal lattice; ionic radii are expressed in angstrom units of length (Å).

Ionization energy: the ionization energy is the energy required to remove an electron completely from its atom, molecule, or radical.

Ionization potential: the energy required to remove a given electron from its atomic orbital; the values are given in electron volts (eV).

Irreversible reaction: A reaction in which the reactant(s) proceed to product(s), but there is no significant backward reaction:

$nA + mB \rightarrow \text{Products}$

In this reaction, the products do not recombine or change to form reactants in any appreciable amount.

Isomers: Compounds that have the same number and types of atoms – the same molecular formula – but differ in the structural formula, i.e., the manner in which the atoms are combined with each other.

Isotope: a variant of a chemical element which differs in the number of neutrons in the atom of the element; all isotopes of a given element have the same number of protons in each atom and different isotopes of a single element occupy the same position on the periodic table of the elements.

IUPAC: International Union of Pure and Applied Chemistry; the organization that establishes the system of nomenclature for organic and inorganic compounds using prefixes and suffixes, developed in the late 19th Century.

Joule: Metric unit of energy, equivalent to the work done by a force of one Newton applied over distance of one meter (= 1 kg m^2/s^2); one joule (J) = 0.239 calories (1 calorie = 4.187 J).

Kelvin: The SI Unit of temperature. It is the temperature in degrees Celsius plus 273.15.

Ketone: An organic compound that contains a carbonyl group (R_1COR_2).

Kerosene: A light middle distillate that in various forms is used as aviation turbine fuel or for burning in heating boilers or as a solvent, such as white spirit.

Kilovolt (kV): 1 000 volts. The amount of electric force carried through a high-voltage transmission line is measured in kilovolts.

Kilowatt (kW): A measure of electrical power equal to 1 000 watts. 1 kW = 3,413 Btu/hr = 1,341 horsepower.

Kilowatt: (kW): A measure of electrical power equal to 1,000 watts. 1 kW = 3412 Btu/hr = 1.341 horsepower.

Kilowatt hour: (kWh): A measure of energy equivalent to the expenditure of one kilowatt for one hour. For example, 1 kWh will light a 100-watt light bulb for 10 hours. 1 kWh = 3412 Btu.

Lag phase: the growth interval (adaption phase) between microbial inoculation and the start of the exponential growth phase during which there is little or no microbial growth.

Landfill gas: A type of biogas that is generated by decomposition of organic material at landfill disposal sites. Landfill gas is approximately 50 percent methane. See also biogas.

Lignin: Structural constituent of wood and (to a lesser extent) other plant tissues, which encrusts the walls and cements the cells together.

Limiting reactant: The reactant that is present in the smallest stoichiometric amount and which determines the maximum extent to which a reaction can proceed; if the reaction is 100% complete then all of the limiting reactant is consumed and the reaction can proceed no further.

Limnology: the branch of science dealing with characteristics of freshwater, including biological properties as well as chemical and physical properties.

Lipophilic: F-loving; applied to molecular entities (or parts of molecular entities) tending to dissolve in fatlike (e.g., hydrocarbon) solvents.

Lipophilicity: The affinity of a molecule or a moiety (portion of a molecular structure) for a lipophilic (fat soluble) environment. It is commonly measured by its distribution behavior in a biphasic system, either liquid-liquid (e.g., partition coefficient in octanol/water).

Lithosphere: The part of the geosphere consisting of the outer mantle and the crust that is directly involved with environmental processes through contact with the atmosphere, the hydrosphere, and living things; varies from (approximately) 40 60 miles in thickness; also called the terrestrial biosphere.

Loading rate: The amount of a chemical that can be absorbed on soil on a per volume of soil basis.

LTU: Land Treatment Unit; a physically delimited area where contaminated land is treated to remove/minimize contaminants and where parameters such as moisture, pH, salinity, temperature and nutrient content can be controlled.

Live cull: A classification that includes live cull trees; when associated with volume, it is the net volume in live cull trees that are 5.0 inches in diameter and larger.

Logging residues: The unused portions of growing-stock and non-growing-stock trees cut or killed logging and left in the woods.

M85: An alcohol fuel mixture containing 85 percent methanol and 15 percent gasoline by volume. Methanol is typically made from natural gas, but can also be derived from the fermentation of biomass.

Macromolecule: A large molecule of high molecular mass composed of more than 100 repeated monomers (single chemical units of lower relative mass); a large complex molecule formed from many simpler molecules.

Masking: Occurs when two components have opposite, cancelling effects such that no effect is observed from the combination.

Mass number: The number of protons plus the number of neutrons in the nucleus of an atom.

Matter: Any substance that has inertia and occupies physical space; can exist as solid, liquid, gas, plasma, or foam.

Megawatt: (MW) A measure of electrical power equal to one million watts (1,000 kW).

Melting point: The temperature when matter is converted from solid to liquid.

Mesophilic digestion: Takes place optimally around 37 to 41°C or at ambient temperatures between 20 to 45°C where mesophiles are the primary micro-organism present.

Metabolic by-product: a product of the reaction between an electron donor and an electron acceptor; metabolic by-products include volatile fatty acids, daughter products of chlorinated aliphatic hydrocarbons, methane, and chloride.

Metabolism: the physical and chemical processes by which foodstuffs are synthesized into complex elements, complex substances are transformed into simple ones, and energy is made available for use by an organism; thus all biochemical reactions of a cell or tissue, both synthetic and degradative, are included; the sum of all of the enzyme-catalyzed reactions in living cells that transform organic molecules into simpler compounds used in biosynthesis of cellular components or in extraction of energy used in cellular processes.

Metabolize: a product of metabolism.

Methane (CH_4): A flammable, explosive, colorless, odorless, tasteless gas that is slightly soluble in water and soluble in alcohol and ether; boils at -161.6°C (-259°F) and freezes at -182.5ºC (-297ºF); formed in marshes and swamps from decaying organic matter, and is a major explosion hazard underground; a major constituent (up to 97%) of natural gas, and is used as a source of petrochemicals and as a fuel.

Methane loss: The amount of methane that does not end up in the biomethane stream; usually expressed as amount lost per total methane gas treated.

Methane number: A measure of the gas resistance to knocking in an internal combustion engine; by definition, methane has a methane number of 100.

Methane slip: Methane that is lost to the atmosphere in the upgrading process; usually expressed as amount lost per total methane gas treated.

Methanogens: strictly anaerobic archaebacteria, able to use only a very limited spectrum of substrates (for example, molecular hydrogen, formate, methanol, methylamine, carbon monoxide or acetate) as electron donors for the reduction of carbon dioxide to methane.

Methanogenic: the formation of methane by certain anaerobic bacteria (methanogens) during the process of anaerobic fermentation.

Methanol: A fuel typically derived from natural gas, but which can be produced from the fermentation of sugars in biomass.

Micelles: a spherical cluster formed by the aggregation of soap molecules in water.

Microbe: the shortened term for microorganism.

Microclimate: a highly localized climatic conditions; the climate that organisms and objects on the surface are exposed to close to ground, under rocks, and surrounded by vegetation and it often quite different form the surrounding macroclimate.

Microcosm: a diminutive, representative system analogous to a larger system in composition, development, or configuration.

Microorganism (micro-organism): an organism of microscopic size that is capable of growth and reproduction through biodegradation of food sources, which can include hazardous contaminants; microscopic organisms including bacteria, yeasts, filamentous fungi, algae, and protozoa; a living organism too small to be seen with the naked eye; includes bacteria, fungi, protozoans, microscopic algae, and viruses.

Micro-turbine: A small combustion turbine with an output of 25 to 500 kW; composed of a compressor, combustor, turbine, alternator, recuperator and generator. Relative to other technologies for small-scale power generation, micro-turbines offer a number of advantages, including: a small number of moving parts, compact size, light weight, greater efficiency, lower emissions, lower electricity costs, potential for low cost mass production, and opportunities to utilize waste fuels.

Million: 1×10^6

Mill residue: Wood and bark residues produced in processing logs into lumber, plywood, and paper.

Mineralization: the biological process of complete breakdown of organic compounds, whereby organic materials are converted to inorganic products (e.g., the conversion of hydrocarbons to carbon dioxide and water); the release of inorganic chemicals from organic matter in the process of aerobic or anaerobic decay.

Mixed waste: any combination of waste types with different properties or any waste that contains both hazardous waste and source, special nuclear, or byproduct material; as defined by the US EPA, mixed Waste contains both hazardous waste (as defined by RCRA and its amendments) and radioactive waste.

Modified/unmodified diesel engine: Traditional diesel engines must be *modified* to heat the oil before it reaches the fuel injectors in order to handle straight vegetable oil. Modified, any diesel engine can run on veggie oil; without modification, the oil must first be converted to biodiesel.

Moiety: A term generally used to signify part of a molecule, e.g., in an ester R^1COOR^2, the alcohol moiety is R^2O.

Moisture content: (MC): The weight of the water contained in wood, usually expressed as a percentage of weight, either oven-dry or as received.

Moisture content, dry basis: Moisture content expressed as a percentage of the weight of oven-wood, i.e.: [(weight of wet sample - weight of dry sample) / weight of dry sample] x 100.

Moisture content, wet basis: Moisture content expressed as a percentage of the weight of wood as-received, i.e.: [(weight of wet sample - weight of dry sample) / weight of wet sample] x 100.

Molality (m): The gram moles of solute divided by the kilograms of solvent).

Molar: A term expressing molarity, the number of moles of solute per liters of solution.

Molarity (M): The gram moles of solute) divided by the liters of solution).

Mole: A collection of 6.022×10^{23} number of objects. Usually used to mean molecules.

Mole fraction: The number of moles of a particular substance expressed as a fraction of the total number of moles.

Molecular weight: The mass of one mole of molecules of a substance.

Molecule: The smallest unit in a chemical element or compound that contains the chemical properties of the element or compound.

Mole fraction: the number of moles of a component of a mixture divided by the total number of moles in the mixture.

Monoaromatic: aromatic hydrocarbons containing a single benzene ring.

Monosaccharide: A simple sugar such as fructose or glucose that cannot be decomposed by hydrolysis; colorless crystalline substances with a sweet taste that have the same general formula, $C_nH_{2n}O_n$.

MTBE: Methyl tertiary butyl ether is highly refined high octane light distillate used in the blending of petrol.

Municipal solid waste (MSW): All types of solid waste generated by a community (households and commercial establishments), usually collected by local government bodies.

Native fauna: the native and indigenous animal of an area.

Native flora: the native and indigenous plant life of an area.

Native flora: the native and indigenous flora of an area.

Natural organic matter (NOM): An inherently complex mixture of polyfunctional organic molecules that occurs natural in the environment and is typically derived from the decay of floral and faunal remains; although they do occur naturally, the fossil fuels (coal, crude oil, and natal gas) are usually not included in the term *natural organic matter*.

Nitrate enhancement: a process in which a solution of nitrate is sometimes added to groundwater to enhance anaerobic biodegradation.

Nitrogen fixation: The transformation of atmospheric nitrogen into nitrogen compounds that can be used by growing plants.

Nitrogen oxides (NOx): Products of combustion that contribute to the formation of smog and ozone.

Non-forest land: Land that has never supported forests and lands formerly forested where use of timber management is precluded by development for other uses; if intermingled in forest areas, unimproved roads and non-forest strips must be more than 120 feet wide, and clearings, etc., must be more than 1 acre in area to qualify as non-forest land.

Non-attainment area: Any area that does not meet the national primary or secondary ambient air quality standard established (by the Environmental Protection Agency) for designated pollutants, such as carbon monoxide and ozone.

Non-industrial private - An ownership class of private lands where the owner does not operate wood processing plants.

Non-point source pollution: pollution that does not originate from a specific source. Examples of non-point sources of pollution include the following: (i) sediments from construction, forestry operations and agricultural lands, (ii) bacteria and micro-organisms from failing septic systems and pet wastes, (iii) nutrients from fertilizers and yard debris, (iv) pesticides from agricultural areas, golf courses, athletic fields and residential yards, oil, grease, antifreeze, and metals washed from roads, parking lots and driveways, (v) toxic chemicals and cleaners that were not disposed of correctly, and (vi) litter thrown onto streets, sidewalks and beaches, or directly into the water by individuals. See Point Source Pollution.

Normality (N): The gram equivalents of solute divided by the liters of solution).

Nucleophile: a chemical reagent that reacts by forming covalent bonds with electronegative atoms and compounds.

Nuclide: a nucleus rather than to an atom - isotope (the older term) it is better known than the term nuclide, and is still sometimes used in contexts where the use of the term nuclide might be more appropriate; identical nuclei belong to one nuclide, for example each nucleus of the carbon-13 nuclide is composed of 6 protons and 7 neutrons.

Nutrients: major elements (for example, nitrogen and phosphorus) and trace elements (including sulfur, potassium, calcium, and magnesium) that are essential for the growth of organisms.

Oceanography: the science of the ocean and its physical and chemical characteristics.

Octane: A flammable liquid (C_8H_{18}) found in petroleum and natural gas; there are 18 different octane isomers which have different structural formulas but share the molecular formula C_8H_{18}; used as a fuel and as a raw material for building more complex organic molecules.

Octanol-water partition coefficient (K_{ow}): the equilibrium ratio of a chemical's concentration in octanol (an alcoholic compound) to its concentration in the aqueous phase of a two-phase octanol-water system, typically expressed in log units (log K_{ow}); K_{ow} provides an indication of a chemical's solubility in fats (lipophilicity), its tendency to bioconcentrate in aquatic organisms, or sorb to soil or sediment.

Off-gas: The gas that is left when methane has been separated from the biogas in the upgrading process; contains mostly carbon dioxide but, in general, may also contain small amounts of methane.

Oil equivalent: The tonne of oil equivalent (toe) is a unit of energy: the amount of energy released by burning one tonne of crude oil, approx. 42 GJ.

Oil from tar sand: synthetic crude oil (q.v.).

Oil mining: application of a mining method to the recovery of bitumen.

Oleophilic: oil seeking or oil loving (e.g., nutrients that stick to or dissolve in oil).

Open-loop biomass: Biomass that can be used to produce energy and bioproducts even though it was not grown specifically for this purpose; include agricultural livestock waste, residues from forest harvesting operations and crop harvesting.

Order of reaction: a chemical rate process occurring in systems for which concentration changes (and hence the rate of reaction) are not themselves measurable, provided it is possible to measure a chemical flux.

Organic: Compounds that contain carbon chemically bound to hydrogen; often contain other elements (particularly O, N, halogens, or S); chemical compounds based on carbon that also contain hydrogen, with or without oxygen, nitrogen, and other elements.

Organic carbon (soil) partition coefficient (K_{oc}): the proportion of a chemical sorbed to the solid phase, at equilibrium in a two-phase, water/soil or water/sediment system expressed on an organic carbon basis; chemicals with higher K_oc values are more strongly sorbed to organic carbon and, therefore, tend to be less mobile in the environment.

Organic chemistry: The study of compounds that contain carbon chemically bound to hydrogen, including synthesis, identification, modelling, and reactions of those compounds.

Organic liquid nutrient injection: an enhanced bioremediation process in which an organic liquid, which can be naturally degraded and fermented in the subsurface to result in the generation of hydrogen. The most commonly added for enhanced anaerobic bioremediation include lactate, molasses, hydrogen release compounds (HRCs¨), and vegetable oils.

Organic material: dead plant and animal tissues that originates from living sources such as plants, insects, and microbes.

Organochlorine compounds (chlorinated hydrocarbons): Organic pesticides that contain chlorine, carbon, and hydrogen (such as DDT); these pesticides affect the central nervous system.

Organo-metallic compounds: compounds that include carbon atoms directly bonded to a metal ion.

Organophosphorus compound: A compound containing phosphorus and carbon; many pesticides and most nerve agents are organophosphorus compounds, such as Malathion.

Osmotic potential: expressed as a negative value (or zero), indicates the ability of the soil to dissolve salts and organic molecules; the reduction of soil water osmotic potential is caused by the presence of dissolved solutes.

Outer-sphere adsorption complex: Sorption of an ion or molecule to a solid surface where waters of hydration are interposed between the sorbate and sorbent.

Oven dry: the weight of a soil after all water has been removed by heating in an oven at a specified temperature (usually in excess of 100°C, 212°F) for water; temperatures will vary if other solvents have been used.

Oxidation: the transfer of electrons away from a compound, such as an organic contaminant; the coupling of oxidation to reduction (see below) usually supplies energy that microorganisms use for growth and reproduction. Often (but not always), oxidation results in the addition of an oxygen atom and/or the loss of a hydrogen atom.

Oxidation number: A number assigned to each atom to help keep track of the electrons during a redox-reaction.

Oxidation reaction: A reaction where a substance loses electrons.

Oxidation-reduction-reaction: A reaction involving the transfer of electrons; a reaction (redox reactions) that involves oxidation of one reactant and reduction of another.

Oxidize: The transfer of electrons away from a compound, such as an organic contaminant. The coupling of oxidation to reduction (see below) usually supplies energy that microorganisms use for growth and reproduction. Often (but not always), oxidation results in the addition of an oxygen atom and/or the loss of a hydrogen atom.

Oxygen enhancement with hydrogen peroxide: an alternative process to pumping oxygen gas into groundwater involves injecting a dilute solution of hydrogen peroxide. Its chemical formula is H_2O_2, and it easily releases the extra oxygen atom to form water and free oxygen. This circulates through the contaminated groundwater zone to enhance the rate of aerobic biodegradation of organic contaminants by naturally occurring microbes. A solid peroxide product [e.g., oxygen releasing compound (ORC)] can also be used to increase the rate of biodegradation.

Oxygenate: A substance which, when added to gasoline, increases the amount of oxygen in that gasoline blend; includes fuel ethanol, methanol, and methyl tertiary butyl ether (MTBE).

Ozone (O_3): A form of oxygen containing three atoms instead of the common two (O_2); formed by high-energy ultraviolet radiation reacting with oxygen.

PAHs: polycyclic aromatic hydrocarbons. Alkylated *PAHs* are *alkyl group* derivatives of the parent *PAHs*. The five target alkylated *PAHs* referred to in this report are the alkylated naphthalene, phenanthrene, dibenzothiophene, fluorene, and chrysene series.

Particulate: A small, discrete mass of solid or liquid matter that remains individually dispersed in gas or liquid emissions.

Pathogen: an organism that causes disease (e.g., some bacteria or viruses).

Pay zone thickness: the depth of a tar sand deposit from which bitumen (or a product) can be recovered.

Permeability: The capability of the soil to allow water or air movement through it. The quality of the soil that enables water to move downward through the profile, measured as the number of inches per hour that water moves downward through the saturated soil.

Permeable reactive barrier (PRB): a subsurface emplacement of reactive materials through which a dissolved contaminant plume must move as it flows, typically under natural gradient and treated water exits the other side of the permeable reactive barrier.

Pesticide: A chemical that is designed and produced to control for pest control, including weed control.

pH: An expression of the intensity of the alkaline or acidic strength of water; values range from 0-14, where 0 is the most acidic, 14 is the most alkaline and 7 is neutral.

Photic zone: The upper layer within bodies of water reaching down to about 200 meters, where sunlight penetrates and promotes the production of photosynthesis; the richest and most diverse area of the ocean.

Photocatalysis: The acceleration of a photoreaction in the presence of a catalyst in which light is absorbed by a substrate that is typically adsorbed on a (solid) catalyst.

Photocatalyst; A material that can absorb light, producing electron–hole pairs that enable chemical transformations of the reaction participants and regenerate its chemical composition after each cycle of such interactions.

Photosynthesis: Process by which chlorophyll-containing cells in green plants concert incident light to chemical energy, capturing carbon dioxide in the form of carbohydrates.

Phototrophs: organisms or chemicals that utilize light energy from photosynthesis.

Phytodegradation: the process in which some plant species can metabolize VOC contaminants. The resulting metabolic products include trichloroethanol, trichloroacetic acid, and dichloracetic acid; mineralization products are probably incorporated into insoluble products such as components of plant cell walls.

Phytovolatilization; the process in which VOCs are taken up by plants and discharged into the atmosphere during transpiration.

PM_{10}: Particulate matter below 10 microns in diameter; this corresponds to the particles inhalable into the human respiratory system, and its measurement uses a size selective inlet.

$PM_{2.5}$: Particulate matter below 2.5 microns in diameter; this is closer to, but slightly finer than, the definitions of respirable dust that have been used for many years in industrial hygiene to identify dusts which will penetrate the lungs.

pOH: A measure of the basicity of a solution; the negative log of the concentration of the hydroxide ions.

Point emissions: Emissions that occur through confined air streams as found in stacks, ducts, or pipes.

Point source pollution: any single identifiable source of pollution from which pollutants are discharged, such as a pipe. Examples of point sources include: (i) discharges from wastewater treatment plants, (ii) operational wastes from industries, and (iii) combined sewer outfalls. See Non-point Source Pollution.

Polar compound: an organic compound with distinct regions of positive and negative charge. *Polar compounds* include alcohols, such as sterols, and some *aromatics*, such as monoaromatic-steroids. Because of their polarity, these compounds are more soluble in polar solvents, including water, compared to non-polar compounds of similar molecular structure.

Pollutant: Either (i) a non-indigenous chemical that is present in the environment or (ii) an indigenous chemical that is present in the environment in greater than the natural concentration. Both types of pollutants are the result of human activity and have an overall detrimental effect upon the environment or upon something of value in that environment.

Polymer: A large molecule made by linking smaller molecules (monomers) together.

Positional isomers: Compounds which differ only in the position of a functional group; 2-pentanol and 3-pentanol are positional isomers.

Pour point: the lowest temperature at which oil will pour or flow when it is chilled without disturbance under definite conditions.

Precipitation: Formation of an insoluble product that occurs via reactions between ions or molecules in solution.

Primary substrates: The electron donor and electron acceptor that are essential to ensure the growth of microorganism; these compounds can be viewed as analogous to the food and oxygen that are required for human growth and reproduction.

Process heat: Heat used in an industrial process rather than for space heating or other housekeeping purposes.

Producer gas: Fuel gas high in carbon monoxide (CO) and hydrogen (H2), produced by burning a solid fuel with insufficient air or by passing a mixture of air and steam through a burning bed of solid fuel.

Producers: organisms or chemicals that utilize light energy and store it as chemical energy.

Prokaryotes: microorganisms that lack a nuclear membrane so that their nuclear genetic material is more diffuse in the cell.

Propagule: any part of a plant (e.g. bud) that facilitates dispersal of the species and from which a new plant may form.

Protease: An enzyme that helps proteolysis: protein catabolism by hydrolysis of peptide bonds.

Protozoa: microscopic animals consisting of single eukaryotic cells.

Pulpwood: Round wood, whole-tree chips, or wood residues that are used for the production of wood pulp.

Putrescible waste: Solid waste that contains organic matter capable of being decomposed by microorganisms and of such a character and proportion as to cause obnoxious odors and to be capable of attracting or providing food for birds or animals.

Pyrolysis: The thermal decomposition of biomass at high temperatures (greater than 205°C, or 400°F) in the absence of air; the end product of pyrolysis is a mixture of solids (char), liquids (oxygenated oils), and gases (methane, carbon monoxide, and carbon dioxide) with proportions determined by operating temperature, pressure, oxygen content, and other conditions.

Quad: One quadrillion Btu (10^{15} Btu) = ca. 172 million barrels of oil equivalent.

Rate: a derived quantity in which time is a denominator quantity so that the progress of a reaction is measured with time.

Rate constant, k; See *Order of reaction*.

Rate-controlling step (rate-limiting step, rate-determining step): the elementary reaction having the largest control factor exerts the strongest influence on the rate; a step having a control factor much larger than any other step is said to be rate-controlling.

Raw gas: Untreated gas.

Reaction rate: The change in concentration of the starting chemical in given time interval.

Reaction (irreversible): A reaction in which the reactant(s) proceed to product(s), but there is no significant backward reaction:

$$nA + mB \rightarrow \text{Products}$$

In this reaction, the products do not recombine or change to form reactants in any appreciable amount.

Receptor: an object (animal, vegetable, or mineral) or a locale that is affected by the pollutant.

Redox (reduction-oxidation reactions): oxidation and reduction occur simultaneously; in general, the oxidizing agent gains electrons in the process (and is reduced) while the reducing agent donates electrons (and is oxidized).

Refractory lining: A lining, usually of ceramic, capable of resisting and maintaining high temperatures.

Refuse-derived fuel (RDF): Fuel prepared from municipal solid waste; non-combustible materials such as rocks, glass, and metals are removed, and the remaining combustible portion of the solid waste is chopped or shredded.

Relative density: the density of a gas divided by the density of air.

Releases: On-site discharge of a toxic chemical to the surrounding environment; includes emissions to the air, discharges to bodies of water, releases at the facility to land, as well as contained disposal into underground injection wells.

Releases (to air, point and fugitive air emissions): All air emissions from industry activity; point emissions occur through confined air streams as found in stacks, ducts, or pipes; fugitive emissions include losses from equipment leaks, or evaporative losses from impoundments, spills, or leaks.

Releases (to land): Disposal of toxic chemicals in waste to on-site landfills, land treated or incorporation into soil, surface impoundments, spills, leaks, or waste piles. These activities must occur within the boundaries of the facility for inclusion in this category.

Release (to underground injection): A contained release of a fluid into a subsurface well for the purpose of waste disposal.

Releases (to water, surface water discharges): Any releases going directly to streams, rivers, lakes, oceans, or other bodies of water: any estimates for storm water runoff and non-point losses must also be included.

Renewable energy: see Bioenergy

Renewable resource: A resource that can replace itself; virtually inexhaustible in duration, but limited in the amount of energy that is available per unit of time. Some (such as geothermal and biomass) may be stock-limited in that stocks are depleted by use, but on a time scale of decades, or perhaps centuries, they can probably be replenished. Renewable energy resources include: biomass, hydro, geothermal, solar and wind. In the future they could also include the use of ocean thermal, wave, and tidal action technologies. Utility renewable resource applications include bulk electricity generation, onsite electricity generation, distributed electricity generation, non-grid-connected generation, and demand-reduction (energy efficiency) technologies.

Residues: Bark and woody materials that are generated in primary wood-using mills when round wood products are converted to other products.

Residuum (pl. residua, also known as resid or resids): the non-volatile portion of petroleum that remains as residue after refinery distillation; hence, atmospheric residuum, vacuum residuum.

Respiration: the process of coupling oxidation of organic compounds with the reduction of inorganic compounds, such as oxygen, nitrate, iron (III), manganese (IV), and sulfate.

Reversible reaction: A reaction in which the products can revert to the starting materials (A and B). Thus:

$nA + MB \leftrightarrow$ Products

Rhizodegradation: the process whereby plants modify the environment of the root zone soil by releasing root exudates and secondary plant metabolites. Root exudates are typically photosynthetic carbon, low molecular weight molecules, and high molecular weight organic acids. This complex mixture modifies and promotes the development of a microbial community in the rhizosphere. These secondary metabolites have a potential role in the development of naturally occurring contaminant-degrading enzymes.

Rhizosphere: the soil environment encompassing the root zone of the plant.

Rotation: Period of years between establishment of a stand of timber and the time when it is considered ready for final harvest and regeneration.

Round wood products - Logs and other round timber generated from harvesting trees for industrial or consumer use.

Sandstone: a sedimentary rock formed by compaction and cementation of sand grains; can be classified according to the mineral composition of the sand and cement.

Saturated steam: Steam at boiling temperature for a given pressure.

Secondary oil recovery: application of energy (e.g., water flooding) to recovery of crude oil from a reservoir after the yield of crude oil from primary recovery diminishes.

Secondary wood processing mills: A mill that uses primary wood products in the manufacture of finished wood products, such as cabinets, moldings, and furniture.

Silage: Fermented high-moisture stored fodder which can be fed to cattle, sheep, and other such ruminants (cud-chewing animals) or used as a feedstock for anaerobic digesters.

Single displacement reactions: reactions where one element trades places with another element in a compound. These reactions come in the general form of:

$A + BC \rightarrow AC + B$

Examples include:

(i) magnesium replacing hydrogen in water to make magnesium hydroxide and hydrogen gas:

$Mg + 2H_2O \rightarrow Mg(OH)_2 + H_2$

(ii) the production of silver crystals when a copper metal strip is dipped into silver nitrate:

$Cu(s) + 2AgNO_3(aq) \rightarrow 2Ag(s) + Cu(NO_3)_2(aq)$

Slow pyrolysis: A process in which biomass is converted at approximately 500°C (930°F) into a char. The char is pulverized and directed to the entrained flow gasifier. Since the pyrolysis also produces an excess of gas, the gas needs to be utilized efficiently. See Flash Pyrolysis.

Sludge: Bio-solids separated from liquids during processing; may contain up to 97% v/v water by volume.

Soil organic matter: Living and partially decayed (non-living) materials as well as assemblages of biomolecules and transformation products of organic residue decay known as humic substances.

Solubility: the amount of a substance (solute) that dissolves in a given amount of another substance (solvent); a measure of the solubility of an inorganic chemical in a solvent, such as water; generally, ionic substances are soluble in water and other polar solvents while the non-polar, covalent compounds are more soluble in the non-polar solvents; in sparingly soluble, slightly soluble or practically insoluble salts, degree of solubility in water and occurrence of any precipitation process may be determined from the solubility product, Ksp, of the salt – the smaller the Ksp value, the lower the solubility of the salt in water.

Soluble: capable of being dissolved in a solvent.

Solute: Any dissolved substance in a solution.

Solution: Any liquid mixture of two or more substances that is homogeneous.

Solvolysis: generally, a reaction with a solvent, involving the rupture of one or more bonds in the reacting solute: more specifically the term is used for substitution, elimination, or fragmentation reactions in which a solvent species is the nucleophile; hydrolysis, if the solvent is water or alcoholysis if the solvent is an alcohol.

Sorbate: Sometimes referred to as adsorbate, is the solute that adsorbs on the solid phase.

Sorbent (adsorbent): The solid phase or substrate onto which the sorbate sorbs; the solid phase may be more specifically referred to as an absorbent or adsorbent if the mechanism of removal is known to be absorption or adsorption, respectively.

Sorption: A general term that describes removal of a solute from solution to a contiguous solid phase and is used when the specific removal mechanism is not known.

Sorption isotherm: Graphical representation of surface excess (i.e., the amount of substance sorbed to a solid) relative to sorptive concentration in solution after reaction at fixed temperature, pressure, ionic strength, pH, and solid-to-solution ratio.

Sorptive: Ions or molecules in solution that could potentially participate in a sorption reaction.

Specific gravity: the mass (or weight) of a unit volume of any substance at a specified temperature compared to the mass of an equal volume of pure water at a standard temperature.

Stable: as applied to chemical species, the term expresses a thermodynamic property, which is quantitatively measured by relative molar standard Gibbs energies; a chemical species A is more stable than its isomer B under the same standard conditions.

Stand (of trees): A tree community that possesses sufficient uniformity in composition, constitution, age, spatial arrangement, or condition to be distinguishable from adjacent communities.

Starch: A polysaccharide containing glucose (long-chain polymer of amylose and amylopectin) that is the energy storage reserve in plants.

Steam turbine: A device for converting energy of high-pressure steam (produced in a boiler) into mechanical power which can then be used to generate electricity.

Stereochemistry: The branch of organic chemistry that deals with the three dimensional structure of molecules.

Stereogenic carbon (asymmetric carbon): A carbon atom which is bonded to four different groups or atoms; a chiral molecule must contain a stereogenic carbon, and therefore has no plane of symmetry and is not superimposable on its mirror image.

Stereoisomers: Isomers which have the same bonding connectivity but have a different three-dimensional structure; examples would be cis-2-butene and trans-2-butene (geometric isomers), and the left and right handed forms of 2-butanol (enantiomers).

Straight vegetable oil (SVO): Any vegetable oil that has not been optimized through the process of transesterification.

Stratosphere: the portion of the atmosphere of the Earth where ozone is formed by the reaction of ultraviolet light on dioxygen molecules.

Strong acid: an acid that releases H^+ ions easily – examples are hydrochloric acid and sulfuric acid.

Strong base: a basic chemical that accept and hold proton tightly – an example is the hydroxide ion.

Superheated steam: Steam which is hotter than boiling temperature for a given pressure.

Structural formula: A convention used to represent the structures of organic molecules in which not all the valence electrons of the atoms are shown.

Structural isomerism: The relationship between two compounds which have the same molecular formula, but different structures; they may be further classified as functional, positional, or skeletal isomers. This relation is also called constitutional isomerism.

Substrate: A chemical species of particular interest, of which the reaction with some other chemical reagent is under observation (e.g., a compound that is transformed under the influence of a catalyst); also the component in a nutrient medium, supplying microorganisms with carbon (C-substrate), nitrogen (N-substrate) as food needed to grow.

Sustainable: An ecosystem condition in which biodiversity, renewability, and resource productivity are maintained over time.

Sustainable development: Development and economic growth that meets the requirements of the present generation without compromising the ability of future generations to meet their needs; a strategy seeking a balance between development and conservation of natural resources.

Sustainable enhancement: an intervention action that continues until such time that the enhancement is no longer required to reduce contaminant concentrations or fluxes.

Synthetic crude oil (syncrude): a hydrocarbon product produced by the conversion of coal, oil shale, or tar sand bitumen that resembles conventional crude oil; can be refined in a petroleum refinery.

Synthetic ethanol: Ethanol produced from ethylene, a petroleum by-product.

Tar sand (bituminous sand): A formation in which the bituminous material (bitumen) is found as a filling in veins and fissures in fractured rocks or impregnating relatively shallow sand, sandstone, and limestone strata; a sandstone reservoir that is impregnated with a heavy, extremely viscous, black hydrocarbonaceous, petroleum-like material that cannot be retrieved through a well by conventional or enhanced oil recovery techniques; (FE 76-4): The several rock types that contain an extremely viscous hydrocarbon which is not recoverable in its natural state by conventional oil well production methods including currently used enhanced recovery techniques.

Terrestrial biosphere: The part of the geosphere consisting of the outer mantle and the crust that is directly involved with environmental processes through contact with the atmosphere, the hydrosphere, and living things; varies from (approximately) 40 60 miles in thickness; also called the lithosphere.

Thermochemical conversion: Use of heat to chemically change substances from one state to another, e.g. to make useful energy products.

Thermodynamic equilibrium: The thermodynamic state that is characterized by absence of flow of matter or energy.

Thermodynamics: The study of the energy transfers or conversion of energy in physical and chemical processes' defines the energy required to start a reaction or the energy given out during the process.

Total suspended particulate matter: The mass concentration determined by filter weighing, usually using a specified sampler which collects all particles up to approximately 20 microns depending on wind speed.

Toxicity: a measure of the toxic nature of a chemical; usually expressed quantitatively as LD_{50} (median lethal dose) or LC_{50} (median lethal concentration in air) – the latter refers to inhalation toxicity of gaseous substances in air; both terms refer to the calculated concentration of a chemical that can kill 50% of test animals when administered.

Toxicological chemistry: the chemistry of toxic substances with emphasis upon their interactions with biologic tissue and living organisms.

Triglyceride: An ester of glycerol and three fatty acids; the fatty acids represented by 'R' can be the same or different:

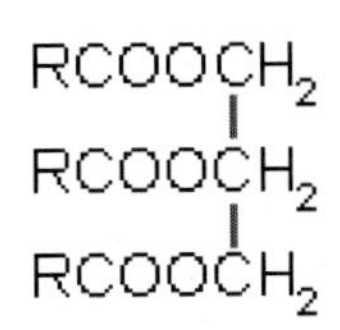

Trophic: the trophic level of an organism is the position it occupies in a food chain.

Troposphere: the portion of the atmosphere of the Earth that is closest to the surface.

Turbine: A machine for converting the heat energy in steam or high temperature gas into mechanical energy. In a turbine, a high velocity flow of steam or gas passes through successive rows of radial blades fastened to a central shaft.

Timberland: Forest land that is producing or is capable of producing crops of industrial wood, and that is not withdrawn from timber utilization by statute or administrative regulation.

Tipping fee: A fee for disposal of waste.

Ton (short ton): 2,000 pounds.

Tonne (Imperial ton, long ton, shipping ton): 2,240 pounds; equivalent to 1,000 kilograms or in crude oil terms about 7.5 barrels of oil.

Topping and back pressure turbines: Turbines which operate at exhaust pressure considerably higher than atmospheric (non-condensing turbines); often multistage with relatively high efficiency.

Topping cycle: A cogeneration system in which electric power is produced first. The reject heat from power production is then used to produce useful process heat.

Torrefaction: A mild heat treatment at 250 to 300°C (480 to 570°F) that efficiently turns solid biomass into a brittle, easy-to-pulverize material (often referred to as *bio-coal*) that can be treated (handled) as coal.

Total solids (dry solids): The residue remaining when water is evaporated away from the residue and dried under heat.

Transesterification: The chemical process in which an alcohol reacts with the triglycerides in vegetable oil or animal fats, separating the glycerin and producing biodiesel.

Traveling grate: A type of furnace in which assembled links of grates are joined together in a perpetual belt arrangement. Fuel is fed in at one end and ash is discharged at the other.

Trillion: 1×10^{12}

Turbine: A machine for converting the heat energy in steam or high temperature gas into mechanical energy. In a turbine, a high velocity flow of steam or gas passes through successive rows of radial blades fastened to a central shaft.

Turn down ratio- The lowest load at which a boiler will operate efficiently as compared to the boiler's maximum design load.

Ultraviolet radiation (UV radiation): An electromagnetic radiation with a wavelength from 10 nm to 400 nm, shorter than the wavelength of visible light but longer than the wavelength of X-rays. UV radiation is present in sunlight constituting about 10% of the total light output of the Sun.

Vacuum distillation: A secondary distillation process which uses a partial vacuum to lower the boiling point of residues from primary distillation and extract further blending components.

Vadose zone: the zone between land surface and the water table within which the moisture content is less than saturation (except in the capillary fringe) and pressure is less than atmospheric; soil pore spaces also typically contain air or other gases; the capillary fringe is included in the vadose zone.

Vapor pressure: the pressure exerted by a solid or liquid in equilibrium with its own vapor; depends on temperature and is characteristic of each substance' the higher the vapor pressure at ambient temperature, the more volatile the substance.

Viscosity: a measure of the ability of a liquid to flow or a measure of its resistance to flow; the force required to move a plane surface of area 1 square meter over another parallel plane surface 1 meter away at a rate of 1 meter per second when both surfaces are immersed in the fluid; the higher the viscosity, the slower the liquid flows.

Volatile organic compounds (VOCs): Name given to light organic hydrocarbons which escape as vapor from fuel tanks or other sources, and during the filling of tanks. VOCs contribute to smog.

Volatile solids (VS): Those solids in water or other liquids that are lost on ignition of the dry solids at 550°C.

Volatile fatty acids (VFA): These are acids that are produced by microbes in the silage from sugars and other carbohydrate sources. By definition they are volatile, which means that they will volatilize in air, depending on temperature. These are the first degradation product of anaerobic digestion prior to methane creation.

Volts: A unit of electrical pressure; a measure of the force of electricity. Volts represent pressure, correspondent to the pressure of water in a pipe. A volt is the unit of electromotive force or electric pressure analogous to water pressure in pounds per square inch. It is the electromotive force which, if steadily applied to a circuit having a resistance of one ohm, will produce a current one ampere.

Waste streams - Unused solid or liquid by-products of a process.

Waste vegetable oil (WVO): Grease from the nearest fryer which is filtered and used in modified diesel engines, or converted to biodiesel through the process of transesterification and used in any ol' diesel car.

Water-cooled vibrating grate: A boiler grate made up of a tuyere grate surface mounted on a grid of water tubes interconnected with the boiler circulation system for positive cooling; the structure is supported by flexing plates allowing the grid and grate to move in a vibrating action; ash is automatically discharged.

Watershed: The drainage basin contributing water, organic matter, dissolved nutrients, and sediments to a stream or lake.

Water solubility: the maximum amount of a chemical that can be dissolved in a given amount of pure water at standard conditions of temperature and pressure; typical units are milligrams per liter (mg/L), gallons per liter (g/L), or pounds per gallon (lbs/gall.

Watt: The common base unit of power in the metric system; one watt equals one joule per second, or the power developed in a circuit by a current of one ampere flowing through a potential difference of one volt. One Watt = 3.412 Btu/hr. The term 'kW' stands for "kilowatt" or 1 000 watts. The term 'MW' stands for "Megawatt" or 1 000 000 watts.

Weak Acid: an acid that does not release H^+ ions easily – an example is acetic acid.

Weak base: a basic chemical that has little affinity for a proton – an example is the chloride ion.

Weathering: Processes related to the physical and chemical actions of air, water and organisms after oil spill. The major weathering processes include evaporation, dissolution, dispersion, photochemical oxidation, water-in-oil *emulsification*, microbial degradation, adsorption onto suspended particulate materials, interaction with mineral fines, sinking, sedimentation, and formation of tar balls.

Wet deposition: The term used to describe pollutants brought to ground either by rainfall or by snow; this mechanism can be further subdivided depending on the point at which the pollutant was absorbed into the water droplets.

Wet digestion: The customary (but arbitrary) term to refer to a digestion process when using energy crops with a dry matter content of up to approximately 12% w/w in the digester because the digester contents are generally still pumpable with this water content; see Dry digestion.

Wheeling: The process of transferring electrical energy between buyer and seller by way of an intermediate utility or utilities.

Whole-tree harvesting: A harvesting method in which the whole tree (above the stump) is removed.

Wobbe index: The calorific value of a gas divided by the square root of the relative density; as with the calorific value, there is an upper and lower Wobbe index. See Calorific Value.

Xylenes: The term that refers to all three types of xylene isomers (meta-xylene, ortho-xylene, and para-xylene); produced from crude oil; used as a solvent and in the printing, rubber, and leather industries as well as a cleaning agent and a thinner for paint and varnishes; a major component of JP-8 fuel.

Yarding: The initial movement of logs from the point of felling to a central loading area or landing.

Yield: The mass (or moles) of a chosen final product divided by the mass (or moles) of one of the initial reactants.

Zwitterion: A particle that contains both positively charged and negatively charged groups; for example, amino acids ($H_2NHCHRCO_2H$) can form zwitterions ($^+H_3NCHRCOO^-$).

ABOUT THE AUTHOR

Dr. James G. Speight, PhD

Dr. James G. Speight has doctorate degrees in Chemistry, Geological Sciences, and Petroleum Engineering and is the author of more than 80 books in petroleum science, petroleum engineering, and environmental sciences.

Dr. Speight has fifty years of experience in areas associated with (i) the properties, recovery, and refining of reservoir fluids, conventional petroleum, heavy oil, and tar sand bitumen, (ii) the properties and refining of natural gas, gaseous fuels, (iii) the production and properties of petrochemicals, (iv) the properties and refining of biomass, biofuels, biogas, and the generation of bioenergy, and (v) the environmental and toxicological effects of fuels. His work has also focused on safety issues, environmental effects, remediation, and safety issues as well as reactors associated with the production and use of fuels and biofuels. He is the author of more than 80 books in petroleum science, petroleum engineering, biomass and biofuels, environmental sciences.

Although he has always worked in private industry which focused on contract-based work, he has served as Adjunct Professor in the Department of Chemical and Fuels Engineering at the University of Utah and in the Departments of Chemistry and Chemical and Petroleum Engineering at the University of Wyoming. In addition, he was a Visiting Professor in the College of Science, University of Mosul, Iraq and has also been a Visiting Professor in Chemical Engineering at the following universities: University of Missouri-Columbia, the Technical University of Denmark, and the University of Trinidad and Tobago.

In 1996, Dr. Speight was elected to the Russian Academy of Sciences and awarded the Gold Medal of Honor that same year for outstanding contributions to the field of petroleum sciences. In 2001, he received the Scientists without Borders Medal of Honor of the Russian Academy of Sciences and was also awarded Dr. Speight the Einstein Medal for outstanding contributions and service in the field of Geological Sciences. In 2005, the Academy awarded Dr. Speight the Gold Medal - Scientists without Frontiers, Russian Academy of Sciences, in recognition of Continuous Encouragement of Scientists to Work Together Across International Borders. In 2007 Dr. Speight received the Methanex Distinguished Professor award at the University of Trinidad and Tobago in recognition of excellence in research. In 2018 he was awarded the American Excellence Award for *Excellence in Client Solutions* by the United States Institute of Trade and Commerce. Washington, DC.

Dr. James G. Speight
2476 Overland Road,
Laramie, WY 82070-4808, USA
E-mail: JamesSp8@aol.com
Web page: https://www.drjamesspeight.com

INDEX

D

E

F

G

Z